NOVEL STRATEGIES TO IMPROVE SHELF-LIFE AND QUALITY OF FOODS

Quality, Safety, and Health Aspects

Innovations in Agricultural and Biological Engineering

NOVEL STRATEGIES TO IMPROVE SHELF-LIFE AND QUALITY OF FOODS

Quality, Safety, and Health Aspects

Edited by

Santosh K. Mishra, PhD

Megh R. Goyal, PhD

First edition published 2021

Apple Academic Press Inc.
1265 Goldenrod Circle, NE,
Palm Bay, FL 32905 USA

4164 Lakeshore Road, Burlington,
ON, L7L 1A4 Canada

CRC Press
6000 Broken Sound Parkway NW,
Suite 300, Boca Raton, FL 33487-2742 USA

2 Park Square, Milton Park,
Abingdon, Oxon, OX14 4RN UK

Library and Archives Canada Cataloguing in Publication

Title: Novel strategies to improve shelf-life and quality of foods : quality, safety, and health aspects / edited by Santosh K. Mishra, PhD, Megh R. Goyal, PhD.

Names: Mishra, Santosh K., editor. | Goyal, Megh Raj, editor.

Series: Innovations in agricultural and biological engineering.

Description: Series statement: Innovations in agricultural and biological engineering | Includes bibliographical references and index.

Identifiers: Canadiana (print) 20200308416 | Canadiana (ebook) 20200308556 | ISBN 9781771888844 (hardcover) | ISBN 9781003010272 (ebook)

Subjects: LCSH: Food—Preservation. | LCSH: Food—Quality. | LCSH: Food—Biotechnology.

Classification: LCC TP371.2 .N68 2021 | DDC 664/.028—dc23

Library of Congress Cataloging-in-Publication Data

CIP data on file with US Library of Congress

ISBN: 978-1-77188-884-4 (hbk)
ISBN: 978-1-00301-027-2 (ebk)

ABOUT THE LEAD EDITOR

Santosh K. Mishra, PhD
Assistant Professor, Department of Dairy Microbiology, College of Dairy Science and Technology, Guru Angad Dev Veterinary and Animal Sciences University, Ludhiana, Punjab, India

Santosh K. Mishra, PhD, is Assistant Professor in the Department of Dairy Microbiology at the College of Dairy Science and Technology, Guru Angad Dev Veterinary and Animal Sciences University, Ludhiana, Punjab, India. He received his BTech degree in Dairy Technology from Maharashtra Animal and Fisheries Sciences University, Nagpur, India; and his MSc and PhD degrees from the National Dairy Research Institute, Karnal, Haryana. He is presently working on areas of functional foods and dairy products incorporating live probiotics and the technology of functional lactic cultures for fermented and non-fermented dairy products. He also served the dairy industry as Quality Assurance Executive at Mother Dairy, New Delhi. He also handles externally funded projects by the Indian Department of Science and Technology (DST), Indian Ministry of Food Processing Industries (MoFPI), and Indian University Grants Commission (UGC) as principal investigator or co-principal investigator. He has received several awards for best papers and posters/presentations. He is the recipient of junior and senior research fellowships during his master's and doctoral programs at the National Dairy Research Institute, Karnal, Haryana. Recently he has received the Award of Honor at the International Conference sponsored by Partap College of Education, Ludhiana, in association with the International Professionals Development Association, UK. He is a member of various scientific societies, including life member of SASNET-Fermented Foods, Anand; member of the Indian Dairy Associations, New Delhi. He has published several research, review, and popular articles in national and international journals, and also published several book chapters and teaching reviews in various training programs. He has recently completed a young scientist project by DST, SEED Department, Govt. of India, New Delhi, on isolation and characterization of novel oxalate degrading lactic acid bacteria (LAB) for potential probiotic management of kidney stone. Readers may contact him at: *skmishra84@gmail.com*

ABOUT THE SENIOR EDITOR-IN-CHIEF

Megh R. Goyal, PhD, PE

Retired Professor in Agricultural and Biomedical Engineering, University of Puerto Rico, Mayaguez Campus; Senior Acquisitions Editor, Biomedical Engineering and Agricultural Science, Apple Academic Press, Inc.

Megh R. Goyal, PhD, PE, is Retired Professor in Agricultural and Biomedical Engineering in the General Engineering Department at the College of Engineering, University of Puerto Rico–Mayaguez Campus; and Senior Acquisitions Editor and Senior Technical Editor-in-Chief in Agriculture and Biomedical Engineering for Apple Academic Press, Inc. He has worked as a Soil Conservation Inspector and as a Research Assistant at Haryana Agricultural University and Ohio State University.

During his professional career of 50 years, Dr. Goyal has received many prestigious awards and honors. He was the first agricultural engineer to receive the professional license in Agricultural Engineering in 1986 from the College of Engineers and Surveyors of Puerto Rico. In 2005, he was named the "Father of Irrigation Engineering in Puerto Rico for the Twentieth Century" by the American Society of Agricultural and Biological Engineers (ASABE), Puerto Rico Section, for his pioneering work in micro irrigation, evapotranspiration, agroclimatology, and soil and water engineering. The Water Technology Centre of Tamil Nadu Agricultural University in Coimbatore, India, recognized Dr. Goyal as one of the experts "who rendered meritorious service for the development of micro irrigation sector in India" by bestowing the Award of Outstanding Contribution in Micro Irrigation. This award was presented to Dr. Goyal during the inaugural session of the National Congress on "New Challenges and Advances in Sustainable Micro Irrigation" held at Tamil Nadu Agricultural University. Dr. Goyal received the Netafim Award for Advancements in Microirrigation: 2018 from the American Society of Agricultural Engineers at the ASABE International Meeting in August 2018.

A prolific author and editor, he has written more than 200 journal articles and several textbooks and has edited over 80 books. He is the editor of three book series published by Apple Academic Press: Innovations in Agricultural & Biological Engineering, Innovations and Challenges in Micro Irrigation, and Research Advances in Sustainable Micro Irrigation. He is also instrumental in the development of the new book series Innovations in Plant Science for Better Health: From Soil to Fork.

Dr. Goyal received his BSc degree in engineering from Punjab Agricultural University, Ludhiana, India; his MSc and PhD degrees from Ohio State University, Columbus; and his Master of Divinity degree from Puerto Rico Evangelical Seminary, Hato Rey, Puerto Rico, USA.

ABOUT THE BOOK SERIES: INNOVATIONS IN AGRICULTURAL AND BIOLOGICAL ENGINEERING

Under this book series, Apple Academic Press Inc. is publishing book volumes over a span of 8–10 years in the specialty areas defined by the American Society of Agricultural and Biological Engineers (www.asabe.org). Apple Academic Press Inc. aims to be a principal source of books in agricultural and biological engineering. We welcome book proposals from readers in areas of their expertise.

The mission of this series is to provide knowledge and techniques for agricultural and biological engineers (ABEs). The book series offers high-quality reference and academic content on agricultural and biological engineering (ABE) that is accessible to academicians, researchers, scientists, university faculty and university-level students, and professionals around the world.

Agricultural and biological engineers ensure that the world has the necessities of life, including safe and plentiful food, clean air and water, renewable fuel and energy, safe working conditions, and a healthy environment by employing knowledge and expertise of the sciences, both pure and applied, and engineering principles. Biological engineering applies engineering practices to problems and opportunities presented by living things and the natural environment in agriculture.

ABE embraces a variety of the following specialty areas (www.asabe.org): aquaculture engineering, biological engineering, energy, farm machinery and power engineering, food, and process engineering, forest engineering, information, and electrical technologies, soil, and water conservation engineering, natural resources engineering, nursery, and greenhouse engineering, safety, and health, and structures and environment.

For this book series, we welcome chapters on the following specialty areas (but not limited to):

1. Academia to industry to end-user loop in agricultural engineering.
2. Agricultural mechanization.
3. Aquaculture engineering.
4. Biological engineering in agriculture.

5. Biotechnology applications in agricultural engineering.
6. Energy source engineering.
7. Farm to fork technologies in agriculture.
8. Food and bioprocess engineering.
9. Forest engineering.
10. GPS and remote sensing potential in agricultural engineering.
11. Hill land agriculture.
12. Human factors in engineering.
13. Impact of global warming and climatic change on agriculture economy.
14. Information and electrical technologies.
15. Irrigation and drainage engineering.
16. Micro-irrigation engineering.
17. Milk engineering.
18. Nanotechnology applications in agricultural engineering.
19. Natural resources engineering.
20. Nursery and greenhouse engineering.
21. Potential of phytochemicals from agricultural and wild plants for human health.
22. Power systems and machinery design.
23. Robot engineering and drones in agriculture.
24. Rural electrification.
25. Sanitary engineering.
26. Simulation and computer modeling.
27. Smart engineering applications in agriculture.
28. Soil and water engineering.
29. Structures and environment engineering.
30. Waste management and recycling.
31. Any other focus areas.

For more information on this series, readers may contact:

Megh R. Goyal, PhD, PE
Book Series Senior Editor-in-Chief:
Innovations in Agricultural and Biological Engineering
E-mail: goyalmegh@gmail.com

OTHER BOOKS ON AGRICULTURAL AND BIOLOGICAL ENGINEERING BY APPLE ACADEMIC PRESS, INC.

Management of Drip/Trickle or Micro Irrigation
Megh R. Goyal, PhD, PE, Senior Editor-in-Chief

Evapotranspiration: Principles and Applications for Water Management
Megh R. Goyal, PhD, PE, and Eric W. Harmsen, Editors

Book Series: Research Advances in Sustainable Micro Irrigation
Senior Editor-in-Chief: Megh R. Goyal, PhD, PE
Volume 1: Sustainable Micro Irrigation: Principles and Practices
Volume 2: Sustainable Practices in Surface and Subsurface Micro Irrigation
Volume 3: Sustainable Micro Irrigation Management for Trees and Vines
Volume 4: Management, Performance, and Applications of Micro Irrigation Systems
Volume 5: Applications of Furrow and Micro Irrigation in Arid and Semi-Arid Regions
Volume 6: Best Management Practices for Drip Irrigated Crops
Volume 7: Closed Circuit Micro Irrigation Design: Theory and Applications
Volume 8: Wastewater Management for Irrigation: Principles and Practices
Volume 9: Water and Fertigation Management in Micro Irrigation
Volume 10: Innovation in Micro Irrigation Technology

Book Series: Innovations and Challenges in Micro Irrigation
Senior Editor-in-Chief: Megh R. Goyal, PhD, PE
Volume 1: Management of Drip/Trickle or Micro Irrigation
Volume 2: Sustainable Micro Irrigation Design Systems for Agricultural Crops
Volume 3: Principles and Management of Clogging in Micro Irrigation
Volume 4: Performance Evaluation of Micro Irrigation Management
Volume 5: Potential Use of Solar Energy and Emerging Technologies in Micro Irrigation

Volume 6: Micro Irrigation Management: Technological Advances and Their Applications
Volume 7: Micro Irrigation Engineering for Horticultural Crops
Volume 8: Micro Irrigation Scheduling and Practices
Volume 9: Engineering Interventions in Sustainable Trickle Irrigation
Volume 10: Management Strategies for Water Use Efficiency and Micro Irrigated Crops
Volume 11: Fertigation Technologies in Micro Irrigation

Book Series: Innovations in Agricultural & Biological Engineering
Senior Editor-in-Chief: Megh R. Goyal, PhD, PE

- Dairy Engineering: Advanced Technologies and Their Applications
- Developing Technologies in Food Science: Status, Applications, and Challenges
- Emerging Technologies in Agricultural Engineering
- Engineering Interventions in Agricultural Processing
- Engineering Interventions in Foods and Plants
- Engineering Practices for Agricultural Production and Water Conservation: An Interdisciplinary Approach
- Engineering Practices for Management of Soil Salinity: Agricultural, Physiological, and Adaptive Approaches
- Engineering Practices for Milk Products: Dairyceuticals, Novel Technologies, and Quality
- Evapotranspiration
- Field Practices for Wastewater Use in Agriculture
- Flood Assessment: Modeling and Parameterization
- Food Engineering: Emerging Issues, Modeling, and Applications
- Food Process Engineering: Emerging Trends in Research and Their Applications
- Food Technology: Applied Research and Production Techniques
- Handbook of Research on Food Processing and Preservation Technologies: Volume 1: Nonthermal and Innovative Food Processing Methods
- Modeling Methods and Practices in Soil and Water Engineering
- Nanotechnology and Nanomaterial Applications in Food, Health and Biomedical Sciences
- Nanotechnology Applications in Dairy Science: Packaging, Processing, and Preservation

- Novel Dairy Processing Technologies: Techniques, Management, and Energy Conservation
- Novel Strategies to Improve Shelf-Life and Quality of Foods
- Processing of Fruits and Vegetables: From Farm to Fork
- Processing Technologies for Milk and Milk Products: Methods, Applications, and Energy Usage
- Scientific and Technical Terms in Bioengineering and Biological Engineering
- Soil and Water Engineering: Principles and Applications of Modeling
- Soil Salinity Management in Agriculture: Technological Advances and Applications
- State-of-the-Art Technologies in Food Science: Human Health, Emerging Issues and Specialty Topics
- Sustainable Biological Systems for Agriculture: Emerging Issues in Nanotechnology, Biofertilizers, Wastewater, and Farm Machines
- Technological Interventions in Dairy Science: Innovative Approaches in Processing, Preservation, and Analysis of Milk Products
- Technological Interventions in Management of Irrigated Agriculture
- Technological Interventions in the Processing of Fruits and Vegetables
- Technological Processes for Marine Foods, From Water to Fork: Bioactive Compounds, Industrial Applications, and Genomics

CONTENTS

Contributors .... *xvii*

Abbreviations .... *xxi*

Preface .... *xxvii*

Part I: Quality and Health Aspects of Food Preservation 1

1. **Preservation of Bovine Colostrum: Immunotherapeutic Agents to Promote Health 3**
   Krishan K. Mishra
2. **Essential Oil-Based Combined Processing Methods to Pursue Safer Foods 21**
   Manju Gaare, Shaik Abdul Hussain, and Santosh K. Mishra
3. **Potential Use of Herbs in Milk and Milk Products 53**
   Pravin D. Sawale, Writdhama Prasad, Shaik Abdul Hussain, Veena Nagarajappa, and Santosh K. Mishra

Part II: Applications of Novel Biocompounds in Quality and Safety of Foods 71

4. **Antifungal Lactic Acid Bacteria (LAB): Potential Use in Food Systems 73**
   Saurabh Kadyan and Diwas Pradhan
5. **Applications of Antimicrobial Enzymes in Foods 95**
   Veena Nagarajappa, Pravin D. Sawale, Surendra N. Battula, and Venus Bansal
6. **Health Benefits of Antimicrobial Peptides 111**
   Sunita Meena, Kapil Singh Narayan, and Sandeep Kumar
7. **Applications of Bdellovibrio Bacteria as a Biocontrol Agent: Food Safety and Mitigating Clinical Pathogens 143**
   Valerie D. Zaffran, Gabrielle Kirshteyn, and Prashant Singh
8. **Safety Aspects of Novel Bacteriocins 161**
   Tejinder P. Singh and Shalini Arora

Part III: Potential of Novel Technologies for Food Preservation 185

9. Potential of Nonthermal Plasma Technology in Food Preservation...... 187

Sujit Das and Subrota Hati

10. Potential of High Hydrostatic Pressure Technology in Food Preservation and Food Safety 211

Rekha Chawla, Venus Bansal, S. Sivakumar, Narender K. Chandla, and Santosh K. Mishra

11. Application of Pulsed Light Technology in Microbial Safety and Food Preservation 227

Venus Bansal, Narender K. Chandla, Rekha Chawla, Veena Nagarajappa, and Santosh K. Mishra

12. Novel Packaging Systems for Food Preservation.................................. 249

Narender K. Chandla, Venus Bansal, Gopika Talwar, Santosh K. Mishra, and Sunil K. Khatkar

13. Role of Biosensors in Quality and Safety of Dairy Foods....................... 275

H. V. Raghu, Ajeet Singh, and Naresh Kumar

14. Potential Use of Lactic Acid Bacteria (LAB): Protective Cultures in Food Biopreservation 287

Manju Gaare and Santosh K. Mishra

Index.. 307

CONTRIBUTORS

Shalini Arora
Assistant Professor, Department of Dairy Technology, College of Dairy Science and Technology, Lala Lajpat Rai University of Veterinary and Animal Sciences (LUVAS), Hisar – 125001, Haryana, India, Phone: +91-7988425439, E-mail: shaliniarora.luvas@gmail.com

Venus Bansal
Assistant Professor, Department of Dairy Technology, College of Dairy Science and Technology, Guru Angad Dev Veterinary and Animal Sciences University (GADVASU), Ludhiana – 141004, Punjab, India, Phone: +91-9478476400, E-mail: venus3b3@gmail.com

Surendra N. Battula
Principal Scientist, Dairy Chemistry Division, National Dairy Research Institute (NDRI), SRS, Adugodi, Bengaluru – 560030, Karnataka, India, Phone: +91-9449028628, E-mail: bsn_ndri@yahoo.com

Narender K. Chandla
Assistant Professor, Department of Dairy Engineering, College of Dairy Science and Technology, Guru Angad Dev Veterinary and Animal Sciences University (GADVASU), Ludhiana – 141004, Punjab, India, Phone: +91-9464116738, E-mail: chandla84@gmail.com

Rekha Chawla
Assistant Professor, Department of Dairy Technology, College of Dairy Science and Technology, Guru Angad Dev Veterinary and Animal Sciences University (GADVASU), Ludhiana – 141004, Punjab, India, Phone: +91-7589145459, E-mail: mails4rekha@gmail.com

Sujit Das
PhD Candidate, Department of Rural Development and Agricultural Production, North-Eastern Hill University-Tura Campus, Tura – 794001, Meghalaya, India, Phone: +91-7908567743, E-mail: sujitdas557@gmail.com

Manju Gaare
Assistant Professor, Department of Dairy Microbiology, GN Patel College of Dairy Technology, Sardar Dantiwada Agriculture University (SDAU), Village Dantiwada, Sardarkrushinagar – 385506, Gujarat, India, Phone: +91-8971769001, E-mail: manjugdsc@gmail.com

Megh R. Goyal
PE, Retired Professor in Agricultural and Biomedical Engineering from College of Engineering at University of Puerto Rico-Mayaguez Campus, and Senior Technical Editor-in-Chief in Agricultural and Biomedical Engineering for Apple Academic Press Inc., PO Box 86, Rincon-PR – 006770086, USA, E-mail: goyalmegh@gmail.com

Subrota Hati
Assistant Professor, Department of Dairy Microbiology, Anand Agricultural University, Anand – 388110, Gujarat, India, Phone: +91-9409669561, E-mail: subrota_dt@yahoo.com

Shaik Abdul Hussain
Scientist, Dairy Technology Division, National Dairy Research Institute (NDRI), Karnal – 132001, Haryana, India, Phone: +91-9896668983, E-mail: abdulndri@gmail.com

Saurabh Kadyan
Scientist, Dairy Microbiology Division, ICAR-National Dairy Research Institute (NDRI), Karnal – 132001, Haryana, India, Phone: 91-9466564543, E-mail: kadyan.saurabh3@gmail.com

Sunil K. Khatkar
Assistant Professor, Department of Dairy Technology, College of Dairy Science and Technology, Guru Angad Dev Veterinary and Animal Sciences University (GADVASU), Ludhiana – 141004, Punjab, India, E-mail: absuneelkhatkar@gmail.com

Gabrielle Kirshteyn
Undergraduate Student, Biological Science Department, Florida State University (FSU), Tallahassee – 32306-2400, Florida, USA, Phone: 850-644-4796, E-mail: gak17@my.fsu.edu

Naresh Kumar
Principal Scientist, National Referral Center, Dairy Microbiology Division, ICAR-National Dairy Research Institute (NDRI), Karnal – 132001, Haryana, India, Phone: +91-8901023594, E-mail: nrshgoyal@yahoo.com

Sandeep Kumar
PhD Research Scholar, Animal Biochemistry Division, National Dairy Research Institute (NDRI), Karnal – 132001, Haryana, India, E-mail: sandeepvermma@gmail.com

Sunita Meena
Scientist, Animal Biochemistry Division, National Dairy Research Institute (NDRI), Karnal – 132001, Haryana, India, Phone: 91-8930938362, E-mail: sunitameena1188@gmail.com

Krishan K. Mishra
Associate Professor and Head, Department of Veterinary Medicine, College of Veterinary Science and Animal Husbandry, Nanaji Deshmukh Veterinary Science University (NDVSU), Rewa – 486001, Madhya Pradesh, India, Phone: +91-8966888486, E-mail: drmishra79@gmail.com

Santosh K. Mishra
Assistant Professor, Department of Dairy Microbiology, College of Dairy Science and Technology, Guru Angad Dev Veterinary and Animal Sciences University (GADVASU), Ludhiana – 141004, Punjab, India, Phone: +91-9464995049, E-mail: skmishra84@gmail.com

Veena Nagarajappa
Assistant Professor, Department of Dairy Chemistry, College of Dairy Science and Technology, Guru Angad Dev Veterinary and Animal Sciences University (GADVASU), Ludhiana – 141004, Punjab, India, Phone: +91-9855886831, E-mail: veena.ndri@gmail.com

Kapil Singh Narayan
PhD Research Scholar, Animal Biochemistry Division, National Dairy Research Institute (NDRI), Karnal – 132001, Haryana, India, E-mail: kapilnsnk@gmail.com

Diwas Pradhan
Scientist, Dairy Microbiology Division, ICAR-National Dairy Research Institute (NDRI), Karnal – 132001, Haryana, India, Phone: 91-8053319936, E-mail: zawidprd@gmail.com

Writdhama Prasad
Scientist, Dairy Technology Division, National Dairy Research Institute (NDRI), Karnal, Haryana, India, Phone: +917206282166, E-mail: wgprasad.ndri@gmail.com

H. V. Raghu
Scientist, National Referral Center, Dairy Microbiology Division, ICAR-National Dairy Research Institute (NDRI), Karnal – 132001, Haryana, India, Phone: +91-9466963599, E-mail: 4rvsy.dmndri@gmail.com

Pravin D. Sawale
Assistant Professor, College of Dairy Technology, Maharashtra Animal and Fishery Sciences University (MAFSU), Warud (Pusad) – 445204, Maharashtra, India, Phone: 91-8469367101,
E-mail: pravins92@gmail.com

Ajeet Singh
Senior Research Fellow, National Referral Center, Dairy Microbiology Division,
ICAR-National Dairy Research Institute (NDRI), Karnal – 132001, Haryana, India,
Phone: +91-8791539165, E-mail: ajeetchoudharygkv@gmail.com

Prashant Singh
Assistant Professor, Department of Nutrition, Food and Exercise Sciences,
Florida State University (FSU), Tallahassee – 32306-2400, Florida, USA,
Phone: 850-644-4796, E-mail: psingh2@fsu.edu

Tejinder P. Singh
Assistant Professor, Department of Dairy Microbiology, College of Dairy Science and Technology,
Lala Lajpat Rai University of Veterinary and Animal Sciences (LUVAS), Hisar – 125001, Haryana,
India, Phone: +91-8929970050, E-mail: 88tejindersingh@gmail.com

S. Sivakumar
Assistant Scientist, Department of Dairy Technology, College of Dairy Science and Technology,
Guru Angad Dev Veterinary and Animal Sciences University (GADVASU), Ludhiana – 141004,
Punjab, India, Phone: +91-8556065594, E-mail: drsiva2003@yahoo.com

Gopika Talwar
Assistant Professor, Department of Dairy Engineering, College of Dairy Science and Technology,
Guru Angad Dev Veterinary and Animal Sciences University (GADVASU), Ludhiana – 141004,
Punjab, India, E-mail: engg_gopika@yahoo.co.in

Valerie D. Zaffran
PhD Research Scholar, Department of Nutrition, Food and Exercise Sciences,
Florida State University (FSU), Tallahassee – 32306-2400, Florida, USA,
Phone: 561-628-7114, E-mail: vdz10@my.fsu.edu

ABBREVIATIONS

AC	ascorbic acid
AFPs	antifungal peptides
AIDS	acquired immune deficiency syndrome
AMPs	antimicrobial peptides
AMSdb	antimicrobial sequence database
APD	antimicrobial peptide database
APP	atmospheric pressure plasma
APPJ	atmospheric-pressure plasma jet
ARB	antibiotic-resistant bacteria
AU	arbitrary units
BALO	*Bdellovibrio* and like organisms
BC	bovine colostrum
BCCs	bovine colostrum concentrates
BHQ	butyl hydroquinone
BHV	bovine herpes virus
BOPA	biaxially oriented nylon
BoPET	biaxially oriented PE terephthalate
C_6H_5-COOH	benzoic acid
CAMPs	cationic antimicrobial peptides
CAS	controlled atmospheric storage
CAS/P	controlled atmospheric storage/packaging
CAT	catalase
C-C	carbon-carbon
CDPs	cyclic dipeptides
CDs	corona discharges
CFSAN	Center for Food Safety and Applied Nutrition
CFU	colony-forming unit
CH_3-CH(OH)-COOH	lactic acid
CH_3-CH_2-COOH	propionic acid
CH_3-COOH	acetic acid
CTI	critical temperature-based indicators
CTTI	critical temperature time-based indicators
CVD	cardiovascular diseases

DBD	dielectric barrier discharge
DC	direct current
DE	dextrose equivalent
DNA	deoxyribonucleic acid
DRS-S4	dermaseptin-S4
DS	destructurized starch
E. coli	*Escherichia coli*
EDTA	ethylenediamine tetra-acetic acid
ELISA	enzyme-linked immune sorbent assay
EMP	Embden-Meyerhof-Parnas
EOs	essential oils
EPA	Environment Protection Agency
ESBL	extended-spectrum ß-lactamase
ETEC	enterotoxigenic *Escherichia coli*
EVA	ethylene vinyl alcohol
EVOH	ethylene vinyl alcohol copolymer
FAO	Food and Agriculture Organization
FBD	food-borne diseases
FDA	Food and Drug Administration
FFDCA	Federal Food, Drug, and Cosmetic Act
FIFO	first-in first-out
FIFRA	Federal Insecticide, Fungicide, and Rodenticide Act
FMIA	federal meat inspection act
FSIS	Food Safety and Inspection Services
FSSAI	Food Safety and Standards Authority of India
GAD	glutamate decarboxylase
GALT	gut-associated lymphoid tissues
GI	gastrointestinal
GIT	gastrointestinal tract
GOX	glucose oxidase
GRAS	generally regarded as safe
GST	glutathione S-transferase
H. pylori	*Helicobacter pylori*
H_2O_2	hydrogen peroxide
HBC	hyperimmune bovine colostrum
HCMV	human cytomegalovirus
HHP	high hydrostatic pressure
HI	host-independent

HPA	hydroxy propionaldehyde
HPP	high pressure processing
IFN	interferon
IgA	immunoglobulin A
IGF	insulin-like growth factor
IgG	immunoglobulin G
IgM	immunoglobulin M
IL	interleukin
ILP	intense light pulses
IPL	intense pulsed light
IR	infrared
LAB	lactic acid bacteria
LDPE	low density polyethylene
Lf	lactoferricin
LIFO	last-in first-out
LOD	limit of detection
LP	lactoperoxidase
LPS	lactoperoxidase system
LPS	lipopolysaccharide
LTAs	lipoteichoic acids
LYZ	lysozyme
Man-PTS	mannose phosphotransferase system
MAP	modified atmosphere packaging
MBC	minimum bactericidal concentration
MDA	malondialdehyde
MDR	multiple drug resistant
MF	microfiltration
MIC	minimal inhibitory concentration
mRNA	messenger ribonucleic acid
MRS	De Man, Rogosa, and Sharpe
MW	molecular weight
NAs	nucleic acids
NCR	nodule specific cysteine rich
NK	natural killer
NPN	non-protein-nitrogen
NTP	non-thermal plasma
OH	ohmic heating
$OSCN^-$	hypothiocyanite ion
PA	polyamide

PAA	peracetic acid
PBS	phosphate buffer saline
PC	protective cultures
PDB	protein data bank
PE	polyethylene
PEF	pulsed electric field
PET	polyethylene terephthalate
PL	pulsed light
PLA	phenyl lactic acid
PLT	pulsed light technology
PP	polypropylene or polypropene
PPA	phenylpyruvic acid
PPIA	poultry products inspection act
PRPs	proline rich polypeptides
PUFA	polyunsaturated fatty acids
PVOH	polyvinyl alcohol
QCM	quartz crystal microbalance
QPS	qualified presumption of safety
RCT	rennet coagulation time
RFID	radio frequency identification
RFs	radio frequencies
RIF	radio frequency indicators
RNA	ribonucleic acid
ROS	reactive oxygen species
RRS	radiation on radiosensitivity
S. dublin	*Salmonella dublin*
SCN-	thiocyanate
SDBC	spray dried bovine colostrum
SELEX	systematic evolution of ligands by exponential
SEMI	scanning electron microscope imaging
SH	sulphydryl
SI	self-immunity
SPR	surface plasma resonance
SPs	surface plasmons
TAs	teichoic acids
TGF ß	transforming growth factor ß
TGF	tissue growth factors
TII	toxin identification indicators
TPS	thermoplastic starch

TSB	tryptic soy broth
TTIs	time temperature indicators
UHT	ultra high temperature
URTI	upper respiratory tract infection
USA	United States of America
USDA	United States Department of Agriculture
US-FDA	United States Food and Drug Administration
UV	ultraviolet
WHO	World Health Organization

PREFACE

The tropical climate average ambient temperature throughout the year is very high in most parts of the world, especially in developing nations, which is very favorable for the growth of spoilage microorganisms. On the other hand, due to the acute power shortage, it is difficult to maintain the cold chain right from food production at crucial farms to food processing plants and finally to distribution outlets of foods. Therefore, food preservation prior to distribution and sale is a major problem in the tropical climates of most developing nations. In order to assure the consumer that the product is safe for human consumption, due importance is given to the quality and safety part of the production, processing, and distribution.

By searching the literature, one can find volumes of books and specialized publications on augmentation of shelf-life and quality of food products. Unfortunately, most of these publications have dealt with theoretical aspects of these strategies and technologies with little emphasis on real application in consumer and food products.

This book volume attempts to illustrate various aspects of enhancement of shelf-life along with quality and safety applications. This book has several potential users. It can be a reference book for those students who are taking a college- or university-level food safety or quality assurance course for the first time. The objective of editors was to compile information that dairy and food science students are expected to be familiar with as part of their college or university program before they seek career positions in the food industry. This book will be further useful to food industry quality practitioners or employees who need to become familiar with updated information pertaining to their routine work. This book is organized in such a way that each chapter treats one major application of food safety and quality enhancement through various means.

The book contains three main parts such as (1) Health and quality aspects of food preservation: the concepts of health aspects of bovine colostrum (BC) along with their preservation techniques and application of natural herbs and essential oils (EOs) in different types of food have been discussed. (2) Applications of novel biomolecules in quality and safety of foods: the concepts of novel compounds from different natural resources, their antimicrobial properties, different hurdle formulations with each other, mechanisms of action, applications in foods, etc., and their health promotional aspects are highlighted. (3) Novel research techniques in food bio-preservation: the application of novel technologies, formulations with other biomolecules to form hurdle concepts for the improvement of food shelf-life and safety concepts are elucidated.

We introduce this book volume under book series *Innovations in Agricultural & Biological Engineering* (www.appleacademicpress.com). This book volume is a treasure house of information and excellent reference material for researchers, scientists, students, growers, traders, processors, industries, and others for quality control and safety of food products during production, processing, and transportation in any food industry and boosts their confidence in the area of safety and quality aspects of food products.

This book has surpassed our vision and expectations due to the contributions of all cooperating authors to this book volume, which have been most valuable in the compilation. Their names are mentioned in each chapter and in the list of contributors. We are grateful to all of them for their expertise, commitment, and dedication. We hope that this book will prove itself a useful source for the preservation of easily perishable food products with a wide range of food and consumer product applications.

We will like to thank editorial and production staff at Apple Academic Press, Inc., for making every effort to publish this book when all are concerned with health and food issues.

We request the reader to offer your constructive suggestions that may help to improve the next edition. Also, we would like to thank our families who have taught us the importance of working hard, having clear goals, and standing for what we believe is right. It is a lesson that guides us in everything we do. Last but not least, we wish to thank our wives, Anamika and Subhadra, for their understanding and patience throughout this project.

As educators, there is a piece of advice to one and all in the world: "*Permit that our Almighty God, our Creator, provider of all and excellent Teacher, feed our life with Healthy Food Products and His Grace…; and Get married to your profession…*"

—Santosh Kumar Mishra, PhD
Megh R. Goyal, PhD

Part I:
Quality and Health Aspects of Food Preservation

CHAPTER 1

PRESERVATION OF BOVINE COLOSTRUM: IMMUNOTHERAPEUTIC AGENTS TO PROMOTE HEALTH

KRISHAN K. MISHRA

ABSTRACT

Bovine colostrum (BC) is a unique and immuno therapeutically valuable by-product of large scale dairy production units. The uses of BC in both humans and as well as animals are related largely to the historical concept of "immune milk," which has been found appropriate for transferring passive immunity. Besides immunoglobulins at a much higher level than the ordinary milk, colostrum contains a plethora of other bioactive immune constituents [such as tissue growth factors (TGF) and proline-rich polypeptides (PRP)] and antimicrobial fractions (such as lactoferrin, antioxidants, cytokines, lymphokines, etc.) that stimulate various immunological mechanisms, thus providing prophylaxis against various infectious and non-infectious diseases. BC has the potential to act like a matchless source of biomolecules and treasured raw material for the manufacture of nutraceuticals. Various techniques are employed for increasing the keeping quality of BC by processing and preservation for prolonged use, like, freezing, spray drying, lyophilization, pasteurization, etc. for retaining a reasonable proportion of bioactive components especially immunoglobulins. Novel techniques like membrane processing, microfiltration (MF), and high pressure treatment have been advocated for better efficacy and retaining biomolecules post-treatment by employing various manipulation methods for the formulation of a specific composition. This chapter primarily discusses various immunotherapeutic roles of BC along with different storage techniques for its extended safe use along with the influence of various preservation methods on immune-biological or nutritional qualities.

1.1 INTRODUCTION

Bovine colostrum (BC) is called "initial" milk that is produced by cows post-parturition, which has an immunological composition and nutrient profile differing considerably from regular milk produced later on. Neonates are supplied with immune factors like antibodies by their dam either before parturition, at the time of birth, or just after parturition, which helps in protecting the neonates against various types of disease etiologies, until they institute their own immune mechanism for the identification and clearance of pathogens.

In the majority of farm animals, the majority of the immunoglobulins are acquired by the neonates through the ingestion of colostrum, through the gut but human beings on the contrary acquire the bulk of immunoglobulins, and the IgGs (especially immunoglobulin G) especially, by placental passage prior to birth. In the present scenario, colostrums-based products are being produced at an industrial level in many countries like the USA (United States of America), China, Australia, New Zealand, India, etc. and are being promoted as a functional food supplement for promoting health, after regulating their IgG contents [11].

The feasting of the colostrum by humans and the post-birth of newborns of dairy animals is an established custom in several customary societies. The application of BC for the prevention of infectious diseases is correlated to the ancient concept of "immune milk" being able to convey passive immunity. The ruminant colostrum is predominantly richer in immunoglobulins especially IgG as opposed to IgA (immunoglobulin A), which is the chief class of immunoglobulins present in human milk. In addition, BC contains Igs at extremely higher levels compared to regular bovine milk [31], but also contains a plethora of physiologically bioactive components such as growth-promoting factors for the growth of muscles, repair of sites of inflammation, prevention of bacterial translocation and stimulation of gut immunity, healing of gastrointestinal tract (GIT) ulcers, etc. [44, 75] besides many antimicrobial portions comprising lactoferrin, lactoperoxidase (LP), and lysozymes (LYZ) [35, 39, 63]. Similarly, the content of many fat-soluble vitamins (like retinol, beta carotene, and tocopherol) and water-soluble vitamins (like Vitamin B_1, B_2 and B_{12}, Pyridoxine, and niacin) are also very high in colostrum as compared to normal milk in addition to lactoferrin, which is considered as a non-specific antibacterial factor [28].

This chapter explores the capability of BC as a natural immunotherapeutic agent and its usefulness in promoting human and animal health along with various methods for preservation and storage of BC for its extended use.

1.2 COLOSTRUM VERSUS IMMUNITY

Multiple immune-enhancing factors (like immunoglobulin recognized permeability fraction and fractions comprising enzyme proteins and unusual peptides, etc.) have been reported to be present in colostrum [2]. Immunomodulatory functions for BC are gradually being discovered. For example, cytokine transforming growth factor ß (TGF ß) has been reported to be present in colostrum [14] and it can suppress human lymphocyte proliferation [70].

Proline-rich polypeptides (PRPs) present in BC are recognized to ignite the debilitated immune system and pacify the overactive immune system due to allergies and autoimmune diseases, hence acting as immunomodulators [75] and promote T-cell development from early thymic pioneers, apart from affecting the and task of mature T-cells. PRPs are also known to suppress T-cells in a typical T-cell-independent humoral immune response [78]. Cytokines and lymphokines existing in the colostrum have been thought to swift the acute response for chemotaxis and protein synthesis and function as immune-modulators by virtue of their modulation of duration and strength of immune response [5].

Colostral leucocytes have shown to augment lymphocyte response to specific antigens and nonspecific mitogens along with their capability to increase antigen-presenting capacity [56, 57]. Cytokines present in colostrum have also anti-viral and anti-tumor activity [75]. Apart from this, the consumption of colostrum is deliberated to improve T-cell mediated response and develop B-cell immunity in neonatal animals [49, 50]. It encourages *in vitro* phagocytic activity of leukocytes from cattle [73] and humans [38]. Leucocytes present in BC motivate the production of interferon (IFN) and decrease the viral reproduction in addition to inhibiting the cellular wall permeation [34]. Lactalbumins in colostrum has shown action against many forms of cancer and viruses and are known to elevate the serotonin levels while decreasing cortisol levels [15].

Colostrum contains fat-soluble vitamins (vitamin A, D, and E), which serve in the development of specific immune functions in calves during the first week post-partum. The ß-carotene in colostrum improves lymphocytic blastogenic response, cytotoxic actions of natural killer (NK) cells, and cytokine production of macrophages [12]. Insulin-like growth factor (IGF)-I, due to its anabolic effects, is accountable for arbitrating the biological promotion of the activity of growth hormone. In a similar manner, IGF-II (although existing in low concentration in bovine milk as well as colostrum when compared to IGF-I) works in similar anabolic fashion as that of IGF-I thus reducing catabolism in food-deprived animals.

Colostrum is an amusing source of many other growth factors, which are generally produced in higher concentrations than the human colostrum [69] and is comparatively steady in extreme thermal and acidic settings and it is therefore capable of surviving the unadorned conditions like gastric acidity and milk-processing thus preserving their natural action. The BC has a plethora of other immunoregulatory and inflammatory biomolecules, such as interleukins (IL-1 beta, IL-2, and IL-6, interferon-gamma), tumor necrosis factor in addition to a wide range of non-antimicrobial amalgamations, which help in regulating infections and inflammations via various biological measures at the molecular level.

BC is a harmless and nominal nutraceutical that is needed for the prophylaxis and therapy of a number of transmissible diseases and immunity-related ailments. The literature has confirmed that the prophylactic role of colostrum for neo-natal animals especially calves is well established [23, 68]. BC has been in use as an established and effective nutraceutical for the enrichment of immune function in a miscellaneous range of animal species including cattle [64, 77], horse [33], pig [1], sheep [13], cats [65], mice [26], hamsters [13], ferrets [4], and lizards [25]. Nonetheless, the defensive and therapeutic role of colostrum especially in the young ones and adult animals is yet being recognized by many researchers and has much scope for further research. Some workers have found feeding of colostrum to be beneficial even after the absorptive phase, i.e., up to 12 hours since birth, because of local activity of colostral immunoglobulins in the intestine against *E. coli* [20, 37, 47].

Supplementary colostrum from day four of birth until weaning has shown to reduce mortality in calves with clinical diarrhea [40]. Colostrum also contains trypsin inhibitor [75] and the unaffected colostrum passes down the digestive tract thereby helping in preserving the well-being of epithelial lining and immune system and is thus defensive against many gastrointestinal (GI) disorders [2, 6] and enteric infections (like rotavirus, *Salmonella*, *E. coli*, *Shigella* spp. and *Clostridium* spp.). Colostrum and its constituents are also effective against a wide range of other community pathogens, including Streptococcus spp. *Staphylococcus aureus*, *Candida* spp., Feline immunodeficiency virus, and *Helicobacter pylori*.

Biodynamic oligosaccharides may be imperative in defense against pathogens and in endorsing the growth of favorable microflora in the colon. Pathogenic bacteria are invited and binded to oligosaccharides and glycoconjugate sugars of BC, which inhibits their access to the mucosal lining [46].

Disease precise hyperimmune bovine colostrum (HBC) having to counteract action against rotavirus, cryptosporidia, *Shigella*, and *H. pylori* can be useful adjunctive utility in circumstances concomitant with these infectious organisms [29]. It is efficient in handling hemorrhagic and other diarrheas in infants and curtailing the probability of disease progressing to hemolytic uremic syndrome [55]. Studies, with standardized colostrum preparations in animal models of human disease, have advocated the use of BC in human preclinical trials [71].

1.3 PROTECTIVE ROLE OF BOVINE COLOSTRUM (BC)

Colostrum has a protective role [6] in neonatal GI immunity due to its efficacy for stimulation and development of gut-associated lymphoid tissues (GALT) in infants, which is accountable for augmenting the juvenile gut immunity [10, 30]. Many studies have shown the effects of ingesting regular colostrum on immune function and battle against infections [8].

Secretory IgA, present in milk and colostrum, is the primary immunoglobulin for providing immunity to the mucosal membranes like intestines and this may be a contributory factor towards the shielding effects of these secretions having antimicrobial effects, such as virus neutralization and microbial agglutination. Additionally, inhibition of adherence and invasion of mucosal epithelial cells helps to produce a non-inflammatory extracellular and intracellular immune defense [7]. Bacterial enterotoxins have been reported to be neutralized by attachment and entry of secretory IgA into intestinal epithelial cells [19].

The clinical utility of BC has been substantiated by a plethora of clinical trials moderately supported by huge data comprising clinical case presentations and circumstantial outcomes. Till today, no effective commercialized vaccine is available against enterotoxigenic *Escherichia coli* (ETEC), which is the main cause of GI upsets in children, travelers as well as military persons in underdeveloped countries. This constraint along with hardships against the development of the ETEC vaccine has a compelling quest for an alternative and unconventional protective product including nutraceuticals like BC.

The use of hyperimmune BC (HBC), which is rich in immunoglobulins directed against GI infectious agents and microbes, has an added advantage of being natural and consumable having the least capacity towards disturbance of gut microflora [71, 72]. The hyperimmune colostrum produced by pre-partum vaccination of cows with the specific agent can be useful in the treatment and prophylaxis of wide range of diseases like diarrhea [69] by virtue of its constituents like lactoferrin, cytokines, and many growth factors, which promote growth and tissue repair in the GI tract apart from the provision of the immune barrier. The specific antibodies (chiefly IgG) against the important disease-causing dynamics have the ability to impede the motility of infectious agents like ETEC along with the endorsement of complement arbitrated damage. Oral provision of IgG boosted colostrum has supported to relieve insulin resistance and liver injury, which can be mainly attributed to the modifications in the activity of T-cells especially NK ones.

Nucleotides, nucleosides, and nucleobases relate to the non-protein-nitrogen (NPN) portion of the milk. Dietary nucleotides induce immune modulation (e.g., augmentation of antibody responses in case of infants and along with nucleosides help in iron absorption in the GI tract) and to affect desaturation and elongation rates in fatty acid synthesis, especially long-chain polyunsaturated fatty acids (PUFA) in initial phases of life. Ribonucleotides are also known to modulate the *in vitro* cell proliferation and apoptosis by altered constituents using human cell culture models [66]. The polyvalent concentrate from the BC has neutralizing abilities against lipopolysaccharide (LPS) endotoxins produced by the infection by gram-negative microbes. They also retard the endotoxemia in various experimentation models with relevant usefulness in various clinical practices and protocols.

1.4 THERAPEUTIC ROLE OF BOVINE COLOSTRUM (BC)

Bovine colostrum concentrates (BCCs) have been extensively used as a significant nutraceutical and therapeutic agents against GI pathogens. The decrease in the influx of LPS from the gut may be due to the abundant existence of bioactive constituents in BCC [72]. There is a common association among various types of endotoxemia developing during operative procedures and the ensuing surge in mediators of the acute phase reaction in surgical patients. Prophylactic application of colostrum is likely to reduce the incursion of lipo-polysaccharides originating from Gram-negative microbial pathogens across the gut, thereby inhibiting enterogenic endotoxaemia in patients [48]. Although *Cryptosporidium parvum* (an omnipresent enteric pathogen) is self-limiting in the healthy human adult, yet the infection may be life intimidating for those incapable to mount a satisfactory immune response. Many colostrum derivatives like long-chain fatty acids are naturally anti-cryptosporidium and may have utility for both, scrutinizing the mechanism of fatty acid facilitated inhibition of sporozoite-host cell adhesion and nutritional treatment of veterinary as well as human cases of cryptosporidiosis [67].

Colostrum has thus shown its worthiness in the amelioration of cryptosporidiosis in patients with AIDS (acquired immune deficiency syndrome). BC has been found useful in the prevention of recurring infections of the upper respiratory tract and diarrheal episodes, as it shrinks the total incidences and the need for hospitalization arising out of these respiratory, enteric or other GI ailments and thus has been recommended to be delivered as a beneficial therapeutic alternative in young ones suffering from frequent URTI (upper respiratory tract infection) and diarrheal episodes [43, 58]. Colostral immunoglobulins have also been reported to decrease the level of serum cholesterol in subjects suffering from hyper-cholesterolemia, commonly cited as a hazard issue for many diseases related to the cardiovascular system [16].

1.5 PRESERVATION AND STORAGE OF BOVINE COLOSTRUM (BC)

With the expansion in the quest for healthier nutrition and lifestyle, people are increasingly becoming conscious about the role of nutrition in the occurrence of various lifestyle-related disorders. More number of people is impressively interested in foodstuffs and nutraceuticals having healthier paybacks. During recent times, colostrum-based products have been gaining much attention of customers to adapt to functional foods with additional health benefits, especially those related to the prevention of nutrition-related disorders [42]. The process of production of harmless and unwavering colostrum products to be safely consumed by people is highly complicated owing to the extremely inconstant bacteriological quality of uncooked colostrum and its heat sensitivity. At present, the colostrum-based products are available in the market in two forms: spray-dried powder or lyophilized powder. These types of dehydration processes facilitate consequent handling, processing, prolongation of the keeping quality, and cost reduction [11].

The stability of immunoglobulins is an important factor, which needs to be evaluated under different processing and storage conditions for the specific use of BC as nutraceuticals. Many reports exist, which describe the steadiness of immunoglobulins in liquid milk or colostrum after various types of processing such as homogenization, heat, or ultrasonic application [9, 22, 36]. It is also of great significance to select the appropriate storage environments and packaging materials to keep food qualities and protracted shelf-life post drying. Apart from packaging material, the keeping quality of any food powder is highly dependent on many other extrinsic factors like temperature and relative humidity.

More recently, plastic film laminates and aluminum foil have been accustomed as substitutes for the old-fashioned cans made of tin [74]. Stretchable packaging also presents itself as an alternative to conventional one in addition to assisting in cutting down the costs related to packaging and logistics and thereby minimizing wastages [76].

1.5.1 HEAT TREATMENT

On the farm, heat treatment of colostrum at 60°C for 60 min may lower colostrum bacterial load however preserving the IgG content of the BC. Calves receiving colostrum after heat treatment present higher serum concentration of total serum proteins and IgG level in addition to the higher ability of immunoglobulin absorption when compared to calves, which are provided with colostrum in raw form without heat application [27]. There is no reported consequence of heat application on serum concentrations of other constituents like beta-carotene, vitamin A and E : cholesterol ratio, and immunoglobulins like IgA and IgM (immunoglobulin M), etc. On the contrary, better IgG absorption has been endorsed to the decreased bacterial load in the colostrum post-thermal treatment. Calves offered colostrum with low bacterial load also suffer from scarcer episodes of clinical conditions like diarrhea and respiratory disorders like pneumonia thus detailing the convincing evidence for the profits of curtailing microbial load in the colostrum meant for calves as compared to raw colostrum. The IgM seems to be the most sensitive of the Igs to processing while IGF-1 and IGF-ß2 remain unaltered by heat application or lyophilization [18].

1.5.2 PASTEURIZATION

While on one side, colostrum is thought to be a significant nutritional source having health and immunity benefits, it is also an important niche of disease-causing pathogenic microbes like *Mycobacterium* spp., *Mycoplasma* spp., *Salmonella* spp., coliforms, and bovine leukemia virus. Pasteurization of colostrum and milk may significantly cut morbidity and mortality in calves compared with calves receiving non-pasteurized colostrum and milk (frozen at –20°C and previously reheated up to 40°C), respectively, during the first three weeks of life [3]. In addition, it also reduces the occurrences of many diseases such as paratuberculosis.

Earlier methods for pasteurization of colostrum utilized similar high-temperature methods as those used for pasteurization of milk, which often produced undesirable outcomes like denatured IgGs and escalation in viscosity. However, the latest studies have indicated that these issues can be sidestepped by applying a lower temperature with extended time for pasteurization of colostrum. It may be quite appropriate to heat colostrum up to 60°C for one hour for maintenance of IgG and other attributes while simultaneously decreasing pathogenic microorganisms. [41].

1.5.3 CHEMICAL PRESERVATION

Chemical preservatives are not reputed for preserving colostrum satisfactorily. In contrast, freezing has been attributed to the preferred method.

1.5.4 FREEZING AND FREEZE DRYING/LYOPHILIZATION

Freezing of colostrum after pasteurization has not been reported to cause alterations in the concentration of IgG and lactoferrin. On the other hand, freeze-drying of pasteurized colostrum has also without any deleterious effect on IgG, IgM, IGF-1, and lactoferrin concentrations. Similarly, storing colostrum at minus 20°C for three months has no reported effect on IgG and IgM levels, which are probably decreased in freeze-dried colostrum when it is kept above 7°C storage temperature for three months [17]. Freeze drying hence has been advocated for the retention of a greater percentage of bioactive immunoglobulins.

1.5.5 SPRAY DRYING

The external factors (which affect attributes of BC powder produced by spray drying, during storage like flavor, color) and bioactive components are of concern to food manufactures and customers. Browning, development of a hostile musty flavor due to Maillard reaction [32, 51] along with oxidation of lipids [32, 45], lumpiness, and loss of bioactive components owing to protein denaturation [62, 74] are common quality deterioration of SDBC (spray-dried bovine colostrum) powder by unsuitable packing and storage conditions. For the assessment of processing ability, steadiness, and keeping quality forecasts, the concept of water activity together with the concept of glass transition may be undertaken [52–54, 60–62]. The controlled drying conditions during spray drying can be modified to preserve a practically superior percentage of immunoglobulins.

1.5.6 HIGH PRESSURE PROCESSING (HPP)

The high pressure processing (HPP) of colostrum has been established as an encouraging conservation technique for the colostral Igs among various novel technologies available in the market. The processing by utilizing membrane can be engaged to affect the constituents of various preparations. HPP of BC is able to conserve an acceptable IgG level while reducing viral (bovine herpesvirus (BHV) type 1, feline calicivirus) and bacterial counts (*E. coli* and *S. dublin*) and increased colostrum viscosity, which often requires dilution with water when intended for calf-feeding [21]. There is no difference between serum IgG levels in calves fed colostrum post-pressure-processing and those fed colostrum post-heating apart from having similar but lower efficiency of absorption. Additional research is needed to evaluate this technology for on-farm use.

1.5.7 MICROFILTRATION (MF)

Because of the inactivation of valuable bioactive components by the conventional heat treatment methods, the microfiltration (MF) by utilizing cross-flow at mild temperatures may be an encouraging opportunity for the colostral treatment, thereby arriving at a product with extended shelf-life along with sufficiently viable bioactivity. Multiple cross-flow MF experiments have established that cross-flow MF can be utilized for processing of BC and by tolerating protein denaturation to some magnitude, an amalgamation of choice between pore sizes and application of mild heat could be a rational choice.

1.5.8 MICROFILTRATION (MF) TREATMENT IN COMBINATION WITH HIGH PRESSURE PROCESSING (HPP)

This technique is a unique and combined process encompassing HPP succeeding MF, where skimmed colostrum soiled with innumerable bacterial species is exposed to cross-flow MF at specific pore size followed by decreasing the saturates of MF having leftover bacterial load to untraceable amounts by HPP. This process in general results in the reduction of IgG. While utilizing this method for colostrum with high bacterial loads, this HPP has proved to cause lesser damages to IgG content than the thermal treatment when performed at different variable conditions providing comparable efficacy of decreasing the bacterial load in the treated colostrum [24].

1.5.9 COLOSTRUM SILAGE

It is a fermentation process utilizing an anaerobic approach for the conservation of colostrum produced in excess at the farm, which can be fed to calves as a milk replacement later on and a suitable substitute to storage by the freezing method and future delivery to the animal. This type of anaerobic fermentation method serves as a worthy substitute for the storage of surplus colostrum without wastage and decomposition while keeping viable immunoglobulins capable of transferring passive immunity in the neonatal calves [59]. The fermentation process leads to the development of favorable microorganisms like lactic acid bacteria (LAB) and a reduction in pH, which helps in preserving colostrum at ambient temperature. Faster drop in pH and lesser average rates, followed by greater titratable acidity and concentration of lactic acid results when higher storage temperature during fermentation is present.

The rapidity and intensity of development of microbial inhabitant and degradation of the nutritional considerations, like lactose and casein, are right away subjective to the temperature at which the colostrum is fermented while the changes in the components during fermentation under chiller temperatures are more even. Non-protein nitrogen (NPN) portion of the treated colostrum upsurge across the fermentation time, with a superior surge at upper-temperature levels representing higher proteolytic action. Anaerobically fermented colostrum (colostrum silage) as a milk replacer has not been reported appropriate for dairy calves, especially when the fermentation process has been performed under warm climatic conditions.

1.6 SUMMARY

The BC contains various antibodies, antioxidants apart from an excess of bioactive immune constituents, such as tissue growth factors (TGF) and PRP, antimicrobial fractions like lactoferrin, cytokines, lymphokines, etc., which have shown activities against various infectious and non-infectious diseases. BC is increasingly being used for the treatment of various diseases like pneumonia, diarrhea, septicemia, gastroenteritis, etc. This chapter may be useful in exploring the capability of BC as a natural immunotherapeutic agent and its usefulness in promoting human and animal health along with various methods for preservation and storage of BC for its extended use.

KEYWORDS

- **bovine colostrum**
- **colostrum silage**
- **heat treatment**
- **high pressure processing**
- **immune milk**
- **lyophilization**
- **prophylaxis**
- **spray drying**

REFERENCES

1. Alexander, A. N., & Carey, B. V., (1999). Oral IGF-I enhances nutrient and electrolyte absorption in neonatal piglet intestine. *Am. J. Physiol., 271*, 619–625.
2. Antonius, C. M., Van, H., Kussendrager, K. D., & Steijns, J. M., (2000). *In vitro* antimicrobial and antiviral activity of components in bovine mild and colostrums involved in non-specific defense. *Br. J. Nutr., 84*(1), 127–134.
3. Armengol, R., & Fraile, L., (2016). Colostrum and milk pasteurization improve health status and decrease mortality in neonatal calves receiving appropriate colostrum ingestion. *J. Dairy Sci.,* *99*(6), 4718–4725. doi: 10.3168/jds.2015-10728.
4. Bitzan, M. M., Gold, B. D., & Phil-Pott, D. J., (1998). Inhibition of *Helicobacter pylori* and *Helicobacter mustelae* binding to lipid receptors by bovine colostrum. *J. Infect. Dis., 177*, 955–961.
5. Bocc, V., Von, B. K., & Corradeschi, F., (1991). What is the role of cytokines in human colostrums? *J. Bio. Regulat. Homeo. Agents, 3*, 121–124.
6. Bogstedt, A., Johansen, K., Hatta, H., Kim, M., Casswall, T., Sevensson, L., & Hammarstrom, L., (1996). Passive immunity against diarrhea. *Acta. Paediatr., 85*, 125–128.

7. Brandtzaeg, P., & Johansen, F. E., (2007). IgA and intestinal homeostasis. In: Kaetzel, C. S., (ed.), *Mucosal Immune Defense: Immunoglobulin A* (pp. 221–268). Springer, New York, USA.
8. Brinkworth, G. D., & Buckley, J. D., (2003). Concentrated bovine colostrum protein supplementation reduces the incidence of self-reported symptoms of upper respiratory tract infections in adult males. *Eur. J. Nutr., 42*(4), 228–232.
9. Cao, J., Wang, X., & Zheng, H., (2007). Comparative studies on thermo resistance of protein G-binding region and antigen determinant region of immunoglobulin G in acidic colostral whey. *Food Agric Immunol., 18*(1), 17–30. doi: 10.1080/09540100701220267.
10. Carbonare, S. B., Silva, M. L., Palmeira, P., & Carmeiro-Sampaio, M. M., (1997). Human colostrum IgA antibodies reacting to enteropathogenic *Escherichia coli* antigens and their persistence in the feces of breast-fed infant. *J. Diarr. Dis. Res., 15*(2), 53–58.
11. Chelack, B. J., Paul, S., Morley, & Haines, D. M., (1993). Evaluation of methods for dehydration of bovine colostrum for total replacement of normal colostrum in calves. *Can. Vet. J., 34*(7), 407–412.
12. Chew, B. P., (1993). Role of carotenoids in the immune response. *J. Dairy Sci., 76*, 2804.
13. Clarkson, M. J., Faull, W. H., & Kerry, J. B., (1985). Passive transfer of clostridial antibodies from bovine colostrum to lambs. *Vet. Rec., 116*, 467–469.
14. Cross, M. L., & Gill, H. S., (2000). Immunomodulatory properties of milk. *Br. J. Nutr., 84*, S81–S89.
15. Dichtelmuller, W., & Lissner, R., (1990). Antibodies from colostrum in oral immunotherapy. *J. Clin. Bio. Chem., 28*, 19–23.
16. Earnest, C. P., Jordan, A. N., Safir, M., Weaver, E., & Church, T. S., (2005). Cholesterol-lowering effects of bovine serum immunoglobulin in participants with mild hypercholesterolemia. *Am. J. Clin. Nutr., 81*, 792–798.
17. El-Fattah, A. M. A., Hassan, F., Rabo, R., Samia, M., & El-Kashef, H. A., (2014). Preservation methods of buffalo and bovine colostrum as a source of bioactive components. *International Dairy Journal, 39*(1), 24–27.
18. Elfstrand, L., Mansson, H. L., Paulsson, M., Nyberg, L., & Akesson, B., (2002). Immunoglobulins, growth factors and growth hormone in bovine colostrum and the effects of processing. *International Dairy Journal, 12*(11), 879–887. https://doi.org/10.1016/S0958-6946(02)00089-4.
19. Fernandez, M. I., Pedron, T., Tournebize, R., & Olivo-Marin, J. C., (2003). Anti-inflammatory role for intracellular dimeric immunoglobulin a by neutralization of lipopolysaccharide in epithelial cells. *Immunity, 18*, 739–749. doi: 10.1016/S1074-7613(03)00122-5.
20. Fisher, E. W., & Martinez, A. A., (1976). Colibacillosis in calves. *Vet. Annual, 16*, 22–29.
21. Foster, D. M., Poulsen, K. P., Sylvester, H. J., Jacob, M. E., Casulli, K. E., & Farkas, B. E., (2016). Effect of high-pressure processing of bovine colostrum on immunoglobulin G concentration, pathogens, viscosity, and transfer of passive immunity to calves. *J. Dairy Sci., 99*(11), 8575–8588. doi: 10.3168/jds.2016-11204.
22. Fukumoto, L. R., Li-Chan, E., Kwan, L., & Nakai, S., (1994). Isolation of immunoglobulins from cheese whey using ultra filtration and immobilized metal affinity chromatography. *Food Res. Int., 27*(4), 335–348. doi: 10.1016/0963-9969(94)90189-9.
23. Gay, C. C., Mc Guire, T. C., & Parish, S. M., (1983). Seasonal variation in passive transfer of immunoglobulin G1 to new born calves. *J. Am. Vet. Med. Assoc., 183*(5), 566–568.

24. Gosch, T. S., Apprich, W., Kneifel, S., & Novalin, A., (2014). Combination of microfiltration and high pressure treatment for the elimination of bacteria in bovine colostrum. *International Dairy Journal*, *34*(1), 41–46.
25. Graczyk, T. K., Cranfield, M. R., & Bastwick, E. F., (1999). Hyperimmune bovine colostrum treatment of moribund Leopard geckos (*Eubiepharis macularius*) infected with *Cryptosporidium* sp. *Vet. Res.*, *30*, 377–382.
26. Jenkins, M. C., O'Brien, C., Trout, J., Guidry, A., & Fayer, R., (1999). Hyperimmune bovine colostrum specific for recombinant *Cryptosporidium parvum* antigen confers partial protection against cryptosporidiosis in immune suppressed adult mice. *Vaccine*, *17*, 2453–2460.
27. Johnson, J. L., Godden, S. M., Molitor, T., Ames, T., & Hagman, D., (2007). Effects of feeding heat-treated colostrum on passive transfer of immune and nutritional parameters in neonatal dairy calves. *J. Dairy Sci.*, *90*(11), 5189–5198.
28. Kehoe, S. I., Jayarao, B. M., & Heinrichs, A. J., (2007). A survey of bovine colostrum composition and colostrum management practices on Pennsylvania dairy farms. *J. Dairy Sci.*, *90*, 4108–4116.
29. Kelly, G. S., (2003). Bovine colostrum: A review of clinical uses. *Altern. Med. Rev.*, *8*(4), 378–394.
30. Korhonen, H., Marnila, P., & Gill, H. S., (2000). Milk immunoglobulin's and complement factors. *Br. J. Nutr.*, *84*(1), 75–80.
31. Korhonen, H., Syvaoja, E. L., & Ahola-Luttila, H., (1995). Bactericidal effect of bovinenormal and immune serum colostrum and milk against *Helicobacter pylori*. *J. Appl. Bacter.*, *78*, 655–662.
32. Kumar, P., & Mishra, H. N., (2004). Storage stability of mango soy fortified yoghurt powder in two different packaging materials: HDPP and ALP. *J. Food Eng.*, *65*(4), 569–576. doi: 10.1016/j.jfoodeng.2004.02.022.
33. Lavoie, J. P., Spensley, M. S., Smith, B. P., & Mihalyi, J., (1989). Absorption of bovine colostral immunoglobulins G and M in newborn foals. *Am. J. Vet. Res.*, *50*, 1598–1603.
34. Lawton, J. W., Shortstride, K. F., & Wong, R., (1979). Interferon synthesis by human colostral leukocytes. *Arch. Dis. Childhood.*, *54*, 127–130.
35. Levay, P. F., & Viljoen, M., (1995). Lactoferrin: A general review. *Haematologic.*, *80*, 252–267.
36. Li-Chan, E., Kummer, A., Losso, J. N., Kitts, D. D., & Nakai, S., (1995). Stability of bovine immunoglobulins to thermal treatment and processing. *Food Res. Int.*, *28*(1), 9–16. doi: 10.1016/0963-9969(95)93325-O.
37. Logan, E. F., & Penhale, W. J., (1971). Studies on the immunity of the calf to colibacillosis. Part-1: The influence of colostral whey and immunoglobulin fractions on experimental colisepticemia. *Vet. Rec.*, *88*, 222–228.
38. Loimoranta, V., Nuutila, J., Marnila, P., Tenovuo, J., Korhonen, H., & Lilius, E. M., (1999). Colostral proteins from cows immunized with *Streotococcus mutans/S. sobrinus* support the phagocytosis and killing of mutans *Streptococci* by human leucocytes. *J. Med. Microbiol.*, *48*, 917–926.
39. Lonnerdal, B., & Lyer, S., (1995). Lactoferrin: Molecular structure and biological function. *Annl. Rev. Nutr.*, *15*, 93–110.
40. McDougall, S., Cullum, A., & Parkinson, T., (1999). Effect of feeding colostrum of high immunoglobulin content on calf growth rates and disease. In: *Proceedings of the 16th Annual Seminar of the Society of Dairy Cattle Veterinarians of the NZVA* (Vol. 192, pp.

33–45). NZVA conference, Nelson, New Zealand, Publication-Veterinary-Continuing-Education, Massey University.
41. McMartin, S., Godden, S., Metzger, L., & Feirtag, J., (2006). Heat treatment of bovine colostrum. Part-I: Effects of temperature on viscosity and immunoglobulin g level. *J. Dairy Sci.*, *89*, 2110–2118.
42. Mehra, R., Marnila, P., & Korhonen, H., (2006). Milk immunoglobulins for health promotion. *Int. Dairy J.*, *16*(11), 1262–1271. doi: 10.1016/j.idairyj.2006.06.003.
43. Menon, R. P., Lodha, R., & Kabra, S. K., (2010). Bovine colostrum in pediatric respiratory diseases: A systematic review. *Indian J. Pediatr.*, *77*, 108–109.
44. Mero, A., Mikkulainen, H., Rishi, J., Pakkanen, R., Aalto, J., & Takala, T., (1997). Effects of bovine colostum supplementation on serum IGF-I, IgG, hormone and saliva IgA during training. *J. Appl. Physiology*, *83*(4), 1144–1158.
45. Nielsen, B. R., Stapclteldt, H., & Skibsted, L. H., (1997). Differentiation between 15 whole milk powders in relation to oxidative stability during accelerated storage: Analysis of variance and canonical variable analysis. *Int. Dairy J.*, *7*(8–9), 589–599. doi: 10.1016/S0958-6946(97)00046-0.
46. Ogra, S. S., & Ogra, P. L., (1978). Immunologic aspects of human colostrum and milk. *J. Pediatr.*, *92*, 546–549.
47. Okabe, T., (1983). Protective effects of antibody against intestinal invasion by *Escherichia coli. Microbiol. Immunol.*, *27*(4), 303–310.
48. Orth, K., Knoefel, W. T., Van, G. M., & Matuschek, C., (2013). Preventively enteral application of immunoglobulin enriched colostrums milk can modulate postoperative inflammatory response. *Eur. J. Med. Res.*, *18*(1), 50–53. doi: 10.1186/2047-783X-18-50.
49. Pabst, H. F., & Spady, D. W., (1990). Effect of breast feeding on antibody response to conjugate vaccine. *Lancet*, *336*, 269–270.
50. Pabst, H. F., Godel, J., Grace, M., Cho, H., & Spady, D. W., (1989). Effect of breast feeding on immune response to BCG vaccination. *Lancet*, *1*, 295–297.
51. Pereyra, G. A. S., Naranjo, G. B., Leiva, G. E., & Malec, L. S., (2010). Maillard reaction kinetics in milk powder: Effect of water activity at mild temperatures. *Int. Dairy J.*, *20*(1), 40–45. doi: 10.1016/j.idairyj.2009.07.007.
52. Rahman, M. S., (2006). State diagram of foods: Its potential use in food processing and product stability. *Trends Food Sci. Technol.*, *17*(3), 129–141. doi: 10.1016/j.tifs.2005.09.009.
53. Rahman, M. S., (2009). Food stability beyond water activity and glass transition: Macro-micro region concept in the state diagram. *Int. J. Food Prop.*, *12*(4), 726–740. doi: 10.1080/10942910802628107.
54. Rahman, M. S., (2010). Food stability determination by macro-micro region concept in the state diagram and by defining a critical temperature. *J. Food Eng.*, *99*(4), 402–416. doi: 10.1016/j.jfoodeng.2009.07.011.
55. Rawal, P., Gupta, V., & Thapa, B. R., (2008). Role of colostrum in gastrointestinal infections. *Indian J. Pediatr.*, *75*(9), 917–921. doi: 10.1007/s12098-008-0192-5.
56. Reber, A. J., Donovan, D. C., Gabbard, J., & Galland, K., (2008). Transfer of maternal colostral leukocytes promotes development of the neonatal immune system I: Effects on monocyte lineage cells. *Vet. Imm. Immunopathol.*, *123*, 186–196.
57. Reber, A. J., Donovan, D. C., Gabbard, J., & Galland, K., (2008b). Transfer of maternal colostral leukocytes promotes development of the neonatal immune system. Part-II: Effects on neonatal lymphocytes. *Vet. Imm. Immunopathol.*, *123*, 305–313.

58. Saad, K., Abo-Elela, M. G. M., El-Baseer, K. A. A., & Ahmed, A. E., (2016). Effects of bovine colostrum on recurrent respiratory tract infections and diarrhea in children. *Medicine (Baltimore), 95*(37), e4560. doi: 10.1097/MD.0000000000004560.
59. Saalfeld, M. H., Pereira, D. I., Borchardt, J. L., & Sturbelle, R. T., (2014). Evaluation of the transfer of immunoglobulin from colostrum anaerobic fermentation (colostrum silage) to newborn calves. *Anim. Sci. J., 85*(11), 963–967. doi: 10.1111/asj.12229.
60. Sablani, S., Syamaladevi, R., & Swanson, B., (2011). A review of methods, data, and applications of state diagrams of food systems. *Food Eng, Rev., 2*(3), 168–203. doi: 10.1007/s12393-010-9020-6.
61. Sablani, S. S., Al-Belushi, K., Al-Marhubi, I., & Al-Belushi, R., (2007). Evaluating stability of vitamin C in fortified formula using water activity and glass transition. *Int. J. Food Prop., 10*(1), 61–71. doi: 10.1080/10942910600717284.
62. Sablani, S. S., Kasapis, S., & Rahman, M. S., (2007). Evaluating water activity and glass transition concepts for food stability. *J. Food Eng., 78*(1), 266–271. doi: 10.1016/j.jfoodeng.2005.09.025.
63. Sanchez, L., Clavo, M., & Brock, J. H., (1992). Biological role of lactoferrin. *Arch. Dis. Child.*, 67, 657–661.
64. Sasaki, M., Davis, C. L., & Larson, B. L., (1976). Production and turnover of IgG1 and IgG2 immmunoglobulins in the bovine around parturition. *J. Dairy Sci., 59*, 2046–2055.
65. Sato, R., Inanami, O., Tanaka, Y., Takase, M., & Naito, Y., (1996). Oral administration of bovine lactoferrin for treatment of intractable stomatitis in feline immunodeficiency virus (FIV)-positive FIV-negative cats. *Am. J. Vet. Res., 57*, 1443–1446.
66. Schlimme, E., Martin, D., & Meisel, H., (2000). Nucleosides and nucleotides: Natural bioactive substances in milk and colostrum. *Br. J. Nutr., 84*(1), S59–68.
67. Schmidt, J., & Kuhlenschmidt, M. S., (2008). Microbial adhesion of *Cryptosporidium parvum*: Identification of a colostrum-derived inhibitory lipid. *Mol. Biochem. Parasitol., 162*(1), 32–39. doi: 10.1016/j.molbiopara.2008.06.016.
68. Smith, T., & Little, R. B., (1922). Cow serum as a substitute for colostrum in new born calves. *J. Exp. Med., 36,* 453–468.
69. Steele, J., Sponseller, J., Schmidt, D., Cohen, O., & Tzipori, S., (2013). Hyperimmune bovine colostrum for treatment of GI infections: A review and update on *Clostridium difficile. Hum. Vaccine Immunother., 9*, 1565–1568. doi:10.4161/hv.24078.
70. Stoeck, M., Ruegg, C., Miescher, S., Carrel, S., Cox, D., Von, F. V., & Akam, S., (1989). Comparison of the immunosuppressive properties of milk growth factor and transforming growth factors beta 1 and beta 2. *J. Immunol., 143*, 3258–3265.
71. Struff, W. G., & Sprotte, G., (2007). Bovine colostrum as a biologic in clinical medicine: A review. Part I: Biotechnological standards, pharmacodynamic and pharmacokinetic characteristics and principles of treatment. *Int. J. Clin. Pharmacol. Ther., 45*, 193–202.
72. Struff, W. G., & Sprotte, G., (2008). Bovine colostrum as a biologic in clinical medicine: A review. Part II: Clinical studies. *Int. J. Clin. Pharmacol. Ther., 46*, 211–225. doi: 10.5414/CPP46211.
73. Sugisawa, H., Itou, T., & Sasaki, T., (2001). Promoting effect of colostrum on the phagocytic activity of bovine polymorphonuclear leukocytes *in vitro*. *Biol. Neonate, 79*, 140–144.
74. Tehrany, E. A., & Sonneveld, K., (2010). Packaging and the shelf life of milk powders. In: Gordon, L. R., (ed.), *Food Packaging and Shelf Life-a Practical Guide* (pp. 128–141). CRC Press, Boca Raton, Florida.

75. Thapa, B. R., (2005). Health factors in colostrum. *Ind. J. Pediatrics., 74*, 579–581.
76. Twede, D., & Goddard, R., (1998). *Packaging Materials* (p. 2) Surrey: Pira International.
77. Tyler, J. W., Steevens, B. J., Hostetler, D. E., Hotle, J. M., & Denbigh, J. L., (1999). Colostral immunoglobulin concentration in Holstein and Guernsey cows. *Am. J. Vet. Res., 60*, 1136–1139.
78. Zimecki, M., (2008). Proline-rich polypeptide from bovine colostrum: Colostrinin with immunomodulatory activity. *Adv. Exp. Med. Biol., 606*, 241–250.

CHAPTER 2

ESSENTIAL OIL-BASED COMBINED PROCESSING METHODS TO PURSUE SAFER FOODS

MANJU GAARE, SHAIK ABDUL HUSSAIN, and
SANTOSH K. MISHRA

ABSTRACT

The application of essential oils (EOs) in modern food processing has evolved as a promising strategy to meet the growing demand of consumers for natural products without compromising food safety. However, food applications are limited due to the characteristic flavor and sometimes cause an adverse effect on sensory. Therefore, many studies have evaluated the approach of combining EOs with industrial processes over foodborne bacteria. This chapter examines the EOs combined food preservation methods that have the potential to be further optimized as effective food processing treatments to achieve food safety.

2.1 INTRODUCTION

Over the last decades, the processed food industry has experienced substantial growth in demand for convenient, healthy, and minimally-processed food products. On one hand, foodborne illnesses and outbreaks, hospitalizations, and quality losses have also increased owing to improper handling after processing and storage and the emergence of antibiotic resistance pathogens. Heat treatment and modern food preservation technologies are traditionally relied upon to produce a wide range of commercially safe products. However, when applied at the required intensity to guarantee shelf-life and safety of foods, significant damage to appearance, aroma, and destruction of nutrients may take place. In addition, the application of antimicrobials of synthetic nature as food preservatives are not readily accepted by consumers who are shifting towards less or synthetic free and minimally processed food products. Hence, Plant EOs for having derived from natural sources carry the image of natural and the majority with GRAS (generally regarded as safe) status for food application can be promising alternative [5, 14, 71].

Essential oils (EOs) and extracts of herbs, spices, and other plants are sources of bioactive molecules and are widely used traditionally in different applications. Most applications are in enhancing the flavor of foods, perfumery, and aromatherapy. In the last few decades, a growing tendency is to meet consumer demand by replacing synthetic food additives with EOs. They may be used in foods mainly to preserve foods preventing spoilages (food biopreservation) and to control the pathogenic microorganisms (food safety). The considerable studies have been conducted for potential utilization as natural antimicrobials in foods such as fruits, juices, milk products, vegetables, sausages, and other meat products. Even a few EOs-based preservatives 'DMC Base Natural,' DOMCA S.A., Spain, and Protecta™ One and Protecta™ Two, Bacaria Corp. Apopka, FL-USA having GRAS tag is commercially available. Despite multifaceted potentials of EOs, their commercial application in foods is not significant as effective concentrations sometimes result in unacceptable damages to taste and odor [64].

This chapter focuses on the role of EOs along with other preservation methods of foods and natural antimicrobials used in food processing that might help the food industry to attain greater inactivation using the lowest doses, thus to confirm the food safety and final quality.

2.2 ESSENTIAL OIL (EO)

EOs is hydrophobic but ether, alcohol, and fat-soluble, high refractive index, and optimal rotation and relative density are less than one, but exceptions exist. These oils are volatile, generally colorless, and remain liquid at room temperature, possess a typical odor of plant or herb from which these are extracted. EOs is contained within the plant, which are primarily responsible for the characteristic essence that the plant emits. There are more than 2,000 species belonging to 60 different families of plants well-known for bearing EOs. Among these, >300 plant species are commercially important [37]. Practically any part of a plant (viz., flowers (clove, lavender, etc.), leaves (thyme, mint, bay, etc.), grasses (lemongrass), seeds (coriander), rhizomes (ginger), wood (cedar), bulb (garlic), fruits (fennel, anise, citrus, etc.), and bark (cinnamon)) can be used to obtain the EO. Different techniques are known for the extraction of EOs viz., steam-distillation, solvent-extraction, hydrodistillation, ultra-sonic extraction, microwave-assisted extraction, supercritical carbon dioxide extraction, etc. Among these, extraction through steam distillation is most popular for production at commercial scale [146].

2.3 COMPOSITION OF ESSENTIAL OIL (EO)

EOs are composed of more than 120 compounds of low molecular weight (MW) (<500 Da). The compound present in large quantities (>80%) is a major constituent; the compounds in small quantities are minor constituents. The composition of EO is highly variable, which influences the biological activity both qualitatively and quantitatively. This in turn is determined by factors such as plant part, genetics, season, stage of harvesting, geographical area, method of extraction, and the environment factors [69]. EOs mainly has two chemicals: terpenes and phenylpropanoids; both have different precursors and biosynthesized through mevalonate and mevalonate-independent and shikimate pathway, respectively [37]. Terpenes are more abundant and frequent in EOs and so far >1000 compounds belonging to this family are known. Terpene compounds can be subdivided into terpenes with a hydrocarbon structure mainly monoterpene, sesquiterpene, and diterpenes; and terpenoids are enzymatic derivatives of terpenes mainly alcohol (cinnamylalcohol), esters (γ-tepinyl acetate), ketone (menthone) and aldehydes (citronellal). Coriander, mint, orange, rosemary, thyme, peppermint, and sage are examples of such plants (Figure 2.1). Phenylpropanoids are phenols or phenol ethers such as cinnamaldehdye, eugenol, vanillin, isoeugenol; and safrole have been thoroughly studied [26]. The typical examples of such plants are cloves, cinnamon, nutmeg, and fennel [70]. Many of these have been explored as natural preservatives in sweets, juices, beverages, and other food preparations.

2.4 ANTIBACTERIAL ACTIVITY OF ESSENTIAL OILS (EOS)

The EOs being hydrophobic compounds, the primary site of action on bacteria will be cell membranes. They dissolve into the lipid portion of the cell membrane disrupting its fluidity thereby increasing the permeability [112]. This irreversible disruption causes disruption in pH gradient and leakage of divalent cations and other cellular molecules. The leakage of cellular contents beyond critical limit of the bacterial cell can lead to cell death [18, 50, 145].

Moreover, EOs depending upon the composition and outcome of interactions between individual components can act on single or multiple targets simultaneously [32, 50, 129]. Therefore, sometimes the overall activity cannot be attributed to any one constituent. Here many researchers argue on preference to whole EO over purified individual active component [31, 76].

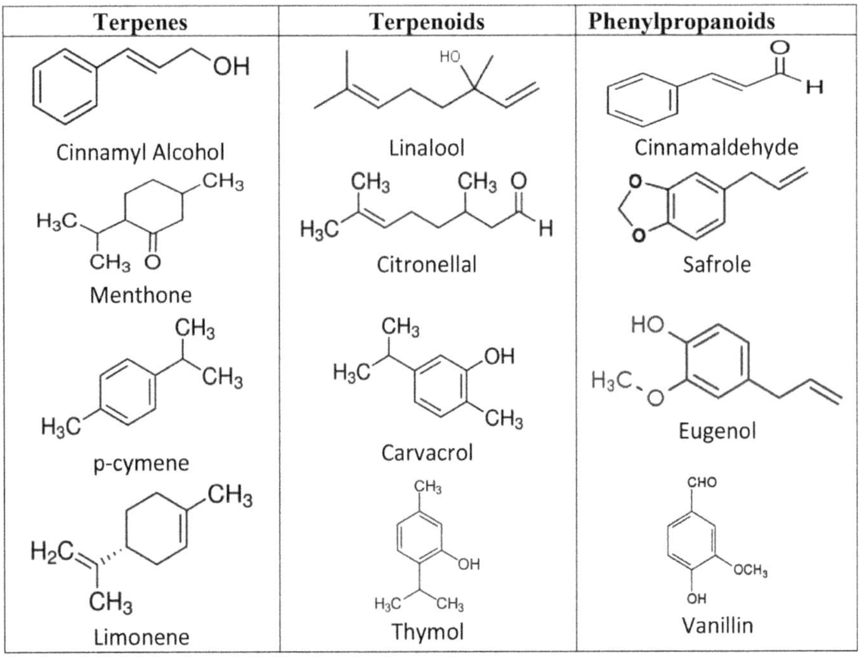

Terpenes	Terpenoids	Phenylpropanoids
Cinnamyl Alcohol	Linalool	Cinnamaldehyde
Menthone	Citronellal	Safrole
p-cymene	Carvacrol	Eugenol
Limonene	Thymol	Vanillin

FIGURE 2.1 Chemical structure of biocomponents in selected essential oils.

The antibacterial activity and their concerning mode of action is believed to be associated with the chemical structure of EO. The hydroxyl group in the phenolic compounds and relative positive and lipophilic character of the hydrocarbon skeleton and functional groups of EOs are important in exerting antibacterial activity of the components [141]. Therefore, EOs characterized by phenols and aldehyde (such as carvacrol, thymol, cinnamaldehyde, and eugenol) generally have high antibacterial activity on a wide range of bacteria comprising not just pathogenic but also spoilage causing [77]. Also, recent studies demonstrated that EOs have significant antimicrobial potential against antibiotic resistant bacteria (ARB) [52]. Other EOs characterized by alcohols, ketones, esters, and hydrocarbons also has vital antibacterial activities [77]. The low susceptibility is linked to obstruction offered by outer membrane for diffusion of EOs [50]. Interestingly, terpineol-4-ol was stronger on *S. enteritidis* and *E. coli* than *L. monocytogenes* [2]. Nevertheless, the antibacterial activity is linked to number and concentrations of low molecular components present in the EO and their interactions. The specific mechanisms and underlying associated molecular aspects are needed to be further explored.

2.5 COMBINED APPROACH BASED ON ESSENTIAL OIL (EO)

The hurdle approach is a combined application of several food preservation treatments intended to enhance the lethal effects. When lethality of combined process is higher than individual processes together, the outcome is synergy whereas lethality of combined processes is less than the sum of separate applications (Figure 2.2). The most successful combinations should be those, which results in higher lethality than separate applications. Nonetheless, in food applications, synergistic effect is preferred over additive effect [88, 122]. In this sense, EOs can be combined with thermal or non-thermal processes or antimicrobials to increase the lethality of processes. The greater inactivation is presumably due to the stress exerted at multiple targets simultaneously causing damages to several functions of the cell [2, 55, 79].

The successful combinations with the resultant effectiveness should be able to reduce the effective concentrations and consequently minimize undesirable change on organoleptic acceptability of foods without compromising the level of inactivation [44]. The strategy of combining EOs with physical treatments (Heat, PEF, HHP, ultrasound, ohmic heating (OH), radiation, etc.) and antimicrobial additives (bacteriocins, EOs, organic acids, etc.) in food processing has been demonstrated in scientific studies (Table 2.1).

2.6 ESSENTIAL OILS (EOS) COMBINED WITH PHYSICAL TREATMENTS

2.6.1 ESSENTIAL OILS (EOS) COMBINED WITH HEAT

Thermal processing is a traditional popular industrial method applied to control foodborne diseases and spoilage. Heat resistance of target bacteria is characterized by D-value and Z-value indices that are used to check death kinetics of any bacteria and status of decimal reduction in bacterial count. D-value (decimal reduction time or decimal reduction dose) is the time or dose) required, at a given condition (e.g., temperature) or set of conditions, to achieve a log reduction, that is, to kill 90% (or 1 log) of relevant microorganisms. Z-value is the reciprocal of the slope resulting from the plot of the logarithm of the D-value versus temperature. EOs increased the bacterial sensitivity to heat, allowing reduction in the intensities of each treatment required to inactivate the same population [73]. Addition of vanillin to TSPYE (pH5) brought down the D_{58} of *C. sakazakii* from 14 min to less than 1 min [148]. Similarly, a five-fold reduction was reported in fruit juice with *T. algeriensis* EO against *E. coli* O157: H7 [3]. In this, EOs, and heat together proved to decrease the requirements of both quantity of EO and thermal intensity. This is a great advantage over the impact of high heat on final quality and cost of foodstuffs such as milk [23], fruit juices [24], and meat-based product [6]. In milk with vanillin 1400 ppm, Cava et al. [23] noted an average 25% reduction in time required for killing of log 4 cells *L. monocytogenes*. Similarly, in apple juice, 200 μl/ml of EO reduced the required temperature by 4.5°C [42].

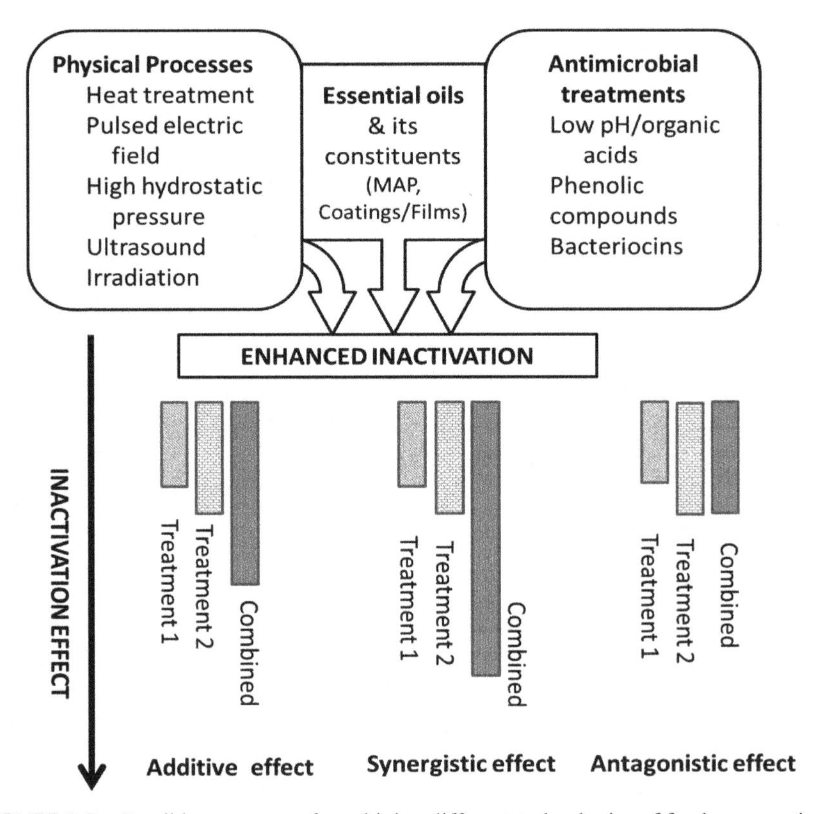

FIGURE 2.2 Possible outcomes of combining different technologies of food preservation.

TABLE 2.1 Microbial Inactivation by Essential Oil-Based Combined Processing Methods

Essential Oil Tested	Combined With the Process or Other Biocompound	Media	Target Microorganism Tested	Log Cycles Reduction Observed	References
(+)-limonene	HHP (175–400 MPa for 20 min)	Buffer	*E. coli* O157:H7 and *L. monocytogenes* EGD-e	4–5	[40]
(+)-limonene,	HHP	Fruit juices	*Escherichia coli* O157:H7	5	[40]
Basil	*Rosemary*	Chicken meat	*S.* Enteritidis	4	[136]
Carvacrol	Nisin	Buffer	*L. mononytogenes* *B. cereus*	3	[120]
Carvacrol (0.2 μL/mL)	Mild heat (54°C/10 min)	Apple juice	*Leuconostoc* sp. *Saccharomyces* sp.	4	[27]
Carvacrol (1% w/w)	Electrolyzed water	Shredded cabbages	Mesophilic and psychrotropic bacteria	<1	[135]
Carvacrol (2.5–3 mM)	HPP (250–300 MPa/20 min)	Laboratory media	*L. monocytogenes*	> 6	[78]
Carvacrol (3 mM)	HPP (300 MPa/20 min)	Milk	*L. monocytogenes*	3.2	[78]
Cinnamon (1–5% w/v)	PEF (10–30 kV/cm, 60–3000 μs)	Skim milk	*Salmonella typhimurium*	>1.96	[117]
Cinnamon bark oil (5 ml/100 ml)	PEF (35 kV/cm, 1700 μs)	Orange, strawberry, apple, and pear juice	*S. enteritidis* *E. coli*	5–6	[105]

TABLE 2.1 *(Continued)*

Essential Oil Tested	Combined With the Process or Other Biocompound	Media	Target Microorganism Tested	Log Cycles Reduction Observed	References
Citral	Heat (54°C)	Phosphate buffer and apple juice	*E. coli* O157:H7	>3	[44]
Cuminum cyminum (15 µL/ml)	*L. acidophilus* (0.5%)	White brined cheese	*S. aureus*	1	[124]
Lemon (200 µL/L)	Mild heat (54–60°C/10 min)	Apple juice	*E. coli* O157:H7	5	[42]
Lemon (200 µL/L)	PEF (25 kV and 100 KJ/Kg) and heat (60°C)	Liquid whole egg	*Salmonella saftenberg* and *L. mononytogenes*	4	[41]
Mandarin (0.05% v/w)	γ-irradiation	Green bean	*L. innocua*	3.3	[128]
Mandarin (0.05% v/w)	UV-C	Green bean	*L. innocua*	3	[128]
Mentha pulegium (0.2–0.86 µL/mL)	Heat (54–60°C)	Apple juice	*Escherichia coli* O157:H7	5	[3]
Metasequoia glyptostroboides (1–2%)	Nisin (62.5–500 IU/ml)	Milk	*L. mononytogenes*	6	[149]
Mint essential oil (0.5 and 1 µL/mL)	HPP (100–300 MPa/3.5 min)	Yogurt	*L. innocua* and *L. monocytogenes*	> 5–6	[47]
Orange, lemon, and mandarin	Heat (54°C/10 min)	–	–	5	[43]

TABLE 2.1 *(Continued)*

Essential Oil Tested	Combined With the Process or Other Biocompound	Media	Target Microorganism Tested	Log Cycles Reduction Observed	References
Oregano (0.01–0.025%v/v)	Ultrasound (26 kHz, 90 µm, 200 W, 14 mm Ø, 300 sec for 5–25 min)	Lettuce	*E. coli* O157:H7	>2	[101]
Oregano (0.2%)	Caprylic acid (0.5%)	Vacuum packed minced meat	*L. mononytogenes*,	2.5	[68]
			LAB,	1.5	
			psychrotrophic bacteria	1.5	
Oregano, lemon grass	Gamma irradiation (0.5 and 1 kGy)	Cauliflower	*L. mononytogenes*	4.5	[138]
	UV-C (5 and 10 kJ/m^2)		*Escherichia coli* O157:H7		
Perilla oil (1 mg/ml)	Nisin (15 µg/ml)	Pasteurized milk	*L. mononytogenes*	>6	[151]
			S. aureus	>2.9	
Rosemary	Thyme	Mozzarella cheese	*L. mononytogenes*	1.7	[66]
Vanillin	*Cinnamon* and *clove*	Milk	*L. mononytogenes*	5	[22]
			E. coli O157:H7		
Vanillin (900–1900 ppm)	Heat (50–28°C)	Sweetened lassi	*E. coli* O157:H7	4	[53]

At respective lethality, the heat sometimes generates sub-lethally injured cells and induces stress responses [114]. The repair of injuries under suitable conditions, especially in food-borne pathogens, is a food safety risk. Besides meeting the inactivation requirements of mild heat, combining with EO may control the regrowth of heat injured or adapted cells [74]. Amalaradjou et al. [6] investigated the efficacy of cooking ground beef (60 and 65°C) coupled with trans-cinnamaldehyde (0, 0.1, and 0.3%) in inactivating *E. coli* O157:H7 and fate of survivors throughout one-week storage at 4 or −18°C. There was a significant reduction in D-values and no increase in cell numbers in combination ($P<0.05$) [6, 132].

Bacterial spore formers are well-known for poor sensitivity towards heat treatment and antimicrobials. The combination approach has also been tested against spore formers. The direct addition of EO to the heating medium was only able to show a small decrease in apparent heat resistance of spore formers and can be useful to control the germination of heat treatment surviving spores [45]. Perigo et al. [116] observed a higher loss of viability of *B. cereus* following mild heat treatment in carrot juice with the addition of carvacrol or thymol (0.3 mmol/L). The chilling of ground turkey with EO components was effective in controlling germination and growth of spores of *Clostridium perfringens* [72, 75]. Further, the overall changes in the final product during the storage should be considered.

Most of the EO and heat combined treatments were carried out by the direct addition of antimicrobials in food media. In this case, the bioavailability of EO depletion is recognized to poor solubility in aqueous media. Some researchers found that nano-emulsions may decrease the thermal resistance of pathogens in liquid foods [92]. Mate et al. [98] found that nano-emulsion of D-limonene (0.5 mM) combined with heat was better than direct addition in increasing the thermal sensitivity of *L. monocytogenes*. The strong lethal effect was probably based on the damages facilitated by heat on the bacterial membranes allowing EOs to penetrate the cells to reach the target sites. However, further studies are required on the mechanism of action. Overall, heat coupled with EO may be an attractive approach upon further standardization of variables such as heating temperature, EO miscibility, and intrinsic factors of foods.

2.6.2 ESSENTIAL OILS (EOS) WITH PULSED ELECTRIC FIELD (PEF)

The PEF technology uses a very high voltage field (10–80 kV/cm for 1–100 μs) and is applied preferably in pumpable foods to be processed, such as juices, milk, fermented beverages. The irreversible damage to cell membranes and pore formation caused by electric pulses leads to leakage of intracellular components [80]. Since EOs are membrane-active compounds, the electroporation of target cells by PEF can facilitate the diffusion across the cell membrane thus damaging various functions leading cells to lose viability [8]. However, the impact of combined treatments of PEF and EO is still to known well. No synergy was reported between PEF and carvacrol in apple juice against *Leuconostoc* sp. and *Saccharomyces* sp [27] and in buffer (pH 4 and 7) against *E. coli* O157:H7 and *L. monocytogenes* [3]. Nevertheless, the effectiveness is determined by factors such as pH, organic acids, food components, added EO, PEF parameters, and other intrinsic and extrinsic factors [119]. Therefore, specific applications of PEF and EO must be evaluated for each target organism and food type under study. The research studies have examined effects combination of PEF and EO on actual foods. The inactivation of *S. typhimurium* in skim milk was synergistically enhanced by adding cinnamon and demonstrated maximum synergy at 30 kV/cm–700 μs and 5% w/v [117]. Incorporation of cinnamon bark oil to apple and pear juice (0.1%) and strawberry juice (0.05%) followed by PEF (1575–1700 μs at 35 kV/cm) was enough to destroy 5 log cells of *S.* enteritidis and *E. coli* O157:H7 [105]. The efficacy of combination of PEF and lemon EO in liquid whole egg was dependent on the sequence of application and heat [41]. Further, the stability of EOs to PEF is another aspect to be studied as they can act as additional hurdle following PEF against the surviving bacteria particularly spore formers [119].

2.6.3 ESSENTIAL OILS (EOS) WITH HIGH HYDROSTATIC PRESSURE (HHP)

High hydrostatic pressure (HHP) is followed for solid or liquid foods as non-thermal process that is applied in either batch or continuous method. In HHP, a static pressure (100 to 1000 MPa at- 20°C to 60°C) is applied to food for a few seconds to minutes [99]. Numerous reports have evaluated combination of EOs and HHP on target bacteria. Allyl-isothiocyanate (AIT: 10 to 80 μg/ml) sensitized the *E. coli* O157 to HPP (200 or 250 MPa) [108]. The combination of AIT and cinnamaldehyde showed synergistic effect with HHP in buffer and meat against pathogenic *E. coli* and *L. monocytogenes* [89]. These EOs showed no effect with thymol and this difference may be attributed to the thiol-reactive compounds (such as cinnamaldehyde and allyl-isothiocyanate) that inhibit the thioredoxin reductase and glutathione reductase activities helping in oxidative stress tolerance and exert synergism with pressure [49, 108, 133].

The lethality is often correlated with temperature and intensity of applied pressure has been categorized into different modalities of treatment, such as cold HHP pasteurization, HPP-assisted pasteurization, or HHP-assisted sterilization. Application of high-intensity pressure or prolonged time requires high cost and can adversely affect the acceptability by consumers. A combination of citral or lemongrass EO (0.75 mg/ml) reduced pressure from 350 to 150 MPa for 30 min required to inactivate *Colletotrichum gloeosporiodes* spore in saline solution [113].

Combination of EO-HHP treatments have been successfully applied in foods enabling microbial inactivation at the requirement of lesser pressure or shorter exposure time [113]. In yogurt drink, combination mint EO with HHP synergistically reduced the pressure required or holding time (210 sec) for 5 decimal reductions of *Listeria*. The pH, a_w, and color of drink remained unchanged and no separation of serum protein was observed [47]. Similarly, in apple and orange juice limonene-HPP acted synergistically on *E. coli* O157 [40]. The carvacrol and HHP displayed the synergistic effect on *L. monocytogenes* in milk (3.2 log reduction). However, it was lowered by two-fold compared to that achieved in buffer (> 6 log reduction) [78]. These observations are in support of the requirement of validation of the results of food models for efficacy in real foods. Citral and lemongrass are likely to be organoleptically acceptable antimicrobial for use in citrus-based beverages and juices owing to pleasing fruity flavor [133].

The high levels of destruction of target bacteria seen in the combination method of EOs and HHP in buffer and food are attributed to simultaneous action on cell membranes through a different mode. During pressurization, irreversible damage to cell membranes due to compression causes altered permeability thus leading to leakage of intracellular ions and ATP molecules. The pressurization can also allow easy penetrability of EOs and conversely weakening of cell wall sensitizes the pressure tolerant bacteria to HHP. However, the precise mechanism is not clearly understood. Also, the tailing effect after HHP, which means a small portion of target population survive and remain viable, is continued to be inhibited by EOs and serves as hurdle thereby preventing cells to recover from sublethal injuries [110]. Perez et al. [115] observed additional 1.38 and 1.8 log cells killing of methicillin resistant *S. aureus* in rice puddings by HHP in the presence of cinnamon or clove, respectively. Following HPP during one-week storage in refrigerator, total bacteria continued to decrease by 1.5 log cycle in HHP-cinnamon treatment. The citral-HHP combination showed synergistic effect on *L. monocytogenes*, conversely resulting in sublethal injury after the treatment. Interestingly, the pressure lethality threshold for butyl hydroquinone (BHQ)-HHP against barotolerant *E. coli* O157:H7 (200 MPa) was much lower than *L. monocyogenes* (>300 MPa) [29]. This difference in sensitivity was probably associated with difference in cell wall composition [94].

Bacterial endospore resistance towards HHP intensities is used presently for food processing and it still remains a challenge. However, the induction of germination by HHP renders them susceptible to other treatments [4]. Corthouts et al. [30] also described impact of EO on pressure-induced germination of spore of psychro-tolerant in *B. cereus*. Use of EO as additional hurdle inhibited induced germination; however, the inactivation was at moderate temperature and pressure. EO combination at high temperatures and pressure may inhibit the release of DPA; and spore rehydration resulted in less inactivation of *B. cereus* [93].

2.6.4 ESSENTIAL OILS (EOS) WITH RADIATION METHOD

Both UV and γ-Irradiation are widely investigated as a non-thermal treatment for assurance in microbiological safety in vegetables and meat products mainly by eliminating foodborne pathogens. The antimicrobial property of radiation is owed to photochemical reactions and DNA damage leading to loss of essential cell functions. The lethality of radiation can be significantly enhanced by EOs. For example, Turgis et al. [140] investigated the addition of more than 40 EOs and its main constituents in ground beef before radiation on radiosensitivity (RRS) of *Salmonella* Typhi (RRS = D_{10} of control sample/D_{10} of antimicrobial treated sample). The authors found that RRS was increased up to more than four times and was more effective. In another study, carvacrol addition to ground beef patties increased RRS of *E. coli and Salmonella* Typhi [25]. Utilization of EOs in vapor form also benefits combining radiation. Exposure of rice grains with vapors of *Ocimum basilicum* EO before radiation enhanced RRS of *Aspergillus niger* and *Penicillium chrysogenum* to 1.4 and 1.6, respectively [67].

A further approach to increase radiosensitivity in food systems was made by combining EOs and bacteriocins. Nisin with savory and carvacrol has shown to improve the radiosensitivity (1 KGy) of *L. monocytogenes* and *Salmonella* Typhimurium in peeled carrot, mini-carrots, and ready-to-eat carrots by 4.19

and 6.31 times, respectively [107]. However, increased radiation tolerance of *B. cereues* accompanied by high numbers of injured cells was observed in carvacrol-nisin than carvacrol alone combined with single and repetitive irradiation (1 kGy) [9]. This means that radiation sensitivity depends on the target microorganism, antimicrobials tested food type and process parameters. Non-thermal pulsed light (PL) treatment (1.2×10^5 J/m^2) with mandarin EO nano-encapsulated in chitosan showed 2 log reduction of *L. innocua.* No synergy was reported during the storage and it rather had slight negative effect on color of green bean samples [39]. Ultraviolet (UV) radiation treatment and *Zataria multiflora* EO treatment showed antagonistic effect on biofilm [137].

Alternatively, EO as hurdle with irradiation at low doses of treatment without compromising the inactivation could minimize undesirable changes to sensory parameters of treated products [86, 111]. For example, addition of EO of Chinese cinnamon and Spanish oregano in ground beef reduced from 1.2 to 0.35 kGy and from 1.4 to 0.5 kGy, respectively for *E. coli* O157:H7 and *S.* Typhi (6 log) [140]. EO can pass through bacterial cell wall mainly acting on membrane [10] and it leads to ATP leakage [20, 21]. Radiation further triggers series of events that damage cellular activities and loss of structure complemented by incorporation of EO makes the cell impossible to repair the damages imposed by the combined treatments [85].

MAP (Modified atmosphere packaging) involves packing of food product in gas-impermeable packaging material, where in gaseous environment is altered. The synergies between EOs and other antimicrobials can enhance the lethality of radiation treatment of packaged ready-to-eat foods. Ground beef with added EO of thyme when irradiated under MAP situation enhanced the radiosensitivity of pathogens [25].

The lethality of ɤ-radiation was more effective on *E. coli* O157:H7 and *S.* Typhimurium in green beans with chitosan-containing carvacrol-based coating. The RRS was increased almost by 25% with reduction of D_{10} from 0.32 to 0.24 kJ/kg [127]. Boumail et al. [17] found that the antimicrobial coating and ɤ-radiation together exerted synergistic effect on *E. coli* along with ready-to-eat cauliflower florets bacteria during storage for 7 days. In another study, coating of fresh broccoli with plant-based antimicrobial containing methyl cellulose increased the RRS of *Salmonella* Typhimurium by 2.4 times [58]. Though, the EO coating decreased sensory scores of the product but not in radiation treated product and it is probably due to dispersion of added EO in the product [10, 111]. Further, Gomes et al. [59] insisted to consider: the contribution of the packaging type; and amount of headspace should be carefully considered, especially irradiation processing when EO combined with edible coating and MAP.

2.6.5 ESSENTIAL OILS (EOS) WITH ULTRASOUND TREATMENT

Ultrasound generated waves of specific intensity and amplitude are being used as non-thermal technology in food processing [123]. During ultrasound, vibrations creates high number of microscopic bubbles or cavities, and an abrupt collapse of generated bubbles cause build-up of localized pressure (up to 50 MPa) and temperature (47000°C). Both the cavitation and sonolysis together cause damage to structure and function of cell components causing lysis of cells [118]. Millan et al. [101] described the outcome of oregano EO (0.025%) and ultrasound treatment as synergistic against decontamination of *E. coli* in lettuce leaves. According to reports when ultrasound is applied separately, the survival curve follows a first-order microbial kinetics. On addition of vanillin as hurdle to ultrasound, the survival curve of *L. innocua* was non-linear with enhanced inactivation [54].

Ferrante et al. [48] reported decrease of *L. monocytogenes* (6 log cells) population in orange juice by thermo-sonication treatment with vanillin (1500 ppm) or citral (100 ppm). The low pH of medium showed no significant effect on the sensitivity of organism to the synergy. Vanillin (500 ppm) and ultrasound (20 kHz for 120 μm) brought down the D_{60} spores of A. flavus spores in broth to less than 0.5 min from 1.20 min. The microbial destruction was increased in EO combination with ultrasound when applied either in a continuous or pulsed mode. Oregano and thyme were synergistic with ultrasound in the decontamination of *Salmonella enterica* Abony and *E. coli* [100]. After the exposure to combination, the texture of leaves was not affected, and the particle size of nano-emulsions had no influence on the inactivation process. The synergy was dependent on treatment time [100]. Together with ultrasound, EOs can sensitize bacteria cell membrane; sudden collapse of membrane can alter membrane permeability and cause seepage of intracellular components [125]. EO assisted ultrasound can save energy as lesser intensities are sufficient and cost effective. Nevertheless, before commercial applications, studies are needed to warrant impact of combinations on texture and other sensorial parameters of foods.

2.6.6 ESSENTIAL OILS (EOS) WITH OHMIC HEATING (OH)

OH or Joule heating or electro heating is based on Ohm's Law, and in OH, food is heated by passing electric current through the electrodes in direct contact with the food [61]. The heating takes place rapidly and in uniform fashion in both liquid and particulates containing foods. The EO-OH showed prospective to enhance the inactivation of pathogenic bacteria. EOs with OH in buffered peptone water resulted in reductions of some foodborne pathogens [81]. When these combinations were applied to Salsa, a synergistic bactericidal activity on the above pathogens and against norovirus surrogate MS-2 bacteriophage was observed [81, 82].

2.7 ESSENTIAL OILS (EOS) COMBINED WITH ANTIMICROBIALS

2.7.1 ESSENTIAL OILS (EOS) WITH BACTERIOCINS

Bacteriocins are peptides, ribosomally synthesized by certain bacteria, which can either inhibit or kill other closely related bacteria. Bacteriocins derived from lactic acid bacteria (LAB) represent majority of bacteriocins applied in food preservations. Currently, nisin and pedocin PA-1 have been accepted as commercial additives available as Nisaplin® and ALTA™ 2341, respectively. Among the preference of bacteriocins for combinations with various EOs, nisin has received interest in the preservation and safety of foods. Several researchers have observed synergistic effect on *Listeria monocytogenes, Listeria innocua, Bacillus cereus,* and *Bacillus subtilis*, when nisin was combined with EOs of oregano, valencia, *Metasequoia glyptostroboides*, *Mentha longifolia*, *Cuminum cyminum*, and thyme including their constituents thymol, carvacrol, and cinnamic acid [46, 116, 120]; while nisin and cinnamic acid had additive effect against *B. subtilis* [109]. Though the mode of action is still to be understood thoroughly: probably both treatments exert stress at different sites simultaneously and inflict damages producing many or larger and stable pores [120, 130]. The combination of nisin with oregano, thyme, and cinnamon in minced sheep meat, minced beef, and apple juice have demonstrated enhanced killing of *E. coli* O157:H7 [62, 131, 150]. The divergicin M35 and garlic extract together was effective to inactivate divergicin resistant *L. monocytogenes* in smoked salmon [153]. Combination of Bacteriocin and EO may prevent regrowth of stressed or adapted cells. Enterocin AS-48 a cyclic antibacterial peptide also reduced the Staphylococci in Carbonara sauce for 30 days storage (22°C), when nisin was combined with hydrocinnamic acid (20 mM) or carvacrol (126 mM) [63]. Nevertheless, nisin, and *Zataria multiflora* Boiss combinations were effective on *S. aureus* at low temperature (8°C) than ambient (25°C) storage [104].

In salad, Enterocin AS-48 with EOs reduced *L. monocytogenes*, and eugenol did not show any anti-listerial effects [7]. Moreover, a reduction in effective concentration of EOs was demonstrated in combination with bacteriocins. The MIC of thyme in broth was reduced to 0.71 from 2.2 μg/ml when applied along with enterocin-A against *E. coli* O157:H7 [57], and it could decrease the negative impact on taste of food. The uniform dispersion of antimicrobials within food matrix has sometimes been debated due to hydrophobicity, interactions, and binding to food components [65]. An improvement in effect of nisin and D-limonene was observed in nano-emulsion resulting antimicrobials requirements in smaller amounts and boarder antimicrobial spectrum [15]. Further utilization in successful combinations on commercial scale should be evaluated warranting further studies in various food matrices.

2.7.2 ESSENTIAL OILS WITH OTHER ESSENTIAL OILS (EOS)

Several studies have reported that mixture of different EOs or active constituents of EO enhance the antibacterial action. Combinations of EOs are likely to act on target bacteria at multiple sites either by same or by different mechanism due to interactions between components [16, 36]. Oregano and thyme EOs exhibited an additive effect of against pathogenic bacteria [64, 65]. The occurrence of improved activity of carvacrol-thymol combination against pathogenic microorganisms *S. aureus* and *P. aeruginosa* than individual applications has been reported [87]. Mixtures of phenolic compounds thymol or carvacrol with cinnamon at certain level have resulted synergistic effect in carrot broth against *Bacillus cereus* at 8°C [143]. Nevertheless, antagonistic susceptibility of *E. coli* has been reported in combination of 1,8-cineole and linalool than the independent applications [12, 121].

According to Burt [19], combinations of whole EO may have better antibacterial efficacies than only individual components due to synergistic effects. The synergism between *p*-cymene and carvacrol on *B. cereus* was hypothesized that weak antimicrobial EO such as *p*-cymene facilitates transportation of strong antimicrobial EO such as carvacrol to the cell by causing better swelling of cell wall [142]. Other mechanism such as the inhibition of protective enzymes and the sequential inhibition of common biochemical pathways has also been proposed by other researchers [35, 95]. To establish the mechanism of synergism, it is required to further understand underlying mechanism on metabolic activities of target bacteria.

Reduction in effective concentration by combining the EOs may be beneficial in the food industry. The required dose can be lowered if used in combinations thereby decreasing the negative effect on acceptability of product. The level of inactivation was uncompromised for the combinations of 25% MIC *O. vulgare* and 25% MIC of *R. officinalis* on different pathogens in experimentally contaminated hand-cut fresh vegetables [34]. Grapes covered with combination *O. vulgare* and *R. officinalis* EO had superior inhibitory effect against *A. flavus* and *A. niger* than individual applications [35].

The Chinese cinnamon and cinnamon bark EO combination was effective against pathogenic and spoilage bacteria; and addition to cooked meat was organoleptically acceptable at P = 0.05% [56]. However, the simultaneous use of basil and rosemary EO were not effective on *Salmonella enteritidis* than the separate application [136]. In Teriyanki sauce, only with 0.5% carvacral or thymol inactivated all *L. monocytogenes* in beef slice within 7 days of storage (7°C) and alone did not reduce the populations of tested organisms [102]. Further, the antimicrobial stability of EO combinations during storage should be carefully considered as some bacteria are less tolerant to antimicrobial combinations at low temperatures; besides interactive outcomes of antimicrobial with high salt and organic acids already present in food product [103].

While the application of cinnamon and clove in vapor form exerted concertation-dependent interaction, i.e., synergistic effect on inhibition of Gram-positive pathogens and antagonistic on *E. coli* [60]. Use of mixture of vapors of EO, the antimicrobial spectrum could be extended. Citrus extracts with carvacrol and thymol (2 mM) have reduced up to 6 log of different pathogens, which were not effectively inactivated alone [28]. The combined addition of garlic and holy basil EO each @ 30 μl had inhibitory activity against all tested bacteria than the separate applications [139]. However, further studies on food products for their efficacies at sensorial acceptable levels of EOs during storage are needed for plant EOs utilization in food products.

2.7.3 ESSENTIAL OILS (EOS) WITH ORGANIC ACIDS

Organic acids are generally used in food preservation as additives or metabolically synthesized by food-grade microorganisms. These have wide applications in fruits and vegetables and meat processing industry [126]. The mechanism of action is associated with the lowering of medium pH and can easily be transported through cell membrane of bacteria and dissociating within the cytoplasm causing cell [13]. The enhanced killing effect particularly foodborne pathogens by combining organic acids with EOs are beneficial in food processing [51]. EOs: alter the functionality of cell membrane, lead to increased permeability, and cause disruption of proton motive force probably making the cells sensitive to acids. Moreover, at low pH these facilitate dissolution of hydrophobic EO into lipids of target bacterial cell membrane. A strong synergistic anti-listerial effect was observed by combining lactic acid (50 μl/ml) with rosemary and thyme EOs at sub-inhibitory concentrations (50–300 μl/ml) [38].

Numerous studies [11, 33, 134, 152] have shown synergistic inactivation of lactic acid, acetic acid *Staphylococcus* sp., *Salmonella* sp. and spoilage bacteria in meat products [106]. Washing chicken breast meat with solution containing thymol and acetic acid was effective for reducing around 3 log *Salmonella enteric* in experimentally contaminated products [91]. Some studies have highlighted acid tolerance of pathogenic bacteria under acidic environment [13, 84]. Interestingly acid tolerant *S.* Typhimurium was sensitized in acid environment (pH 3.8) that was created by acetic, lactic, citric acids following the exposure of target bacteria to guava extracts [90, 152].

2.7.4 ESSENTIAL OILS (EOS) WITH ELECTROLYZED SODIUM CHLORIDE

Mahmoud et al. [97] conducted the study to preserve fish carp fillets by combination of electrolyzed NaCl and EO compounds. It was observed that combination treatment increased shelf-life by four times at 5 and 25°C. In another study, Abou et al. [1] reported treatment of semi fried tuna slices with electrolyzed water and 0.5% eugenol and 0.5% linalool that increased shelf-life by 3 times. Nano-encapsulated carvacrol with acidic electrolyzed water caused <1 log reduction of aerobic bacteria in shredded cabbage than those achieved in water (around 3 Log) alone [135]. The results have shown potentials as alternative to synthetic preservatives and had no negative impact on nutritional quality [96]. However, more studies are needed to evaluate effectiveness of this approach in several real food products.

2.7.5 ESSENTIAL OILS (EOS) WITH OTHER ANTIMICROBIALS

Addition of 0.0025% diglycerol monolaurate to TSB (Tryptic soy broth) reduced the MBC (Minimum bactericidal concentration) of carvacrol, thymol, and eugenol on *L. monocytogenes* by 30%, 20%, and 20%, respectively [147]. In another study, the effectiveness of alone and combination of bacteriophage cocktail, BEC8, and *trans*-cinnamaldehyde was evaluated to inactivate *E. coli* O157:H7 inoculated on intact lettuce and spinach leaves

[144]. The combination was more effective in reducing the bacteria to undetectable levels after 10 min at 4, 8, 23, or 37°C. Impact of combination of fatty acids and EOs was evaluated on flavor and taste of several products. Kim and Rhee [83] observed that individual treatments with caprylic, capric, and lauric acid caused only negligible reduction of *E. coli* O157:H7. However, combinations with trans-cinnamaldehyde, carvacrol, eugenol, thymol, and vanillin resulted in reduction up to 7 log [83].

2.8 SUMMARY

EOs as natural antimicrobials can be used to control foodborne pathogens; however, these can affect the sensory attributes of final food products. Therefore, strategies of EO combined with traditional and emerging physical treatments or antimicrobials or both against several target microorganisms can be a viable alternative. It is necessary to understand the underlying mechanism to design effective hurdle approach. Further validation of observations made in laboratory media in model foods needs to be followed by real foods to understand the factors associated. EO possesses the abnormal flavor if exceeds threshold level. Combination processing methods based on EOs can also facilitate development of novel product formulations and minimal processed foods without compromising the food safety.

KEYWORDS

- **bacteriocins**
- **essential oils**
- **hurdle technology**
- **modified atmosphere packaging**
- **ohmic heating**
- **pulse electric field**

REFERENCES

1. Abou-Taleb, M., & Kawai, Y., (2008). Shelf-life of semi fried tuna slices coated with essential oil compounds after treatment with anodic electrolyzed NaCl solution. *Journal of Food Protection*, *71*(4), 770–774.

2. Ait-Ouazzou, A., Cherrat, L., Espina, L., Lorán, S., Rota, C., & Pagán, R., (2011). The antimicrobial activity of hydrophobic essential oil constituents acting alone or in combined processes of food preservation. *Innovative Food Science and Emerging Technologies*, *12*(3), 320–329.
3. Ait-Ouazzou, A., Espina, L., Cherrat, L., Hassani, M., Laglaoui, A., Conchello, P., & Pagán, R., (2012). Synergistic combination of essential oils from Morocco and physical treatments for microbial inactivation. *Innovative Food Science and Emerging Technologies*, *16*, 283–290.
4. Akhtar, S., Torres, J. A., & Paredes-Sabja, D., (2015). High hydrostatic pressure-induced inactivation of bacterial spores. *Critical Reviews in Microbiology*, *41*(1), 18–26.
5. Allende, A., Tomás-Barberán, F. A., & Gil, M. I., (2006). Minimal processing for healthy traditional foods. *Trends in Food Science and Technology*, *17*(9), 513–519.
6. Amalaradjou, M. A. R., Baskaran, S. A., Ramanathan, R., & Johny, A. K., (2010). Enhancing the thermal destruction of *Escherichia coli* O157: H7 in ground beef patties by trans-cinnamaldehyde. *Food Microbiology*, *27*(6), 841–844.
7. Antonio, C. M., Abriouel, H., López, R. L., Omar, N. B., Valdivia, E., & Gálvez, A., (2009). Enhanced bactericidal activity of enterocin AS-48 in combination with essential oils, natural bioactive compounds, and chemical preservatives against *Listeria monocytogenes* in ready-to-eat salad. *Food and Chemical Toxicology*, *47*(9), 2216–2223.
8. Arroyo, C., Somolinos, M., Cebrián, G., Condon, S., & Pagan, R., (2010). Pulsed electric fields cause sub lethal injuries in the outer membrane of *Enterobacter sakazakii* facilitating the antimicrobial activity of citral. *Letters in Applied Microbiology*, *51*(5), 525–531.
9. Ayari, S., Dussault, D., Jerbi, T., Hamdi, M., & Lacroix, M., (2012). Radio sensitization of *Bacillus cereus* spores in minced meat treated with cinnamaldehyde. *Radiation Physics and Chemistry*, *81*(8), 1173–1176.
10. Ayari, S., Dussault, D., Millette, M., Hamdi, M., & Lacroix, M., (2009). Changes in membrane fatty acids and murein composition of *Bacillus cereus* and *Salmonella typhimurium* induced by gamma irradiation treatment. *International Journal of Food Microbiology*, *135*(1), 1–6.
11. Barros, J. C., Conceição, M. L. D., Gomes, N. N. J., Costa, A. C. V., & Souza, E. L., (2012). Combination of *Origanum vulgare* L. essential oil and lactic acid to inhibit *Staphylococcus aureus* in meat broth and meat model. *Brazilian Journal of Microbiology*, *43*(3), 1120–1127.
12. Bassolé, I. H. N., & Juliani, H. R., (2012). Essential oils in combination and their antimicrobial properties. *Molecules*, *17*(4), 3989–4006.
13. Beales, N., (2004). Adaptation of microorganisms to cold temperatures, weak acid preservatives, low pH, and osmotic stress: A review. *Comprehensive Reviews in Food Science and Food Safety*, *3*(1), 1–20.
14. Behrens, J. H., Barcellos, M. N., Frewer, L. J., Nunes, T. P., Franco, B. D. G. M., Destro, M. T., & Landgraf, M., (2010). Consumer purchase habits and views on food safety: A Brazilian study. *Food Control*, *21*(7), 963–969.
15. Bei, W., Zhou, Y., Xing, X., Zahi, M. R., Li, Y., Yuan, Q., & Liang, H., (2015). Organogel-nanoemulsion containing nisin and D-limonene and its antimicrobial activity. *Frontiers in Microbiology*, *6*(1010), 1–9.
16. Berthold-Pluta, A., Stasiak-Różańska, L., Pluta, A., & Garbowska, M., (2018). Antibacterial activities of plant-derived compounds and essential oils against *Cronobacter* strains. *European Food Research and Technology*, pp. 1–11.

17. Boumail, A., Salmieri, S., & Lacroix, M., (2016). Combined effect of antimicrobial coatings, gamma radiation, and negative air ionization with ozone on *Listeria innocua, Escherichia coli* and mesophilic bacteria on ready-to-eat cauliflower florets. *Postharvest Biology and Technology, 118,* 134–140.
18. Braschi, G., Patrignani, F., Siroli, L., Lanciotti, R., Schlueter, O., & Froehling, A., (2018). Flow cytometric assessment of the morphological and physiological changes of *Listeria monocytogenes* and *Escherichia coli* in response to natural antimicrobial exposure. *Frontiers in Microbiology, 9*(2783), 1–11.
19. Burt, S., (2004). Essential oils: Their antibacterial properties and potential applications in foods – a review. *International Journal of Food Microbiology, 94*(3), 223–253.
20. Caillet, S., & Lacroix, M., (2006). Effect of gamma radiation and oregano essential oil on murein and ATP concentration of *Listeria monocytogenes. Journal of Food Protection, 69*(12), 2961–2969.
21. Caillet, S., Shareck, F., & Lacroix, M., (2005). Effect of gamma radiation and oregano essential oil on murein and ATP concentration of *Escherichia coli* O157: H7. *Journal of Food Protection, 68*(12), 2571–2579.
22. Cava-Roda, R. M., Taboada-Rodríguez, A., Valverde-Franco, M. T., & Marín-Iniesta, F., (2010). Antimicrobial activity of vanillin and mixtures with cinnamon and clove essential oils in controlling *Listeria monocytogenes* and *Escherichia coli* O157:H7 in Milk. *Food Bioprocess Technology, 5, 2120-2131.*
23. Cava-Roda, R. M., Taboada, A., Palop, A., López-Gómez, A., & Marin-Iniesta, F., (2012). Heat resistance of *Listeria monocytogenes* in semi-skim milk supplemented with vanillin. *International Journal of Food Microbiology, 157*(2), 314–318.
24. Char, C., Guerrero, S., & Alzamora, S. M., (2009). Survival of *Listeria innocua* in thermally processed orange juice as affected by vanillin addition. *Food Control, 20*(1), 67–74.
25. Chiasson, F., Borsa, J., Ouattara, B., & Lacroix, M., (2004). Radiosensitization of *Escherichia coli* and *Salmonella Typhi* in ground beef. *Journal of Food Protection, 67*(6), 1157–1162.
26. Chouhan, S., Sharma, K., & Guleria, S., (2017). Antimicrobial activity of some essential oils-present status and future perspectives. *Medicines, 4*(58), 1–21.
27. Chueca, B., Ramírez, N., Arvizu-Medrano, S. M., García-Gonzalo, D., & Pagán, R., (2015). Inactivation of spoiling microorganisms in apple juice by a combination of essential oils' constituents and physical treatments. *Food Science and Technology International, 22*(5), 389–398.
28. Chung, D., Cho, T. J., & Rhee, M. S., (2018). Citrus fruit extracts with carvacrol and thymol eliminated 7-log acid-adapted *Escherichia coli* O157:H7, *Salmonellat yphimurium*, and *Listeria monocytogenes*: A potential of effective natural antibacterial agents. *Food Research International, 107,* 578–588.
29. Chung, Y. K., Malone, A. S., Yousef, A. E., & Peleg, M., (2008). Sensitization of microorganisms to high pressure processing by phenolic compounds. *High Pressure Processing of Foods*, 145–172.
30. Corthouts, J., & Michiels, C. W., (2016). Inhibition of nutrient and high pressure induced germination of *Bacillus cereus* spores by plant essential oils. *Innovative Food Science and Emerging Technologies, 34,* 250–258.
31. Cox, S. D., Mann, C. M., Markham, J. L., Bell, H. C., Gustafson, J. E., & Warmington, J. R., (2000). *The Mode of Antimicrobial Action of the Essential Oil of Melaleuca Alternifolia (Tea Tree Oil)* (pp. 170–175).

32. Cristani, M., D'Arrigo, M., Mandalari, G., Castelli, F., Sarpietro, M. G., Micieli, D., Venuti, V., et al., (2007). Interaction of four monoterpenes contained in essential oils with model membranes: Implications for their antibacterial activity. *Journal of Agricultural and Food Chemistry*, *55*(15), 6300–6308.
33. De Oliveira, C. E. V., Stamford, T. L. M., Neto, N. J. G., & De Souza, E. L., (2010). Inhibition of *Staphylococcus aureus* in broth and meat broth using synergies of phenolics and organic acids. *International Journal of Food Microbiology*, *137*(2), 312–316.
34. De Sousa, J. P., De Azerêdo, G. A., De Araújo, T. R., Da Silva, V. M. A., Da Conceição, M. L., & De Souza, E. L., (2012). Synergies of carvacrol and 1,8-cineole to inhibit bacteria associated with minimally processed vegetables. *International Journal of Food Microbiology*, *154*(3), 145–151.
35. De Souza, E. L., De Azerêdo, G. A., De Sousa, J. P., De Figueiredo, R. C. B. Q., & Stamford, T. L. M., (2013). Cytotoxic effects of *Origanum vulgare* L. and *Rosmarinus officinalis* L. essential oils alone and combined at sublethal amounts on *Pseudomonas fluorescense* in a vegetable broth. *Journal of Food Safety*, *33*(2), 163–171.
36. Delaquis, P. J., Stanich, K., Girard, B., & Mazza, G., (2002). Antimicrobial activity of individual and mixed fractions of dill, cilantro, coriander, and eucalyptus essential oils. *International Journal of Food Microbiology*, *74*(1), 101–109.
37. Dhifi, W., Bellili, S., Jazi, S., Bahloul, N., & Mnif, W., (2016). Essential oil chemical characterization and investigation of some biological activities: A critical review. *Medicines*, *3*(4), 25.
38. Dimitrijević, S. I., Mihajlovski, K. R., Antonović, D. G., Milanović, S. M. R., & Mijin, D. Ž., (2007). A study of the synergistic antilisterial effects of a sub-lethal dose of lactic acid and essential oils from *Thymus vulgaris* L., *Rosmarinus officinalis* L. and *Origanum vulgare* L. *Food Chemistry*, *104*(2), 774–782.
39. Donsì, F., Marchese, E., Maresca, P., Pataro, G., Vu, K. D., Salmieri, S., Lacroix, M., & Ferrari, G., (2015). Green beans preservation by combination of a modified chitosan based-coating containing nanoemulsion of mandarin essential oil with high pressure or pulsed light processing. *Postharvest Biology and Technology*, *106,* 21–32.
40. Espina, L., García-Gonzalo, D., Laglaoui, A., Mackey, B. M., & Pagán, R., (2013). Synergistic combinations of high hydrostatic pressure and essential oils or their constituents and their use in preservation of fruit juices. *International Journal of Food Microbiology*, *161*(1), 23–30.
41. Espina, L., Monfort, S., Álvarez, I., García-Gonzalo, D., & Pagán, R., (2014). Combination of pulsed electric fields, mild heat and essential oils as an alternative to the ultra pasteurization of liquid whole egg. *International Journal of Food Microbiology*, *189,* 119–125.
42. Espina, L., Somolinos, M., AitOuazzou, A., Condón, S., García-Gonzalo, D., & Pagán, R., (2012). Inactivation of *Escherichia coli* O157: H7 in fruit juices by combined treatments of citrus fruits essential oils and heat. *International Journal of Food Microbiology*, *159*(1), 9–16.
43. Espina, L., Somolinos, M., Lorán, S., Conchello, P., García, D., & Pagán, R., (2011). Chemical composition of commercial citrus fruit essential oils and evaluation of their antimicrobial activity acting alone or in combined processes. *Food Control*, *22*(6), 896–902.
44. Espina, L., Somolinos, M., Pagán, R., & García-Gonzalo, D., (2010). Effect of citral on the thermal inactivation of *Escherichia coli* O157: H7 in citrate phosphate buffer and apple juice. *Journal of Food Protection*, *73*(12), 2189–2196.

45. Esteban, M. D., Conesa, R., Huertas, J. P., & Palop, A., (2015). Effect of thymol in heating and recovery media on the isothermal and non-isothermal heat resistance of *Bacillus* spores. *Food Microbiology, 48,* 35–40.
46. Ettayebi, K., El Yamani, J., & Rossi-Hassani, B. D., (2000). Synergistic effects of nisin and thymol on antimicrobial activities in *Listeria monocytogenes* and *Bacillus subtilis. FEMS Microbiology Letters, 183*(1), 191–195.
47. Evrendilek, G. A., & Balasubramaniam, V. M., (2011). Inactivation of *Listeria monocytogenes* and *Listeria innocua* in yogurt drink applying combination of high pressure processing and mint essential oils. *Food Control, 22*(8), 1435–1441.
48. Ferrante, S., Guerrero, S., & Alzamora, S. M., (2007). Combined use of ultrasound and natural antimicrobials to inactivate *Listeria monocytogenes* in orange Juice. *Journal of Food Protection, 70*(8), 1850–1856.
49. Feyaerts, J., Rogiers, G., Corthouts, J., & Michiels, C. W., (2015). Thiol-reactive natural antimicrobials and high pressure treatment synergistically enhance bacterial inactivation. *Innovative Food Science and Emerging Technologies, 27,* 26–34.
50. Fisher, K., & Phillips, C., (2009). The mechanism of action of a citrus oil blend against *Enterococcus faecium* and *Enterococcus faecalis. Journal of Applied Microbiology, 106*(4), 1343–1349.
51. Friedly, E. C., Crandall, P. G., Ricke, S. C., Roman, M., O'Bryan, C., & Chalova, V. I., (2009). *In vitro* antilisterial effects of citrus oil fractions in combination with organic acids. *Journal of Food Science, 74*(2), M67–M72.
52. Friedman, M., Henika, P. R., & Mandrell, R. E., (2002). Bactericidal activities of plant essential oils and some of their isolated constituents against *Campylobacter jejuni, Escherichia coli, Listeria monocytogenes,* and *Salmonella enterica. Journal of Food Protection, 65*(10), 1545–1560.
53. Manju, G., Grover, C. R., & Suman, R. (2015) Antimicrobial efficacy of vanillin in conjunction with mild heat treatment against *Escherichia coli* O157: H7 in sweetened lassi. *Indian Journal of Dairy Science, 68*(3), 411–416.
54. Gastélum, G., Avila-Sosa, R., López-Malo, A., & Palou, E., (2012). *Listeria innocua* multi-target inactivation by thermo-sonication and vanillin. *Food Bioprocess Technology, 5*(2), 665–671.
55. Gayán, E., Torres, J. A., & Paredes-Sabja, D., (2012). Hurdle approach to increase the microbial inactivation by high pressure processing: Effect of essential oils. *Food Engineering Reviews, 4*(3), 141–148.
56. Ghabraie, M., Vu, K. D., Tata, L., Salmieri, S., & Lacroix, M., (2016). Antimicrobial effect of essential oils in combinations against five bacteria and their effect on sensorial quality of ground meat. *LWT-Food Science and Technology, 66,* 332–339.
57. Ghrairi, T., & Hani, K., (2013). Enhanced bactericidal effect of enterocin A in combination with thyme essential oils against *L. monocytogenes* and *E. coli* O157:H7. *Journal of Food Science and Technology*, pp. 1–9.
58. Gibriel, A. Y., Al-Sayed, H. M. A., Rady, A. H., & Abdelaleem, M. A., (2013). Synergistic antibacterial activity of irradiated and nonirradiated cumin, thyme and rosemary essential oils. *Journal of Food Safety, 33*(2), 222–228.
59. Gomes, C., Moreira, R. G., & Castell-Perez, E., (2011). Microencapsulated antimicrobial compounds as a means to enhance electron beam irradiation treatment for inactivation of pathogens on fresh spinach leaves. *Journal of Food Science, 76*(6), E479–E488.

60. Goñi, P., López, P., Sánchez, C., Gómez-Lus, R., Becerril, R., & Nerín, C., (2009). Antimicrobial activity in the vapor phase of a combination of cinnamon and clove essential oils. *Food Chemistry*, *116*(4), 982–989.
61. Goullieux, A., & Pain, J. P., (2014). Ohmic heating. In: Sun, D. W., (ed.), *Emerging Technologies for Food Processing* (2nd edn., pp. 399–426). Academic Press, San Diego.
62. Govaris, A., Solomakos, N., Pexara, A., & Chatzopoulou, P. S., (2010). The antimicrobial effect of oregano essential oil, nisin and their combination against *Salmonella enteritidis* in minced sheep meat during refrigerated storage. *International Journal of Food Microbiology*, *137*(3), 175–180.
63. Grande, M. J., Lopez, R. L., Abriouel, H., Valdivia, E., Omar, N. B., Maqueda, M., Martinez-Canamero, M., & Galvez, A., (2007). Treatment of vegetable sauces with enterocin AS-48 alone or in combination with phenolic compounds to inhibit proliferation of *Staphylococcus aureus*. *Journal of Food Protection*, *70*(2), 405–411.
64. Gutierrez, J., Barry-Ryan, C., & Bourke, P., (2008). The antimicrobial efficacy of plant essential oil combinations and interactions with food ingredients. *International Journal of Food Microbiology*, *124*(1), 91–97.
65. Gutierrez, J., Barry-Ryan, C., & Bourke, P., (2009). Antimicrobial activity of plant essential oils using food model media: Efficacy, synergistic potential and interactions with food components. *Food Microbiology*, *26*(2), 142–150.
66. Han, J. H., Patel, D., Kim, J. E., & Min, S. C., (2015). Microbial inhibition in mozzarella cheese using rosemary and thyme oils in combination with sodium diacetate. *Food Science and Biotechnology*, *24*(1), 75–84.
67. Hossain, F., Follett, P., Dang, K., Salmieri, S., Senoussi, C., & Lacroix, M., (2014). Radio sensitization of *Aspergillus niger* and *Penicillium chrysogenum* using basil essential oil and ionizing radiation for food decontamination. *Food Control*, *45,* 156–162.
68. Hulankova, R., Borilova, G., & Steinhauserova, I., (2013). Combined antimicrobial effect of oregano essential oil and caprylic acid in minced beef. *Meat Science*, *95*(2), 190–194.
69. Hüsnü, K., Başer, C., & Demirci, F., (2007). Chemistry of essential oils. In: Berger, R., (ed.), *Flavors and Fragrances* (pp. 43–86). Springer, Berlin Heidelberg.
70. Hyldgaard, M., Mygind, T., & Meyer, R. L., (2012). Essential oils in food preservation: Mode of action, synergies, and interactions with food matrix components. *Frontiers in Microbiology*, *3,* 12.
71. Ju, J., Xu, X., Xie, Y., Guo, Y., Cheng, Y., Qian, H., & Yao, W., (2018). Inhibitory effects of cinnamon and clove essential oils on mold growth on baked foods. *Food Chemistry*, *240,* 850–855.
72. Juneja, V. K., & Friedman, M., (2007). Carvacrol, cinnamaldehyde, oregano oil, and thymol inhibit *Clostridium perfringens* spore germination and outgrowth in ground turkey during chilling. *Journal of Food Protection*, *70*(1), 218–222.
73. Juneja, V. K., & Friedman, M., (2008). Carvacrol and cinnamaldehyde facilitate thermal destruction of *Escherichia coli* O157: H7 in raw ground beef. *Journal of Food Protection*, *71*(8), 1604–1611.
74. Juneja, V. K., Hwang, C. A., & Friedman, M., (2010). Thermal inactivation and post-thermal treatment growth during storage of multiple *Salmonella* serotypes in ground beef as affected by sodium lactate and oregano oil. *Journal of Food Science*, *75*(1), M1–M6.

75. Juneja, V. K., Thippareddi, H., & Friedman, M., (2006). Control of *Clostridium perfringens* in cooked ground beef by carvacrol, cinnamaldehyde, thymol, or oregano oil during chilling. *Journal of Food Protection*, *69*(7), 1546–1551.
76. Juven, B. J., Kanner, J., Schved, F., & Weisslowicz, H., (1994). Factors that interact with the antibacterial action of thyme essential oil and its active constituents. *Journal of Applied Bacteriology*, *76*(6), 626–631.
77. Kalemba, D., & Kunicka, A., (2003). Antibacterial and antifungal properties of essential oils. *Current Medicinal Chemistry*, *10*(10), 813–829.
78. Karatzas, A. K., Kets, E. P., Smid, E. J., & Bennik, M. H., (2001). The combined action of carvacrol and high hydrostatic pressure on *Listeria monocytogenes* Scott A. *Journal of Applied Microbiology*, *90*(3), 463–469.
79. Karatzas, A. K., Kets, E. P. W., Smid, E. J., & Bennik, M. H. J., (2001). The combined action of carvacrol and high hydrostatic pressure on *Listeria monocytogenes* scott A. *Journal of Applied Microbiology*, *90*(3), 463–469.
80. Kethireddy, V., Oey, I., Jowett, T., & Bremer, P., (2016). Critical analysis of the maximum non-inhibitory concentration (MNIC) method in quantifying sub-lethal injury in *Saccharomyces cerevisiae* cells exposed to either thermal or pulsed electric field treatments. *International Journal of Food Microbiology*, *233,* 73–80.
81. Kim, S. S., & Kang, D. H., (2017). Combination treatment of ohmic heating with various essential oil components for inactivation of food-borne pathogens in buffered peptone water and salsa. *Food Control*, *80,* 29–36.
82. Kim, S. S., & Kang, D. H., (2017). Synergistic effect of carvacrol and ohmic heating for inactivation of *E. coli* O157:H7, *S.* Typhimurium*, L. monocytogenes*, and MS-2 bacteriophage in salsa. *Food Control*, *73,* 300–305.
83. Kim, S. A., & Rhee, M. S., (2016). Highly enhanced bactericidal effects of medium chain fatty acids (caprylic, capric, and lauric acid) combined with edible plant essential oils (carvacrol, eugenol, β-resorcylic acid, trans-cinnamaldehyde, thymol, and vanillin) against *Escherichia coli* O157:H7. *Food Control*, *60,* 447–454.
84. Lachowicz, K. J., Jones, G. P., Briggs, D. R., Bienvenu, F. E., Wan, J., Wilcock, A., & Coventry, M. J., (1998). The synergistic preservative effects of the essential oils of sweet basil (*Ocimum basilicum* L.) against acid-tolerant food microflora. *Letters in Applied Microbiology*, *26*(3), 209–214.
85. Lacroix, M., Caillet, S., & Shareck, F., (2009). Bacterial radio sensitization by using radiation processing in combination with essential oil: Mechanism of action. *Radiation Physics and Chemistry*, *78*(7), 567–570.
86. Lacroix, M., Chiasson, F., Borsa, J., & Ouattara, B., (2004). Radio sensitization of *Escherichia coli* and *Salmonella typhi* in presence of active compounds. *Radiation Physics and Chemistry*, *71*(1), 65–68.
87. Lambert, R. J. W., Skandamis, P. N., Coote, P. J., & Nychas, G. E., (2001). *A Study of the Minimum Inhibitory Concentration (MIC) and Mode of Action of Oregano Essential Oil, Thymol and Carvacrol* (pp. 453–462).
88. Leistner, L., (2000). Basic aspects of food preservation by hurdle technology. *International Journal of Food Microbiology*, *55*(1–3), 181–186.
89. Li, H., & Gänzle, M., (2016). Effect of hydrostatic pressure and antimicrobials on survival of *Listeria monocytogenes* and enterohaemorrhagic *Escherichia coli* in beef. *Innovative Food Science and Emerging Technologies, 38*, 321–327.

90. Lim, S. W., Kim, S. W., Lee, S. C., & Yuk, H. G., (2013). Exposure of *Salmonella typhimurium* to guava extracts increases their sensitivity to acidic environments. *Food Control*, *33*(2), 393–398.
91. Lu, Y., & Wu, C., (2012). Reductions of *Salmonella enterica* on chicken breast by thymol, acetic acid, sodium dodecyl sulfate or hydrogen peroxide combinations as compared to chlorine wash. *International Journal of Food Microbiology*, *152*(1–2), 31–34.
92. Luis-Villaroya, A., Espina, L., García-Gonzalo, D., Bayarri, S., Pérez, C., & Pagán, R., (2015). Bioactive properties of a propolis-based dietary supplement and its use in combination with mild heat for apple juice preservation. *International Journal of Food Microbiology*, *205,* 90–97.
93. Luu-Thi, H., Corthouts, J., Passaris, I., Grauwet, T., Aertsen, A., Hendrickx, M., & Michiels, C. W., (2015). Carvacrol suppresses high pressure high temperature inactivation of *Bacillus cereus* spores. *International Journal of Food Microbiology*, *197,* 45–52.
94. Luz, I. D. S., De Melo, A. N. F., Bezerra, T. K. A., Madruga, M. S., Magnani, M., & deSouza, E. L., (2014). Sublethal amounts of *Origanum vulgare* L. essential oil and carvacrol cause injury and changes in membrane fatty acid of *Salmonella typhimurium* cultivated in a meat broth. *Food Borne Pathogens and Disease*, *11*(5), 357–361.
95. Lv, F., Liang, H., Yuan, Q., & Li, C., (2011). *In vitro* antimicrobial effects and mechanism of action of selected plant essential oil combinations against four food-related microorganisms. *Food Research International*, *44*(9), 3057–3064.
96. Mahmoud, B. S. M., Kawai, Y., Yamazaki, K., Miyashita, K., & Suzuki, T., (2007). Effect of treatment with electrolyzed NaCl solutions and essential oil compounds on the proximate composition, amino acid and fatty acid composition of carp fillets. *Food Chemistry*, *101*(4), 1492–1498.
97. Mahmoud, B. S. M., Yamazaki, K., Miyashita, K., Shin, II., & Suzuki, T., (2006). New technology for fish preservation by combined treatment with electrolyzed NaCl solutions and essential oil compounds. *Food Chemistry*, *99*(4), 656–662.
98. Maté, J., Periago, P. M., & Palop, A., (2016). When nano-emulsified, d-limonene reduces *Listeria monocytogenes* heat resistance about one hundred times. *Food Control*, *59,* 824–828.
99. Matser, A. M., Krebbers, B., Vanden, B. R. W., & Bartels, P. V., (2004). Advantages of high pressure sterilization on quality of food products. *Trends in Food Science and Technology*, *15*(2), 79–85.
100. Millan-Sango, D., Garroni, E., Farrugia, C., Vanimpe, J. F. M., & Valdramidis, V. P., (2016). Determination of the efficacy of ultrasound combined with essential oils on the decontamination of *Salmonella* inoculated lettuce leaves. *LWT-Food Science and Technology*, *73,* 80–87.
101. Millan-Sango, D., McElhatton, A., & Valdramidis, V. P., (2015). Determination of the efficacy of ultrasound in combination with essential oil of oregano for the decontamination of *Escherichia coli* on inoculated lettuce leaves. *Food Research International*, *67,* 145–154.
102. Moon, H., Kim, N. H., Kim, S. H., Kim, Y., Ryu, J. H., & Rhee, M. S., (2017). Teriyaki sauce with carvacrol or thymol effectively controls *Escherichia coli* O157: H7, *Listeria monocytogenes*, *Salmonella typhimurium*, and indigenous flora in marinated beef and marinade. *Meat Science*, *129,* 147–152.

103. Moon, H., & Rhee, M. S., (2016). Synergism between carvacrol or thymol increases the antimicrobial efficacy of soy sauce with no sensory impact. *International Journal of Food Microbiology*, *217*, 35–41.
104. Moosavy, M. H., Basti, A. A., Misaghi, A., & Salehi, T. Z., (2008). Effect of *Zataria multiflora Boiss* essential oil and nisin on *Salmonella typhimurium* and *Staphylococcus aureus* in a food model system and on the bacterial cell membranes. *Food Research International*, *41*(10), 1050–1057.
105. Mosqueda-Melgar, J., Raybaudi-Massilia, R. M., & Martín-Belloso, O., (2008). Non-thermal pasteurization of fruit juices by combining high-intensity pulsed electric fields with natural antimicrobials. *Innovative Food Science and Emerging Technologies*, *9*(3), 328–340.
106. Naveena, B. M., Muthukumar, M., Sen, A. R., Babji, Y., & Murthy, T. R. K., (2006). Improvement of shelf-life of buffalo meat using lactic acid, clove oil and vitamin C during retail display. *Meat Science*, *74*(2), 409–415.
107. Ndoti-Nembe, A., Vu, K. D., Doucet, N., & Lacroix, M., (2013). Effect of combination of essential oils and bacteriocins on the efficacy of gamma radiation against *Salmonella typhimurium* and *Listeria monocytogenes*. *International Journal of Radiation Biology*, *89*(10), 794–800.
108. Ogawa, T., Nakatani, A., Matsuzaki, H., Isobe, S., & Isshiki, K., (2000). Combined effects of hydrostatic pressure, temperature, and the addition of allyl isothiocyanate on inactivation of *Escherichia coli*. *Journal of Food Protection*, *63*(7), 884–888.
109. Olasupo, N. A., Fitzgerald, D. J., Narbad, A., & Gasson, M. J., (2004). Inhibition of *Bacillus subtilis* and *Listeria innocua* by nisin in combination with some naturally occurring organic compounds. *Journal of Food Protection*, *67*(3), 596–600.
110. Oliveira, T. L. C. D., Ramos, A. L. S., Ramos, E. M., Piccoli, R. H., & Cristianini, M., (2015). Natural antimicrobials as additional hurdles to preservation of foods by high pressure processing. *Trends in Food Science and Technology*, *45*(1), 60–85.
111. Ouattara, B., Sabato, S. F., & Lacroix, M., (2001). Combined effect of antimicrobial coating and gamma irradiation on shelf-life extension of pre-cooked shrimp (*Penaeus* spp.). *International Journal of Food Microbiology*, *68*(1), 1–9.
112. Oussalah, M., Caillet, S., & Lacroix, M., (2006). Mechanism of action of Spanish oregano, Chinese cinnamon, and savory essential oils against cell membranes and walls of *Escherichia coli* O157: H7 and *Listeria monocytogenes*. *Journal of Food Protection*, *69*(5), 1046–1055.
113. Palhano, F. L., Vilches, T. T. B., Santos, R. B., Orlando, M. T. D., Ventura, J. A., & Fernandes, P. M. B., (2004). Inactivation of *Colletotrichum gloeosporioides* spores by high hydrostatic pressure combined with lemongrass essential oil. *International Journal of Food Microbiology*, *95*(1), 61–66.
114. Park, H. G., Han, S. I., Oh, S. Y., & Kang, H. S., (2005). Cellular responses to mild heat stress. *Cellular and Molecular Life Sciences CMLS*, *62*(1), 10–23.
115. Pérez-Pulido, R., Toledo, A. J., Grande, B. M. J., & Gálvez, A., (2012). Bactericidal effects of high hydrostatic pressure treatment singly or in combination with natural antimicrobials on *Staphylococcus aureus* in rice pudding. *Food Control*, *28*(1), 19–24.
116. Periago, P. M., Palop, A., & Fernandez, P. S., (2001). Combined effect of nisin, carvacrol and thymol on the viability of *Bacillus cereus* heat-treated vegetative cells. *Food Science and Technology International*, *7*(6), 487–492.

117. Pina-Pérez, M. C., Martínez-López, A., & Rodrigo, D., (2012). Cinnamon antimicrobial effect against *Salmonella typhimurium* cells treated by pulsed electric fields (PEF) in pasteurized skim milk beverage. *Food Research International, 48*(2), 777–783.
118. Piyasena, P., Mohareb, E., & McKellar, R. C., (2003). Inactivation of microbes using ultrasound: A review. *International Journal of Food Microbiology, 87*(3), 207–216.
119. Pol, I. E., Mastwijk, H. C., Slump, R. A., Popa, M. E., & Smid, E. J., (2001). Influence of food matrix on inactivation of *Bacillus cereus* by combinations of nisin, pulsed electric field treatment, and carvacrol. *Journal of Food Protection, 64*(7), 1012–1018.
120. Pol, I. E., & Smid, E. J., (1999). Combined action of nisin and carvacrol on *Bacillus cereus* and *Listeria monocytogenes. Letters in Applied Microbiology, 29*(3), 166–170.
121. Randrianarivelo, R., Sarter, S., Odoux, E., Brat, P., & Lebrun, M., (2009). Composition and antimicrobial activity of essential oils of *Cinnamosma fragrans. Food Chemistry, 114*(2), 680–684.
122. Raso, J., & Barbosa-Cánovas, G. V., (2003). Non-thermal preservation of foods using combined processing techniques. *Critical Reviews in Food Science and Nutrition, 43*(3), 265–285.
123. Rastogi, N. K., (2011). Opportunities and challenges in application of ultrasound in food processing. *Critical Reviews in Food Science and Nutrition, 51*(8), 705–722.
124. Sadeghi, E., Akhondzadeh- Basti, A., Noori, N., Khanjari, A. L. I., & Partovi, R., (2013). Effect of *Cuminum cyminum* L. essential oil and *Lactobacillus acidophilus* (a probiotic) on *Staphylococcus aureus* during the manufacture, ripening and storage of white brined cheese. *Journal of Food Processing and Preservation, 37*(5), 449–455.
125. Sango, D. M., Abela, D., McElhatton, A., & Valdramidis, V. P., (2014). Assisted ultrasound applications for the production of safe foods. *Journal of Applied Microbiology, 116*(5), 1067–1083.
126. Schirmer, B. C., & Langsrud, S., (2010). A dissolving Co_2 headspace combined with organic acids prolongs the shelf-life of fresh pork. *Meat Science, 85*(2), 280–284.
127. Severino, R., Ferrari, G., Dang, K., & Donsì, F., (2015). *Antimicrobial Effects of Modified Chitosan-Based Coating Containing Nano-Emulsion of Essential Oils, Modified Atmosphere Packaging and Gamma Irradiation Against Escherichia coli O157: H7 and Salmonella Typhimurium on Green Beans, 50,* 215–222.
128. Severino, R., Vu, K. D., Donsì, F., Salmieri, S., Ferrari, G., & Lacroix, M., (2014). Antibacterial and physical effects of modified chitosan based-coating containing nano-emulsion of mandarin essential oil and three non-thermal treatments against *Listeria innocua* in green beans. *International Journal of Food Microbiology, 191,* 82–88.
129. Siroli, L., Braschi, G., deJong, A., Kok, J., Patrignani, F., & Lanciotti, R., (2018). Transcriptomic approach and membrane fatty acid analysis to study the response mechanisms of *Escherichia coli* to thyme essential oil, carvacrol, 2-(E)-hexanal and citral exposure. *Journal of Applied Microbiology, 125*(5), 1308–1320.
130. Smith-Palmer, A., Stewart, J., & Fyfe, L., (2001). The potential application of plant essential oils as natural food preservatives in soft cheese. *Food Microbiology, 18*(4), 463–470.
131. Solomakos, N., Govaris, A., Koidis, P., & Botsoglou, N., (2008). The antimicrobial effect of thyme essential oil, nisin and their combination against *Escherichia coli* O157:H7 in minced beef during refrigerated storage. *Meat Science, 80*(2), 159–166.
132. Somolinos, M., García, D., Condón, S., Mackey, B., & Pagán, R., (2010). Inactivation of *Escherichia coli* by citral. *Journal of Applied Microbiology, 108*(6), 1928–1939.

133. Somolinos, M., García, D., Pagán, R., & Mackey, B., (2008). Relationship between sub lethal injury and microbial inactivation by the combination of high hydrostatic pressure and tert-butyl hydroquinone. *Applied and Environmental Microbiology*, *74*(24), 7570–7577.
134. Souza, E. L. D., Barros, J. C. D., Conceição, M. L. D., Gomes, N. N. J., & Costa, A. C. V. D., (2009). Combined application of *Origanum vulgare* L. essential oil and acetic acid for controlling the growth of *Staphylococcus aureus* in foods. *Brazilian Journal of Microbiology*, *40*(2), 387–393.
135. Sow, L. C., Tirtawinata, F., Yang, H., Shao, Q., & Wang, S., (2017). Carvacrol nanoemulsion combined with acid electrolyzed water to inactivate bacteria, yeast *in vitro* and native microflora on shredded cabbages. *Food Control*, *76,* 88–95.
136. Stojanović-Radić, Z., Pejčić, M., Joković, N., & Jokanović, M., (2018). Inhibition of *Salmonella enteritidis* growth and storage stability in chicken meat treated with basil and rosemary essential oils alone or in combination. *Food Control*, *90,* 332–343.
137. Tajik, H., Naghili, H., Ghasemmahdi, H., Moradi, M., & Badali, A., (2015). Effects of *Zataria multiflora boiss* essential oil, ultraviolet radiation and their combination on *Listeria monocytogenes* biofilm in a simulated industrial model. *International Journal of Food Science & Technology*, *50*(9), 2113–2119.
138. Tawema, P., Han, J., Vu, K. D., Salmieri, S., & Lacroix, M., (2016). Antimicrobial effects of combined UV-C or gamma radiation with natural antimicrobial formulations against *Listeria monocytogenes, Escherichia coli* O157: H7, and total yeasts/molds in fresh cut cauliflower. *LWT-Food Science and Technology*, *65,* 451–456.
139. Torpol, K., Wiriyacharee, P., Sriwattana, S., Sangsuwan, J., & Prinyawiwatkul, W., (2018). Antimicrobial activity of garlic (*Allium sativum* L.) and holy basil (*Ocimum sanctum* L.) essential oils applied by liquid vs. vapor phases. *International Journal of Food Science and Technology*, *53*(9), 2119–2128.
140. Turgis, M., Han, J., Millette, M., Salmieri, S., Borsa, J., & Lacroix, M., (2009). Effect of selected antimicrobial compounds on the radiosensitization of *Salmonella* Typhi in ground beef. *Letters in Applied Microbiology*, *48*(6), 657–662.
141. Ultee, A., Bennik, M. H. J., & Moezelaar, R., (2002). The phenolic hydroxyl group of carvacrol is essential for action against the food-borne pathogen *Bacillus cereus. Applied and Environmental Microbiology*, *68*(4), 1561–1568.
142. Ultee, A., Slump, R. A., Steging, G., & Smid, E. J., (2000). Antimicrobial activity of carvacrol toward *Bacillus cereus* on rice. *Journal of Food Protection*, *63*(5), 620–624.
143. Valero, M., & Francés, E., (2006). Synergistic bactericidal effect of carvacrol, cinnamaldehyde or thymol and refrigeration to inhibit *Bacillus cereus* in carrot broth. *Food Microbiology*, *23*(1), 68–73.
144. Viazis, S., Akhtar, M., Feirtag, J., & Diez-Gonzalez, F., (2011). Reduction of *Escherichia coli* O157:H7 viability on leafy green vegetables by treatment with a bacteriophage mixture and trans-cinnamaldehyde. *Food Microbiology*, *28*(1), 149–157.
145. Wang, L. H., Zhang, Z. H., Zeng, X. A., Gong, D. M., & Wang, M. S., (2017). Combination of microbiological, spectroscopic, and molecular docking techniques to study the antibacterial mechanism of thymol against *Staphylococcus aureus*: Membrane damage and genomic DNA binding. *Analytical and Bioanalytical Chemistry*, *409*(6), 1615–1625.
146. Xavier, V. B., Vargas, R. M. F., Cassel, E., Lucas, A. M., Santos, M. A., Mondin, C. A., Santarem, E. R., et al., (2011). Mathematical modeling for extraction of essential

oil from *Baccharis* spp. by steam distillation. *Industrial Crops and Products*, *33*(3), 599–604.

147. Yamazaki, K., Yamamoto, T., Kawai, Y., & Inoue, N., (2004). Enhancement of anti-listerial activity of essential oil constituents by nisin and diglycerol fatty acid ester. *Food Microbiology*, *21*(3), 283–289.
148. Yemi, P. G., Pagotto, F., Bach, S., & Delaquis, P., (2011). Effect of vanillin, ethyl vanillin, and vanillic acid on the growth and heat resistance of *Cronobacter* species. *Journal of Food Protection*, *74*(12), 2062–2069.
149. Yoon, J. I., Bajpai, V. K., & Kang, S. C., (2011). Synergistic effect of nisin and cone essential oil of *Metasequoia glyptostroboides* Miki ex Hu against *Listeria monocytogenes* in milk samples. *Food and Chemical Toxicology*, *49*(1), 109–114.
150. Yuste, J., & Fung, D. Y. C., (2004). Inactivation of *Salmonella typhimurium* and *Escherichia coli* O157:H7 in apple juice by a combination of nisin and cinnamon. *Journal of Food Protection*, *67*(2), 371–377.
151. Zhao, X., Shi, C., Meng, R., Liu, Z., Huang, Y., Zhao, Z., & Guo, N., (2016). Effect of nisin and perilla oil combination against *Listeria monocytogenes* and *Staphylococcus aureus* in milk. *Journal of Food Science and Technology*, *53*(6), 2644–2653.
152. Zhou, F., Ji, B., Zhang, H., Jiang, H., Yang, Z., Li, J., Ren, Y., & Yan, W., (2007). Synergistic effect of thymol and carvacrol combined with chelators and organic acids against *Salmonella typhimurium*. *Journal of Food Protection*, *70*(7), 1704–1709.
153. Zouhir, A. M., Kheadr, E., Tahiri, I., Benhamida, J., & Fliss, I., (2008). Combination with plant extracts improves the inhibitory action of divergicinM35 against *Listeria monocytogenes*. *Journal of Food Quality*, *31*(1), 13–33.

CHAPTER 3

POTENTIAL USE OF HERBS IN MILK AND MILK PRODUCTS

PRAVIN D. SAWALE, WRITDHAMA PRASAD, SHAIK ABDUL HUSSAIN, VEENA NAGARAJAPPA, and SANTOSH K. MISHRA

ABSTRACT

Functional foods contain bioactive compounds, which offer health benefits. These food ingredients basically include vitamins, phytosterols, bioactive peptides, carotenoids, antioxidants, fatty acids, fibers, etc. Several herbaceous plants and extracts of suave and woody vines are considered as herbal medicines that are recognized for their savory and medicinal properties. Several methods have been mentioned in *Ayurveda* with health benefits, such as application in milk as carriers of these herbs. The major part of the milk in India is used for the preparation of dairy products of traditional origin with a well-established market at the domestic level. Incorporation of herbal bioactives into traditional Indian dairy products not only helps the industry to meet the growing consumer demand for these foods but also facilitates in competing with the ever-increasing world market of functional foods. The research reports confirm the addition of polyphenols (herbs) into milk to improve anti-oxidative activity, thermal stability, and alcohol stability but decreased non-enzymatic browning, rennet coagulation time (RCT), etc. These modifications in milk properties are of utmost importance as these may modify the pre-optimized processing parameters for the preparation of dairy products. This chapter focuses on possible aspects of the incorporation of herbs into milk and different milk products.

3.1 INTRODUCTION

Industrialization and rapid urbanization have resulted in dramatic changes in the social and economic life of people living in developing countries and

India is not an exception. The effect of changes in the lifestyle is directly reflected in the susceptibility and occurrence of lifestyle-related diseases in these populations. Busy work schedule tends to affect the lifestyle by replacement of part of the regular diet by junk foods, which are rich in fat, sugar, and calories. This changing scenario of health status has made people more prone to various nutritional disorders, such as obesity, diabetes, cardiovascular diseases (CVD), etc. [39]. The emergence of several lifestyle-related health disorders in the recent times has resulted into an increasing focus on foods added with bio-functional components [26].

Consumers nowadays are more attracted toward foods, which enhance the health beyond basic nutrition. Thus, functional foods are assuming immense popularity and the food markets are being dominated by a variety of functional foods. A market survey has shown high potential for health promoting as well as value-added foods in the Indian market [38]. The study also reported that the growth in functional food segment has increased from 8.5 to 20% per year, whereas growth in other food products is estimated at about 1 to 4% per year.

In India, majority of milk produced is utilized for preparation of different milk products [89]. Processing and thereby converting the liquid milk into dairy products not only preserves the milk solids for longer duration but also results into value addition products. Although the domestic markets are well established for dairy products in India, yet India dairy entrepreneurs must look for new ways to add more and more functionality in their dairy products to compete with the expanding demand of functional foods.

Some bioactive ingredients targeted for functional dairy/food products include: polyphenols, plant sterols, dietary fibers, Omega-3 fatty acids, probiotics, and prebiotics, etc. Demand for herbal remedies is regularly increasing. About 20% of the Indian plant species are used for medicinal purpose. In pharmacologic terms, 'herb' is generally referred to all the components of herb (*viz*., shoot, and roots, flowers, and leaves, fruits, and seeds) that are used for their medicinal attributes. Literature also suggests their application in culinary purposes; in cheese making as both coagulating and flavor ingredients. Herbs have also been used in foods in many regions of present-day Asian and European countries (viz., Persia, Mesopotamia, Greece, and Rome, and Arabia) [12, 81].

Herbs harbor variety of bioactive compounds, such as polyphenols, carotenoids, flavonoids, sterols, lignans, saponins, terpenoids, etc. Present research has been focused on herbs with functional activities (*viz*., hypolipidemic, immune-stimulating), which will act as potential adjuncts in decreasing the risk of cancer and cardio-vascular diseases. In India, herbal products are mostly sold in pharmacological aspect (medicine) to cure certain ailments and thus are categorized as 'medicines.' In order to ensure the success of program to boost herbal crop production, emphasis on value-addition to new product development cannot be neglected.

Ayurveda is recognized as the most ancient document related to health's philosophy and it covers almost every aspect of human well-being [71, 72]. There is vast literature on incorporation of herbs into cheese, yoghurt, and other western dairy products to enhance their functionality. However, there is no consolidated literature available regarding the work undertaken with the objective of supplementation of herbs into different Indian dairy products.

During the recent era, milk products have gained special importance in the research and development institutions to enhance their functionality. Developments in the functionality improvement of different Indian dairy products, especially by herbal supplementation, are discussed in this chapter.

3.2 MEDICINAL PLANTS

Plant-derived phytochemicals protect against diseases and in maintaining well-being [90]. Plants having medicinal properties are provided for cure of different ailments in various indigenous healthcare systems *viz.,* Allopathy (30), Siddha (600), Amchi (600), Ayurveda (700) and Unani (700) [63]. In India, around 20,000 plants with medicinal activities have been documented [9, 21]. However, the data suggests that traditional communities are using only about 7,000–7,500 plants for medicinal purposes [67]. *Ayurvedic* system of medicine is well-established in developed countries including Europe, America, and Asia [84].

3.3 IMPORTANCE AND SCOPE OF HERBAL MEDICINES

Majority of the population in the developing countries depends upon herbal medicines to cure diseases. Utilization of herbal medicines is interestingly rising in developed and developing countries. For instance in France and Germany, many herbal extracts are considered and prescribed as drugs; and their sales in the European Union was estimated to be>$20 billion in 2014. In India, the herbal drug market is approximately $3.89 billion in the domestic market and $145 billion for export [77]. Herbal medicines as nutraceuticals (health foods) have an estimated market of about $ 80–250 billion in the Western counties [36].

3.4 DELIVERY SYSTEM FOR HERBAL BIOACTIVE COMPOUNDS

Dairy products are increasingly being used as a delivery vehicle for different bioactive ingredients [34, 78] including the herbal extracts [30]. When herbs are added directly into milk, organoleptic, and physicochemical attributes are prone to be affected severely, which tends to limit their application. Encapsulation of herbal extracts has been used as a successful tool to restrict their implications on sensory quality thereby enabling the use of food products as a vehicle for delivery of bioactive ingredients and for their higher bioavailability [47].

The European legislation describes the delivery system as modification in the extent or location at which the active component is to be released. Such an alternation is done by employing appropriate materials possessing well-defined protective attributes to manipulate the release of the active compound. Encapsulation can act as an effective method to deliver the herbal bioactive compounds through aqueous food systems. Microcapsules, developed in the encapsulation system, comprises of two phases: (i) an inner phase, which carries the active compound of interest; and (ii) an outer phase, which acts a carrier material and encapsulates the inner phase. Different kinds of encapsulation techniques have been developed, such as [17]:

- Phase separation (coacervation);
- Solvent dispersion/evaporation;
- Co-crystallization;
- Drying polymerization;
- Interfacial polymerization to suitably encapsulate different compounds: minerals, vitamins, antioxidants, polyphenols, amino acids, and enzymes.

From the perspective of food processing, encapsulation not only limits the chemical and enzymatic degradations [95] but also enhances the miscibility of active component in food products without affecting the inherent organoleptic attributes [19].

Literature highlights some materials for encapsulation of active components, such as [92]: synthetic polymers, acacia gums, semi-synthetic cellulose derivatives, and maltodextrin. Soluble polysaccharides (maltodextrin) have been reported as potential encapsulating ingredient; however, its main drawback is its inferior emulsification activity, specifically for volatile compounds that tend to escape rapidly after encapsulation using maltodextrin. Thus, there is urgent need for more efficient and stable encapsulating ingredient for delivering bio-functional agents through different food systems. Gelatin and gum acacia are also considered as efficient encapsulating ingredients for their lower viscosity and high solubility in aqueous solutions and ability to form stable O/W emulsions. Combination of maltodextrin and gum acacia is reported to yield an emulsion with higher encapsulation efficiency and recovery of bioactive component [92].

United States Food and Drug Administration (US-FDA) has defined Maltodextrin as sensorially bland tasting (non-sweet) polysaccharide of α-1, 4-linked D-glucose monomeric units with <20 dextrose equivalent (DE) value [64]. It has also been suggested that low-DE maltodextrins can be potentially used as fat replacers in different fat-rich products [46]. DE value has great relevance with viscosity and Maillard browning ability of maltodextrins [4]. Maltodextrins are utilized in our body and give calories like starch (*viz.*, 4 kcal/g), however when they are used as fat alternatives, their actual concentration is very low and hence their net calorie contribution in the food product is approximately 1 kcal/g [25]. Encapsulates formed by maltodextrins showed very less stability of encapsulated bioactive compounds, which may be due to its poor emulsification ability and stability.

Desobry et al. reported that drum and freeze-dried β-carotene encapsulation from maltodextrin had higher encapsulation efficiency as compared to spray dried encapsulation [20]. Buffo and Reineccius [15] compared spray, tray, freeze, and drum drying techniques for encapsulating orange oil. They reported that highest encapsulation efficiency was obtained with freeze-drying technique. Zheng et al. [41] encapsulated blueberry polyphenols using ethyl cellulose added with lecithin. Finotelli et al. [24] reported complete encapsulation of ascorbic acid (AC) using maltodextrin as coating agent and spray drying as technique for encapsulation.

Gum Arabic is often considered as a suitable ingredient for encapsulation purposes due to its lower viscosity, high solubility, and superior emulsification capacity. It has shown to retain volatile compounds in the capsules, which was not observed for maltodextrin. Also, literature confirms it's functioning as a surface-active agent. However, gum Arabic is used at a very limited level by food companies due its higher cost than maltodextrin, which is further subjected to fluctuations [44, 79, 69].

Encapsulation using a mix of coating ingredients has also been applied to efficiently deliver the bioactive components. Recent report suggests of increased encapsulation efficiency of maltodextrin and gum Arabic with addition of gelatin along with them [43, 45, 93]. Karaca et al. [37] reported higher encapsulation efficiency of a mix of lentil protein isolate and maltodextrin for encapsulating flaxseed oil.

3.5 APPLICATION OF HERBS IN DAIRY PRODUCTS

3.5.1 FAT-RICH DAIRY PRODUCTS

In India, approximately 39% of milk production is converted into ghee and butter. Ghee comprises of 98% triglycerides and 0.3% moisture. For its high calorific content, the consumers are looking for complementary or alternative products, which may either provide the same sensory profile as *ghee* or *ghee* with enhanced functional attributes. Fat, particularly *ghee,* has a high capacity to imbibe the medicinal components of different herbs without losing their own qualities. Exploring this concept, about 55–60 *ghee* varieties with medicinal attributes are mentioned in *Ayurveda* and many of them are used even today for treatment of various diseases [32].

Herbs contain phenolic compounds, which are considered for their high antioxidant properties. Natural antioxidants (herbs) have been utilized for preparing different functional foods, which are specifically formulated for people affected with diseases, particularly the cardiovascular diseases [50]. Anti-oxidative attribute of herbs have resulted into their increased use in various fat-rich dairy products for decreasing the susceptibility to auto-oxidation and thus prolonging their shelf-life. Synthetic antioxidants (*viz.* butylated hydroxyl-toluene and butylated hydroxyl-anisole) are associated with the onset of cancer and other health problems. Moreover, consumers are increasingly demanding for additives derived from natural sources compared to synthetic-origin ingredients in food products [52]. Herbal extracts have many times higher antioxidant activity than their synthetic counterparts (*viz.*, BHA) [23]. There is not enough research evidence on herbal *ghee* compared to ghee with synthetic antioxidants (Table 3.1). It can be clearly observed that the herb fortified *ghee* has superior antioxidant activity compared to synthetic ones.

Żegarska et al. [94] reported that ethanolic extract of sage (@ 0.1–0.2%) addition to sweet cream prior to churning resulted in lower peroxide value of butterfat during storage at ambient and elevated temperatures. Alcoholic extracts obtained from rosemary, sage, and oregano were reported to retard deteriorative processes, like oxidation and lypolisis, in butter [7]. Ozkan et al. [53] reported that incorporation of hydro-distilled extract of *Satureja*

TABLE 3.1 Reported Examples of Herbal Supplemented Ghee

Type of Ghee	Medicinal Plant	Synthetic Antioxidant	References
Buffalo ghee	*Curcumin (Curcuma longa)*	BHA (butylated hydroxyl-anisole)	[42]
Butteroil	Caraway (*Carum carvil* L.);		
Clove (*Syzygium aromaticum* L.); and Coriander (*Coriandum sativum* L.)	BHT (butylated hydroxyl-toluene)		[5]
Cow ghee	Satavari (*Asparagus racemosus)*		[54]
	Ashavgandha (Withania somnifera		[62]
Sheep ghee	Rosemary *(Rosmarinus officinalis)*	Mixture BHA; and BHT (1:1)	[6]
	Ground sage (*Salvia officinalis*);		
Rosemary (*Rosmarinus officinalis*)			[26]

cilicica (@ 0.5, 1.0, and 2.0% levels) in butter resulted in corresponding increase in the antioxidant activity in a linear fashion with respect to the concentration of essential oil (EO). However, no such effects were observed for sour-cream butter [7]. Addition of extracts of clove, caraway, and coriander resulted in an increase in the antioxidant activity in butter oil [5] and *ghee* [54]. Medicated ghee is prepared by mixing one part of herbs with 4 parts of ghee and 16 parts of liquid (water, milk, or extract of herb), and the mixture is boiled till all water is evaporated. Once the boiling is completed, ghee is clarified, cooled to room temperature, and stored in appropriate containers.

All the medicated ghee formulations prescribed in *Ayurveda* are not meant for oral ingestion. Application of medicated ghee either by external or internal mode depends on the type of disease, category of the herb used, and treatment period prescribed. Moreover, many medicated ghee preparations are prescribed to be used for a particular period of time rather than for regular use. Arjuna *ghee* was developed by adding the functional components of *Terminalia arjuna* for exhibiting protective activity against CVD. The product had higher oxidative stability than the *ghee* devoid of herb addition

[60]. Furthermore, Arjuna *ghee* can be used as an alternative to normal ghee in our daily diet, which is not possible for medicated *ghee*.

3.5.2 FERMENTED MILK

Lassi, a fermented milk beverage of Indigenous origin, is prepared by fermenting milk using lactic cultures and served in salted or sweet form with sour flavor. Owing to its acceptability in both salted and sweetened form, *lassi* is also often considered as most appropriate vehicle for herbal bioactives delivery. Hussain et al. [31] prepared functional *lassi* by utilizing *Aloe vera* herb. Combination of *Lactobacillus paracasei* and NCDC-60 cultures were used @ 2% for its preparation. The *lassi* also showed functional activity in terms of immune protective effects on mice during animal feeding trials.

Srikanth et al. [83] added *Aloe vera* juice into *lassi* for functional attributes. Addition of *Aloe vera* juice (@ 15% level) in *lassi* yielded highest sensory attributes. Probiotic *dahi (*curd) added with *Aloe barbadensis* Miller herb was also prepared by Desobry et al. [20]. Surprisingly, *Aloe vera* addition did not limited the activity of *Lactobacillus paracasei* ssp. *paracasei* L in *dahi,* although literature reports antimicrobial activity of herbs. Viability of probiotic bacteria was higher than 7 log cfu/ml on the 12th day of refrigerated storage.

Addition of catechins @100–2000 ppm level maintained the viability of *Bifido-bacteria* in yoghurt [3]. Similarly, aloin, the active component of herb *Aloe vera*, improved the survival of *Bifidobacteria in* yoghurt [61]. Bakirci [8] incorporated different herbs (viz., *Anhriscus* sp., *Thymus* sp., *Allium* sp., and *Ferule* sp.) individually and these mixtures resulted in increased growth of *Lactobacillus bulgaricus* and *Streptococcus thermophilus.* Furthermore, the authors observed a linear relationship between herb level and growth of starter culture and reported an increase in the growth of cultures with corresponding increase in the level of herb. However, no such effects was observed by Sarabi-Jamab and Niazmand [70], who reported significantly no difference ($p>0.05$) among the viability of *Lactobacillus acidophilus* cultures in control yoghurt samples and in samples exposed to different concentrations of extract of *Ziziphora clinopodioides, Mentha piperita* during refrigerated storage at 4°C.

Peng et al. [55] developed a technology for the preparation of de-alcoholized yogurt beverage using *Hoveniadulcis* and *Pueraria lobata* herbs. Sensorially well-acclaimed yoghurt was prepared by Hassan et al. [27] using garlic extract.

Tarakci et al. [85] supplemented different herbs in labneh (a traditional fermented beverage of Indigenous origin) and studied the associated effects on organoleptic and physicochemical attributes during storage. They found that all herbs affected the sensory attributes but to a different extent, i.e., the effect was dependent upon the herb used and each herb having a typical intensity of the effect. The authors reported that highest sensory scores were obtained for labneh added with dill and parsley herb. Others [13, 51] have added extracts of green tea (catechins) to fruit-flavored milk drinks, chewing gum, and biscuits. ChingYun et al. [18] prepared fermented milk using a *wen-fu* soup (a Chinese herbal aqueous solution). Landge et al. [41] prepared *shrikhand* using Ashwagandha herb; and reported that addition of Ashwagandha powder (@ 0.5%) into *shrikhand* yielded sensorially acceptable product up to 52 days of refrigerated storage.

3.5.3 SANDESH

Sandesh is a popular sweetmeat of the Indian sub-continent. It is prepared using *chhana* (high-temperature acid coagulated milk product) analogous to cottage cheese. Herbs (*viz*., coriander, curry leaf, turmeric, spinach, and aonla (*Phyllanthus emblica*)) were individually incorporated @ 10% level into *sandesh* in their aqueous paste form [10]. The herbal *sandesh* samples were evaluated for total anti-oxidative status using *in-vitro* (Randox's) method and were compared among themselves and with TBHQ and combination of BHA: BHT (@ 1:1) @ 100 and 200 ppm levels, respectively. The authors reported that turmeric yielded highest antioxidant activity compared to all other herbs. Further, the authors also recommended use of coriander in food preservation for antimicrobial and antioxidant attributes. Because addition of coriander increased the keeping quality of *sandesh* up to 8 and 30 days at ambient and refrigerated temperature, respectively.

Bandyopadhyay [11] prepared the *chhana* by using two herbal coagulants. Herbal coagulants chhana was found to have different sensory (flavor, body, and texture, color, and appearances) and chemical composition (fat, protein, lactose, and ash) as compared to that conventional chhana.

3.5.4 CHEESE

Cheeses with added herbs in it for flavoring or functional ingredient are prepared traditionally in many countries under different names (in West

Asia: Turkey and Syria) such as Otlu, OtluLor, and OtluCacik [28]. Herbs are suggested to be added after coagulation of milk during cheese making [88]. Besides influencing the appearance and flavor, herbs also increase the keeping quality of cheeses [1] by exhibiting antimicrobial activity against spoilage causing micro-organisms. Herbs improve the development of flavor in cheeses during ripening by accelerating the lipolysis of fat and proteolysis. Besides this, herbs also improve the nutritional quality of cheeses by providing a typical vitamin and mineral profile [28]. About 25 different kinds of herbs (individually or combination) were commonly used to prepare herbal cheeses with addition @ 0.5 to 2% of the cheese curd [87]. A probiotic product (*ROSALACT*) was prepared by Mocanu et al. [48] by incorporating extracts of *Glycyrrhiza glabra* L. and *Rosa canina* L. into the probiotic cultures (*Bifidobacterium* sp., *Lactobacillus acidophilus*, *Streptococcus thermophilus*), which is inoculated and propagated in pasteurized milk to prepare the product. Sadeghi et al. [66] reported antimicrobial activity of EO of *Cuminum cyminum* in *Feta* cheese against *Staphylococcus aureus.* They reported that the synergistic activity of combination of *Cuminum cyminum* EO and *Lactobacillus acidophilus* against pathogens in *Feta* cheese.

Herbal extracts are also being utilized as coagulating ingredients during cheese preparation. Proteolytic activity of extracts of *Cynara humilis* L. and *Cynara cardunculus* L. have been utilized in Western Europe for preparation of farm chesses. Lamasa et al. [40] reported proteolytic activity of extract from *Cynara cardunculus* L. for hydrolysis of whey proteins, which results in increased digestible besides of yielding functional peptides. The application of these extracts in different food formulae has been recommended to improve texture and flavor attributes and decreasing the allergen of whey protein [14].

Dev [22] reported lower firmness for curd prepared using plant-derived milk coagulating agents as compared to firmness of curd prepared by chymosin as the coagulating agent, which the authors related to higher proteolytic activity of plant derive coagulating agents. Similarly, Dev [22] prepared cheese from ultra-filtered milk using *Cynara cardunculus* L. extract as coagulating agent; however, the resultant cheese had undesirable texture [2].

Herbal extracts also have preservative effect and the same has been exploited in different food products and cheese. Smith-Palmer et al. [82] reported antimicrobial activity of EO with herbs against pathogens *viz., Listeria monocytogenes,* and *Salmonella enteritidis* in soft cheeses. Further, the authors reported that their anti-microbial activity was more in cheeses having low fat than their fat counterparts, thereby suggesting an inverse relation between the fat content in soft cheese and antimicrobial activity. Application of herbal extracts in food products is restricted by intense inherent herbal flavors, which have a huge impact on the flavor of the product. To utilize the preservative effect of herbs without compromising with the sensory attributes of products, workers have recommended simultaneous application of other preservation technologies along with decreased levels of herbal extracts to achieve the desired purpose.

Tsiraki and Savvaidis [81] evaluated the preservative activity of basil EO under different packaging conditions on keeping quality of "*Anthotyros*," a Greek whey cheese. They reported that combination of MAP (Modified atmosphere packaging) and basil EO yielded highest preservative effect on increasing the keeping quality of Anthotyrosat refrigerated storage temperature (4 ± 1°C). Phenolic compounds found in herbs have been reported for their antimicrobial activity [75]. Rosenthal et al. [65] reported antimicrobial activity of tea catechins and ferulic acid against pathogenic bacteria with minor effect on the activity of LABs. Bullerman et al. [16] reported the activity of oleuropein to restrict the production of aflotoxins without affecting growth of fungi. Based upon these observations, oleuropein was recommended for use in mold-ripened cheeses to limit the aflatoxin production without affecting the mold growth [33].

3.6 INCORPORATION OF HERBS INTO OTHER MILK PRODUCTS

Peterson et al. [56] reported that tea leaves extract inhibited the production of Maillard browning aroma compounds in UHT (ultra high temperature) milk, both during processing and storage [76]. They recommended the Maillard reaction inhibiting action of herbs in prevention of the formation of toxic carcinogenic compounds in foods.

Sawale et al. [73, 74] observed the effect of *Pueraria tuberosa* on quality attributes of milk. The authors reported that herb addition was sensorially acceptable up to 0.4%level, and herb addition increased antioxidant activity while browning, pH, and heat stability were decreased in the milk samples. *In vivo* study also validated the functional activity (immunomodulatory and anti-oxidative potential) of herb in experimental mice [49]. Kamle et al. [35] reported that pineapple pulp addition improved the sensory acceptability of *burfi*. Prasad et al. [59] reported increase in antioxidant activity with incorporation of herb in powder and essential forms [57] and combination of EO increased the keeping quality of *burfi* [58]. Sandhya et al. [68] reported increase in keeping quality of curd by incorporation of pomegranate peel extract into it.

3.7 SUMMARY

Herbs have bioactive molecules responsible for several health benefits. Nowadays, research work has been focused on determining the functionality of herbal bioactives, their toxic effects, and delivery mechanism for supplementation in foods. International agencies are set to form the standards for these kinds of food products. However, addition of herbs (depending upon the herb type and its concentration) into foods may result into undesirable effects on their organoleptic attributes, which could ultimately decrease their acceptability. At present, most of the herb supplemented food products are having a typical flavor associated with that added herb and sometimes unnatural color also. This alternation in sensory quality restricts the application of herbs in foods products for functionality enhancement. Technological interventions are thus needed to allow successful incorporation of these nutraceuticals into food systems without alteration in the ideal sensory quality of food products.

KEYWORDS

- **cardiovascular diseases**
- **functional foods**
- **herbal medicine**
- **phytosterols**
- **polyphenols**
- **value addition**

REFERENCES

1. Ağaoğlu, S., Dostbil, N., & Alemdar, S., (2005). The antibacterial efficiency of some herbs used in herby cheese. *YYÜ Vet. Fak. Derg, 16*(2), 39–41.
2. Agboola, S. O., Chan, H. H., Zhao, J., & Rehman, A., (2009). Can the use of Australian cardoon (*Cynara cardunculus* L.) coagulant overcome the quality problems associated with cheese made from ultrafiltered milk? *LWT-Food Sci. Tech.*, *42*, 1352–1359.

3. Akahoshi, R., & Takahashi, Y., (1996). *Yoghurt Containing Bifidobacterium and Process for Producing the Same.* PCT-International Patent WO96/37113A1 (cited from the United Kingdom: Food Science Technology Abstracts (FSTA), 1997-08-P0149).
4. Akoh, C. C., (1998). Fat replacer. *Food Technology*, *52*(3), 47–53.
5. Ali, M., (2004). *Text Book of Pharmacognosy* (p. 328). New Delhi: CBS Publishers.
6. Amr, A. S., (1990). Role of some aromatic herbs in extending the stability of sheep ghee during accelerated storage. *Egypt. J. Dairy Sci.*, *18*(2), 335–344.
7. Ayar, A., Özcan, M., Akgül, A., & Akin, N., (2001). Butter stability as affected by extracts of sage, rosemary, and oregano. *J. Food Lipids*, *8*(1), 15–25.
8. Bakirci, I., (1999). The effects of some herbs on the activities of thermophilic dairy cultures. *Food Nahrung*, *43*(5), 333–335.
9. Ballabh, B., & Chaurasia, O. P., (2007). Traditional medicinal plants of cold desert Ladakh: Used in treatment of cold, cough and fever. *Journal of Ethnopharmacology, 112*, 341–346.
10. Bandyopadhyay, M., Chakraborty, R., & Raychaudhuri, U., (2007). Incorporation of herbs into sandesh, an Indian sweet dairy product, as a source of natural antioxidants. *Int. J. Dairy Tech., 60*(3), 228–233.
11. Bandyopadhyay, M., Chakraborty, R., & Raychaudhuri, U., (2008). Antioxidant activity of natural plant source in dairy dessert (*sandesh*) under thermal treatment. *LWT-Food Science and Technology, 41*, 816–825.
12. Basant, B., Lokesh, T., & Jorawar, S., (2018). Herbs: Way to enhance functionality of traditional dairy products. *Dairy and Vet. Sci. J.*, *6*(3), 555–689.
13. Bender, A. E., & Bender, D. A., (1994). *A Dictionary of Food and Nutrition* (5th edn., p. 420). New York: Oxford University Press.
14. Boza, J. J., Martinez-Augustin, O., & Gil, A., (1995). Nutritional and antigenic characterization of an enzymatic protein hydrolysate. *J. Agric. Food Chem.*, *43*, 872–875.
15. Buffo, R. A., & Reineccius, G. A., (2001). Comparison among assorted drying processes for the encapsulation of flavors. *Perfumer and Flavorist*, *26*, 58–67.
16. Bullerman, H., & Gourma, L. B., (1987). Effects of oleuropein on the growth and aflatoxin production by *Aspergillus parasiticus*. *Zeitschrift fur Lebensmittel-Untersuchung und-Forschung (Journal of Food Examination and Research), 20*, 226–228.
17. Champagne, C. P., & Fustier, P., (2007). Microencapsulation for the improved delivery of bioactive compounds into foods. *Current Opinion Biotechnology, 18*, 184–190.
18. ChingYun, K., MiaoLing, W., MeiJen, L., & ChienJung, H., (2009). Studies on the manufacture of functional fermented milk with Chinese herbs. *J. Taiwan Livestock Res.*, *42*(2), 109–120.
19. Chiou, D., & Langrish, T. A. G., (2007). Development and characterization of novel nutraceuticals with spray drying technology. *Journal of Food Engineering, 82*, 84–91.
20. Desobry, S. A., Netto, F. M., & Labuza, T. P., (1998). Comparison of spray-drying, drum-drying and freeze-drying for carotene encapsulation and preservation. *Journal of Food Science*, *62*(6), 1158–1162.
21. Dev, S., (1997). Ethno therapeutic and modern drug development: The potential of Ayurveda. *Curriculum Documents: Science, 73*(11), 909–928.
22. Esteves, C. L. C, Lucey, J. A., & Pires, E. M. V., (2002). Rheological properties of milk gels made with coagulants of plant origin and chymosin. *Int. Dairy J., 12*, 427–434.

23. Estévez, M., Ramirez, R., Ventanas, S., & Cava, R., (2007). Sage and rosemary essential oils versus BHT for the inhibition of lipid oxidative reactions in liver pâté. *LWT: Food Sci. Tech.*, *40*, 58–65.
24. Finotelli, P. V., & Rocha-Leao, M., (2005). *Microencapsulation of Ascorbic Acid in Maltodextrin and Capsule Using Spray Drying. https://www.academia.edu/7580668/microencapsulation_of_ascorbic_acid_in_maltodextrin_and_capsul_using_spray-drying* (accessed on 25 May 2020).
25. Frye, A. M., & Setser, C. S., (1993). Balking agent and fat substitute. In: Altschul, A. M., (ed.), *Low Calories Foods Handbook* (pp. 211–251). New York: Market, Dekker Inc.
26. Hasler, C. M., (1998). Functional foods: Their role in disease prevention and health promotion. *Food Technology*, *52*(11), 63–70.
27. Hassan, F. A. M., Helmy, W. A., Anab, A. K., Bayoumi, H. M., & Amer, H., (2010). Production of healthy yoghurt by using aqueous extract of garlic. *Arab. Univ. J. Agric. Sci.*, *18*(1), 171–177.
28. Hayaloglu, A. A., & Fox, P. F., (2008). Cheeses of Turkey: Three varieties containing herbs or spices. *Dairy Sci. Tech.*, *88*, 245–256.
29. Hazra, T., & Parmar, P., (2014). Natural antioxidant use in ghee: A mini review. *Journal of Food Research and Technology*, *2*(3), 101–105.
30. Hussain, S. A., Sharma, P., & Singh, R. R. B., (2011). Functional dairy foods: An overview. In: *Souvenir of International Conference on Functional Dairy Foods* (pp. 227–256). Karnal, NDRI.
31. Hussain, S. A., Sharma, P., & Singh, R. R. B., (2012). Product diversification opportunities and emerging technologies for the Indian dairy industry. In: *Souvenir of National Seminar on Dairying in Eastern India: Challenges and Opportunities* (pp 52–60). Indian Dairy Association (EZ) Bihar State Chapter, Patna.
32. Hussain, S. A., Patil, G. R., Singh, R. R. B., & Raju, N., (2015). Potential herbs and herbal nutraceuticals: Food applications and their interactions with food components. *Journal Critical Reviews in Food Science and Nutrition*, *55*, 94–122.
33. Jarvis, B., (1983). Mold and mycotoxins in moldy cheeses. *Microbiologie Aliments Nutr.*, *1*, 187–191.
34. Kaefer, C. M., & Milner, J. A., (2009). The role of herbs and spices in cancer prevention. *J. Nutr. Biochem.*, *19*, 347–361.
35. Kamble, K., Kahate, P. A., Chavan, S. D., & Thakare, V. M., (2010). Effect of pineapple pulp on sensory and chemical properties of burfi. *Veterinary World*, *3*(7), 329–331.
36. Kamboj, V. P., (2000). Herbal medicine. *Current Science*, *78*(1), 35–50.
37. Karaca, A. C., Nickerson, M., & Nicholas, H., (2013). Microcapsule production employing chickpea or lentil protein isolates and maltodextrin: Physicochemical properties and oxidative protection of encapsulated flaxseed oil. *Food Chemistry*, *139*, 448–457.
38. Kenyon, M. M., (1995). Modified starch, maltodextrin, and corn syrup solids as wall material for food encapsulation. In: Risch, S. J., & Reineccius, G. A., (eds.), *Encapsulation and Controlled Release of Food Ingredients* (pp. 43–50). ACS Symposium Series 590, Washington D.C. American Chemical Society.
39. Knowler, W. C., Barret-connor, E., Fowler, S. E., Hamman, R. F., Lachin, J. M., & Walker, E. A., (2003). Reduction in the incidence of type 2 diabetes with lifestyle intervention or metafornima. *New England Journal of Medicine, 346*, 393–403.

40. Lamasa, E. M., Barrosa, R. M., Balcao, V. M., & Malcata, F. X., (2001). Hydrolysis of whey proteins by proteases extracted from *Cynara cardunculus* and immobilized onto highly activated supports. *Enzyme and Microbial Technol., 28*, 642–652.
41. Landge, U. B., Pawar, B. K., & Choudhari, D. M., (2011). Preparation of shrikhand using Ashwagandha powder as additive. *Dairying, Foods and Home Sciences, 30*(2), 79–84.
42. Lodh, J., Khamrui, K., & Prasad, W. G., (2018). Optimization of heat treatment and curcumin level for the preparation of anti-oxidant rich ghee from fermented buffalo cream by central composite rotatable design. *Journal of Food Science and Technology, 55*(5), 1832–1839.
43. Luca, A., Cilek, B., Hasirci, V., Sahin, S., & Sumnu, G., (2013). Effect of degritting of phenolic extract from sour cherry pomace on encapsulation efficiency: Production of nano-suspension. *Food Bioprocess Technology, 6*(9), 2494–2502. e-article, doi: 10.1007/s11947-012-0880-z 2013.
44. Mahdavee, K., Jafari, S. M., Ghorbani, M., & Hemmati, K., (2014). Application of maltodextrin and gum Arabic in microencapsulation of saffron petal's anthocyanins and evaluating their storage stability and color. *Carbohydr. Polym., 25*(105), 57–62.
45. Malacridaa, R. C., & Telisa, V. R. N. (2011) *Effect of Different Ratios of Maltodextrin/ Gelatin and Ultrasound in the Microencapsulation Efficiency of Turmeric Oleoresin.* https://pdfs.semanticscholar.org/b095/4207222f68ee01383c2fb4c0203a81bfdf9f.pdf?_ga=2.110100741.140053027.1592366996-798819573.1560248475 (accessed on 25 May 2020).
46. Marshall, C. M., Beeftlink, H., & Trammer, J., (1999). Toward a rational design of commercial maltodextri. *Trend of Food Science Technology, 10*, 345–355.
47. McClements, D. J., Decker, E. A., Park, Y., & Weiss, J., (2009). Structural design principles for delivery of bioactive components in nutraceuticals and functional foods. *Critical Review in Food Science and Nutrition, 49*(6), 577–606.
48. Mocanu, G. D., Rotaru, G., Botez, E., Vasile, A., Gîtin, L., Andronoiu, D., & Nistor, O., (2009). Research concerning the production of a probiotic dairy product with added medicinal plant extracts. *Food Tech., 32*, 37–44.
49. Moharkar, K., Sawale, P. D., Sumit, A., Suman, K., & Singh, R. R. B., (2013). *In vivo* effect of *Withania somnifera* on immunomodulatory and antioxidative potential of milk in mice. *Food and Agricultural Immunology, 25*(3), 443–452.
50. Najgebauer-Lejko, D., Grega, T., Sady, M., & Domagała, J., (2009). The quality and storage stability of butter made from sour cream with addition of dried sage and rosemary. *Biotechnology in Animal Husbandry, 25*(5–6), 53–761.
51. O'Connella, J. E., & Fox, P. F., (2001). Significance and applications of phenolic compounds in the production and quality of milk and dairy products: A review. *Int. Dairy J., 11*, 103–120.
52. Özcan, M., (2003). Antioxidant activity of rosemary, sage and sumac extracts and their combinations on stability of natural peanut oil. *J. Medicinal Food, 6*(3), 267–270.
53. Ozkan, G., Simsek, B., & Kuleasan, H., (2007). Antioxidant activities of *Satureja cilicica* essential oil on butter and *in vitro*. *J. Food Eng., 79*, 1391–1396.
54. Pawar, N., Arora, S., Singh, R. R. B., & Wadhwa, B. K., (2012). The effects of *Asparagus racemosus* (Shatavari) extract on oxidative stability of ghee, in relation to added natural and synthetic antioxidants. *International Journal of Dairy Technology, 65*(2), 293–299.

55. Peng, D., Ya, D. Z., Ying, M., Yuan, G., & Dong, L., (2010). Production and function of dealcohol yogurt beverage. *China Dairy Industry*, *38*(1), 26–28.
56. Peterson, D. G., & Totlani, V. M., (2005). Influence of flavonoids on the thermal generation of aroma compounds. In: Shahidi, F., (ed.), *Phenolics in Foods and Natural Health Products* (pp 143–160). American Chemical Society, Washington, DC. ACS Symposium Series #909.
57. Prasad, W., Khamrui, K., Mandal, S., & Badola, R., (2018). Effect of combination of essential oils on physicochemical and sensorial attributes of burfi in comparison with individual essential oil and BHA. *International Journal of Dairy Technology*, *71*(3), 810–819.
58. Prasad, W., Khamrui, K., & Sheshgiri, S., (2017). Effect of packaging materials and essential oils on the storage stability of burfi. *Journal of Packaging Technology and Research, 1*(3), 181–192.
59. Prasad, W., Khamrui, K., Mandal, S., & Badola, R., (2017). Anti-oxidative, physicochemical and sensory attributes of burfi affected by incorporation of different herbs and its comparison with synthetic anti-oxidant (BHA). *Journal of Food Science and Technology*, *54*(12), 3802–3809.
60. Prasher, R., (1999). Standardization of *Vasa ghrita* and its extract form and their comparative pharmaco-clinical study with special reference to *Swasa roga* (Asthma). *Master in Medicine Thesis* (p. 227). Gujarat Ayurved University, Jamnagar, India.
61. Pszczola, D. E., (1998). ABC's of nutraceutical ingredients. *Food Technology*, *52*(3), 30–37.
62. Purohit, A. V., (2012). Effect of herb extract (*Withania somnifera*/Ashwagandha) incorporation on storage stability of ghee. *M.Tech. Thesis for NDRI* (p. 180). Karnal.
63. Rabe, T., & Staden, J. V., (1997). Antibacterial activity of South African plants used for medicinal purposes. *Journal of Ethnopharmacology, 56*, 81–87.
64. Roller, S., (1996). Starch derived fat mimetics: Maltodextrin. In: Roller, S., & Jones, S. A., (eds.), *Handbook of Fat Replacer* (pp. 99–118). Boca Raton, CRC Press.
65. Rosenthal, I., Bernstein, S., & Nakimbugwe, D. N., (1999). Effects of tea solids on milk. *Milchwissenschaft, 54*, 149–152.
66. Sadeghi, E., Basti, A. A., Misaghi, A., Salehi, T. Z., & Osgoii, S. B., (2010). Evaluation of effects of *Cuminum cyminum* and probiotic on *Staphylococcus aureus* in feta cheese. *J. Medicinal Plants*, *9*(34), 131–141.
67. Samy, P. R., & Ignacimuthu, S., (2000). Antibacterial activity of some folklore medicinal plants used by tribals in Western Ghats of India. *Journal of Ethnopharmacology, 69*, 63–71.
68. Sandhya, S., Khamrui, K., Prasad, W., & Kumar, M. C., (2018). Preparation of pomegranate peel extracts powder and evaluation of its effect on functional properties and shelf life of curd. *LWT*, *92*, 416–421.
69. Sankarikutty, B. M., Sreekumar, C. S., & Mathew, A. G., (1988). Studies on microencapsulation of cardamom oil by spray drying technique. *Journal of Food Science Technology, 25*, 352–358.
70. Sarabi-Jamab, M., & Niazmand, R., (2009). Effect of essential oil of *Mentha piperita* and *Ziziphora clinopodioides* on *Lactobacillus acidophilus* activity as yogurt starter culture. *American-Eurasian J. Agric. Environ Sci.*, *6*(2), 129–131.
71. Sawale, P. D., Patil, G. R., Hussain, S. A., Singh, A. K., & Singh, R. R. B., (2017). Release characteristics of polyphenols from microencapsulated *Terminalia arjuna*

extract: Effects of simulated gastric fluid. *International Journal of Food Properties*, *20*(12), 3170–3178. doi.org/10.1080/10942912.2017.1280677.
72. Sawale, P. D., Pothuraju, R., Hussain, S. A., Anuj, K., Kapila, S., & Patil, G. R., (2016). Hypolipidemic and anti-oxidative potential of encapsulated *Terminalia arjuna* added vanilla chocolate milk in high cholesterol fed rats. *Journal of Science of Food and Agricultural*, *96*(4), 1380–1385. doi: 10.1002/jsfa.7234.
73. Sawale, P. D., Singh, R. R. B., Arora, S., Kapila, S., Rastogi, S., & Rawat, A., (2012). *In vivo* immnomodulatory and antioxidative potential of *Pueraria tuberose*- milk model in mice using milk as the carrier. *International Journal of Dairy Technology*, *66*(2), 202–206. doi: 10.1111/1471-0307.12011.
74. Sawale, P. D., Singh, R. R. B., Arora, S., & Kapila, S., (2014). Stability and quality of *Pueraria tuberosa*-milk model system. *Journal of Food Science and Technology, 52*(2), 1089–1095. doi: 10.1007/s13197-013-1067-y.
75. Schaller, F., Rahalison, L., Islam, N., Potterat, O., Hostettmann, K., Stoeckli-Evans, H., & Mavi, S., (2000). New potent antifungal 'quinone methide' diterpene with a cassane skeleton from *Bobgunnia madagascariensis*. *Helvetica. Chimica. Acta*, *83*, 407–413.
76. Schamberger, G. P., & Labuza, T. P., (2007). Effect of green tea flavonoids on Maillard browning in UHT milk. *LWT-Food Sci. Tech.*, *40*, 1410–1417.
77. Sharma, A., Tyagi, C. L. K., Singh, M., & Rao, C. V., (2008). Herbal medicine for market potential in India: An overview. *Academic Journal of Plant Sciences*, *1*(2), 26–36.
78. Sharma, R., (2005). Market trends and opportunities for functional dairy beverages. *Australian Journal of Dairy Technology, 60*(2), 196–199.
79. Shiga, H., Yoshii, H., & Nishiyama, T., (2001). Flavor encapsulation and release characteristics of spray dried powder by the blended encapsulant of cyclodextrin and gum Arabic. *Drying Technology*, *19*, 1385–1395.
80. Singh, H., (2006). Prospects and challenges for harnessing opportunities in medicinal plants sector in India). *Law, Environment and Development Journal*, pp. 186–196. http://www.lead-journal.org/content/06196.pdf (accessed on 25 May 2020).
81. Singh, R. R. B., & Hussain, S. A., (2011). Application of herbs in development of functional milk and milk products. In: *Compendium of National Training Programme on "Basic and Applied Approaches in Designing of Dairy based Nutraceuticals and Functional Foods"* (pp. 19–24). NDRI, Karnal, India.
82. Smith-Palmer, A., Stewart, J., & Fyfe, L., (2001). The potential application of plant essential oils as natural food preservatives in soft cheese. *Food Microbiol.*, *18*(4), 463–470.
83. Srikanth, K., Kartikeyan, S., Adarsh, K., & Punitha, K., (2016). *Aloe vera* and its application in dairy and food products. *Research Journal of Animal Husbandry and Dairy Science, 7(*2), 84–90.
84. Subhose, V., (2005). Basic principles of pharmaceutical science in Ayurveda. *Bull. Indian Inst. Hist. Med. Hyderabad*, *35*, 83–92.
85. Tarakci, Z., Temiz, H., & Ugur, A., (2011). The effect of adding herbs to labneh on physicochemical and organoleptic quality during storage. *Int. J. Dairy Technol.*, *64*(1), 108–116.
86. Tarakci, Z., (2004). The influence of *Prangos* sp. on characteristics of vacuum-packed Van herby cheese during ripening. *Milchwissenschaft*, *59*, 619–623.
87. Tolve, R., Galgano, F., & Caruso, M., (2016). Encapsulation of health promoting ingredients: Applications in foodstuffs. *International Journal of Food Sciences and Nutrition*, *67*(8), 888–892.

88. Tekinşen, O. C., (1997). *Dairy Products Technology* (pp. 224–226). *S.Ü. Vet. Fak Yay,* Konya s.
89. Trevelyan, J., (1993). Herbal medicine. *Nursing Times*, *89*, 36–38.
90. Veale, D. J. H., Furman, K. I., & Oliver, D. W., (1992). South African traditional herbal medicines used during pregnancy and childbirth. *Journal of Ethno Pharmacology, 36*, 185–191.
91. Wang, Y., Zhaoxin, L., Fengxia, L., & Xiaomei, B., (2009). Study on microencapsulation of curcumin pigments by spray drying. *European Food Research and Technology, 229*, 391–396.
92. Williams, P. A., & Phillips, G. O., (2000). Gum Arabic. In: Phillips, G. O., & Williams, P. A., (eds.), *Handbook of Hydrocolloids* (pp. 155–168). Cambridge: Woodhead Publishing Limited.
93. Yoshii, H., Soottitantawat, A., & Liu, X. D., (2001). Flavor release from spray-dried maltodextrin/gum Arabic or soy matrices as a function of storage relative humidity. *Innovative Food Science and Emerging Technologies, 2*, 55–61.
94. Żegarska, Z., & Rafałowski, R., (1997). *Antioxidant Effect of Sage Extract in Relation to Butter* (pp. 227–229). Scientific Communications, Olsztyn.
95. Zheng, L., Ding, Z., Zhang, M., & Sun, J., (2011). Microencapsulation of bayberry polyphenols by ethyl cellulose: Preparation and characterization. *Journal of Food Engineering, 104*, 89–95.

Part II

Applications of Novel Biocompounds in Quality and Safety of Foods

CHAPTER 4

ANTIFUNGAL LACTIC ACID BACTERIA (LAB): POTENTIAL USE IN FOOD SYSTEMS

SAURABH KADYAN and DIWAS PRADHAN

ABSTRACT

Fungal spoilage in food is one of the principal causes of food deterioration and wastages. Although, considerable advancements have been made in the traditional methods of fungal control, alternative methods using antifungal LAB and its metabolites offer a more natural solution in the wake of demand for minimally processed, natural foods. LAB produces an array of antifungal metabolites (such as organic acids, phenyllactate, fatty acids, proteinaceous compounds, volatiles, reuterin, etc.), which either act individually or in synergism with each other in different food systems. Many research reports have elucidated the desired inhibition of fungal contaminants present in various foods such as dairy, breads, fruits, and vegetables, animal feeds, etc., by LAB. Although, commercial antifungal formulations based on LAB are presently available for application in different food formats, more detailed research is still needed in terms of final food quality, stability during processing, mechanism of action, synergism activities, safety, etc. to fully harness their potential for commercial food exploitation.

4.1 INTRODUCTION

Fungi are common food contaminants that play a pivotal role in the spoilage and reduction of marketable quality, hygiene, and safety of food materials to an unacceptable level. The fungal spoilages are responsible for significant food waste and contribute enormously to the postharvest losses of food products. In addition, fungal contaminants also compromise safety of the foods by generating poisonous secondary metabolites called mycotoxins. Hence, the control and elimination of fungal contaminants has become a major challenge for food producers and processors including scientists, who are continuously seeking for newer and effective remedies to avert or control fungal spoilage in foodstuffs. Apart from the advances in the existing technologies in the control of fungal spoilages, recently, new biopreservation strategies by utilizing bioprotective potentials of Lactic acid bacteria (LAB) and its antifungal metabolites have also gathered a lot of attention.

LAB is a multifarious class of microorganisms characterized by standard criteria of lactic acid production in major amounts as an end-product during carbohydrate fermentation. The group composed of members with distinct characteristics as Gram-positive, aerotolerant, non-motile, non-sporulating having either rod or coccus morphology. At present, about 17 bacterial genera are categorized as LAB [15]. LAB has a complex nutritional requirement due to which they are found mostly in nutrient-rich habitats, naturally occurring in food products like dairy, cereal, meat, vegetables, etc., and many others in the intestinal linings of humans and other animals. Traditionally, the fermentative metabolism of LAB has been a very important attribute of these microbes in yielding variety of fermented foods across the globe.

The LAB causes rapid acidification of food products, which is followed by many changes in the flavor, body, and texture. In addition to the change in the organoleptic properties of the food material, LAB also acts as a biopreservative microorganisms and the final product has an enhanced shelf-life than the raw material. During its growth in the food products, LAB produces variety of antimicrobial metabolites that exert significant antagonistic activity against many related and unrelated microorganisms, thereby provides advantage of *in situ* shelf-life extension of fermented food products.

Nowadays, there is an increase in the demand of minimally-processed natural foods due to which LAB and its biopreservative metabolites have been looked up as a possible alternative to chemical preservatives added to foods. LAB is ideal candidates for commercial exploitation since many of them have been attributed to GRAS and QPS status by US-FDA and EU, respectively. Hence, LAB possess healthy and natural image. Recently, the beneficial health effects of certain LAB have also been recognized by many researchers due to which many LAB are now commonly used as Probiotics.

The biopreservative potential of LAB is well known and many of the earlier reports have exclusively elaborated on the various antimicrobial substances derived from LAB. These metabolites are diverse in nature depending on the producing LAB species and can have a narrow to a wide spectrum of antimicrobial activity. The metabolites of LAB can inhibit both bacterial as well as fungal contaminants, however, in this chapter, we are going to discuss only the antifungal metabolites derived from LAB although the same metabolites may also have antibacterial activity.

This chapter focuses on the antifungal potential of LAB derived *via* specific metabolites, which are secreted through various biochemical pathways, their modes of action, and their applications as bioprotective cultures or metabolic mixes in a wide variety of food materials.

4.2 ANTIFUNGAL COMPOUNDS FROM LACTIC ACID BACTERIA (LAB)

LAB have been able to secrete several antimicrobial compounds through various fermentation pathways including primary and secondary carbon metabolism, bioconversion, or peptide synthesis, which result in the synthesis of simple and complex biomolecules. These biocompounds exhibit strong antibacterial and antifungal properties, acting in synergism in complex food environments [58].

According to Figure 4.1, the inhibition mechanism of some of the single compounds studied neglecting synergistic effects includes membrane destabilization, inhibition of cell wall synthesizing enzymes, proton gradient interference, creation of reactive oxygen species (ROS), and induction of oxidative stress, etc. LAB produces an assortment of antifungal agents; and their antagonistic mechanism considering the additive effects utilizes any of the above interdependent modes of action. The various antifungal compounds along with its respective producing LAB and their mode of action are outlined in Table 4.1.

4.2.1 ORGANIC ACIDS

LAB principally derives their energy by fermentation, either homo-fermentation or hetero-fermentation of preformed organic carbon resulting in the production of one or more organic acids. Majority of these organic acids are originated as end-products of primary carbon metabolism (e.g., lactic, acetic, formic, propionic, and succinic acid), however, some are obtained from secondary carbon metabolism (e.g., salicylic, and benzoic acids) and others by *de novo* synthesis (e.g., vanillic, and p-coumaric acid). The formation of these organic acids is the primary means of preserving several food types, such as fermented milk, pickles, sausages, etc. These weak organic acids display strong antagonism against spoilage and pathogenic microbiota in different fermented foods.

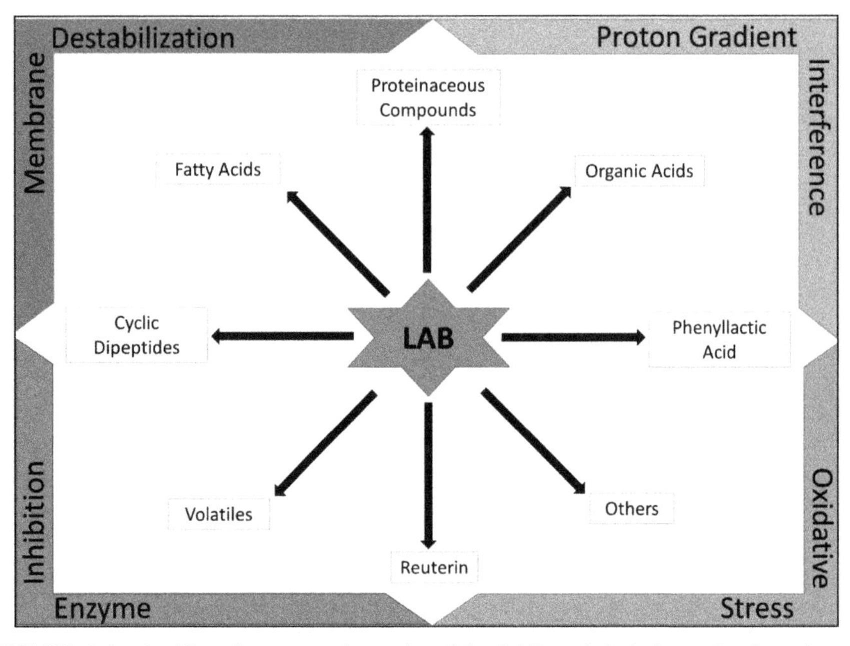

FIGURE 4.1 Antifungal compounds produced by LAB and their interrelated modes of action.

TABLE 4.1 Antifungal Compounds Produced by LAB and Their Inhibitory Mechanisms

Antifungal Compound	LAB	Mode of Action	References
Organic acids	All LAB	Proton Gradient Interference/Cell wall-associated Proton pump inhibition	[12, 65]
1) Primary carbon metabolism	Heterofermentative LAB		[57]
a) Lactic acid	*L. buchneri*, *L. sanfranciscensis*, *L. diolivorans*		[23]
b) Acetic acid			[9]
c) Propionic acid	*L. paracasei* sub sp. *paracasei*		[25]
d) Succinic acid			[51]
2) Secondary Carbon Metabolism/ Bioconversion/Others	*L. casei*, *S. thermophilus*, *L. helveticus*		
a) Benzoic acid	*L. amylovorus*		
b) Salicyclic acid	*L. reuteri*		
c) p-Coumaric acid	*L. plantarum*		
d) Methylcinnamic acid	*L. amylovorus*		

TABLE 4.1 *(Continued)*

Antifungal Compound	LAB	Mode of Action	References
Phenyllactic acid	*L. plantarum, L. alimentarius, P. pentosaceus*	Proton Gradient Interference and enzyme inhibition in synergism with other metabolites	
Fatty acids	*L. plantarum, L. hammesii*	Cell wall instability and Permeability	[6, 59]
Volatiles:	Heterofermentative LAB	Enzyme Inhibition and Membrane Permeability	
1. Carbon Dioxide			
2. Ethanol	Heterofermentative LAB		
3. Diacetyl	*Lactococcus, Leuconostoc, Lactobacillus*, and *Pediococcus* spp.		
Proteinaceous compounds (antifungal peptides)	*L. plantarum, L. fermentum, L. rhamnosus, L. brevis* and *L. paracasei*	Cell wall instability and Permeability	[8, 36, 38, 39]
Cyclic Peptides	*L. plantarum, L. sakei, L. coryniformis, L. casei*	Membrane permeability and proton gradient interference	[28, 41, 61]
Reuterin/Acrolein	*L. reuteri*	Oxidative Stress	[21]
Others	All LAB	Oxidative Stress/ Membrane permeability	[41]
1. Hydrogen peroxide	*L. plantarum*		[63]
2. Mevalonolactone	*L. plantarum*		
3. δ-dodecalactone			

4.2.1.1 ORGANIC ACIDS FROM PRIMARY CARBON METABOLISM

Lactic acid (CH_3-CH(OH)-COOH) is the major metabolite produced by both homo-fermentative and hetero-fermentative LAB. Additionally, hetero-fermentative LAB produces acetic acid (CH_3-COOH) in major and propionic acid (CH_3-CH_2-COOH) in minor amounts, respectively. The antagonistic activity of organic acids against numerous pathogens and spoilage bacteria or fungi can be attributed to low pH conditions making them lipo-soluble so that they can easily diffuse through the cell membrane and reach cell cytoplasm [44, 50]. Inside cytoplasm, acids dissociate with the release of hydrogen ions thereby disturbing intracellular pH homeostasis leading to membrane disruption, obstruction of crucial metabolic reactions, and finally lead to cell death. The pKa value of respective acid and pH of medium influences significantly biocidal activity of organic acids. Higher the pKa more is the acid dissociation in ionic form thereby aggravating intracellular pH changes in the cell cytoplasm and enhanced killing effect due to antifungal activity.

Compared to lactic acid, acetic, and propionic acids possess a higher value of acid dissociation constant pKa values at low pH for dissociation inside the cells. It is believed that organic acids work in synergy with other metabolic compounds in enhancing the biopreservation as evidenced by their low minimal inhibitory concentration (MIC) in combination when compared on individuals [2]. The synergistic approach using acetic acid and lactic acid in curbing fungal growth has already been elucidated [16]. *L. sanfranciscensis* CB1 produced broad spectrum of organic and fatty acids including acetic, butyric, caproic, formic, valeric, and propionic acid with the potential to restrain the growth of a wide array of bread molds [12].

4.2.1.2 ORGANIC ACIDS FROM SECONDARY CARBON METABOLISM/BIOCONVERSION

Benzoic acid (C_6H_5-COOH), a fungistatic agent, is frequently used as a preservative in foods especially dairy-based products. Strains of *L. acidophilus*, *L. helveticus,* and *L. casei* can transform hippuric acid, a component present in milk, into benzoic acid naturally during fermentation. Additional pathways for the production of benzoic acid by LAB include phenylalanine degradation and auto-oxidation of benzaldehyde. Its fungicidal action involves stress response in cells by inhibiting plasma-membrane H^+-ATPase proton pump thereby hampering homeostasis and depletion of available energy required for cell maintenance and growth. Further, various carboxylic acid and their derivatives (acetic, propanoic, vanillic acid, butanoic acids, 4-hydroxy benzoic acid) and other volatile compounds exhibited chao-trophicity (i.e., weakening of non-covalent interactions between macromolecules especially proteins and nucleic acids (NAs)) [13]. Other organic acids (*viz.* p-coumaric and cinnamic acid derivatives) are proposed to be synthesized by bioconversion of extracellular compounds in plant sources.

The cell-free supernatant of *L. buchneri* UTAD 104 containing complex mixture of organic acids inhibited *Pencillum nordicum* and its toxin production [24]. Strains namely *L. reuteri*, *L. amylovorus*, *L. plantarum,* and *Weisella cibaria* produced diverse complex of antifungal carboxylic acids derived *via* secondary carbon metabolism or bioconversion acting in synergism against fungal microflora. Acids include cinnamic acid derivatives, salicylic acid, vanillic, azelaic, p-couramic, and 4-hydroxybenzoic acid [10, 25, 51]. Understanding the complex synergistic mechanisms of organic acids among themselves and other bioprotective compounds play a key role in the shelf-life extension of food products *de novo*.

4.2.2 PROTEINACEOUS ANTIFUNGAL COMPOUNDS

LAB is a predominant producer of proteinaceous compounds in the form of peptides synthesized either ribosomally or non-ribosomally along with enzymatic proteolysis. The structure of ribosomally synthesized peptides differs but mostly they are amphiphilic and possess cationic properties. Antifungal peptides (AFPs) belong to niche area of antimicrobial peptides (AMPs) or bacteriocins shaving specific activity against fungal infections with the added advantage of the development of no resistance by target pathogens. Most of the AFPs are produced *de novo* during ribosomal synthesis by mRNAs (messenger ribonucleic acid), while some are produced by the action of proteases of lactic cultures on various substrate proteins *viz.* wheat, milk, sourdough, etc.

Proteolysis of specific protein-rich substrates by LAB is possible only due to presence of three proteolytic enzyme systems in them: (i) cell envelope associated proteinase responsible for hydrolyzing protein into a pool of peptides; (ii) transfer of theses peptides into the cell cytoplasm via peptide transport systems; and (iii) breakdown of transferred peptides into smaller peptides or amino acids by numerous intracellular peptidases. These compounds largely produced by *Lactobacillus* species along with other genera of *Streptococcus*, *Lactococcus*, and *Pediococcus* are active against a broad spectrum of food associated fungi.

AFPs often lose their activity after treatment with proteolytic enzymes because of their proteolytic nature. AFPs comprising of cationic peptides with high hydrophobicity display antifungal traits by enabling cell wall cellular lysis. Isolation and purification of AFPs has been documented from various food product hydrolysates, e.g., wheat germ, kefir, sour cream, Chinese traditional fermented meat Dong [11, 36, 48].

Number of milk-derived peptides like lactoferricin (Lf) and Isracidin differ from AFPs as they are produced by action of digestive enzymes-chymosin/trypsin on milk proteins (lactoferrin, casein, and whey proteins) that possess antifungal activity. Recently, bioactive peptides SSSEESII and DMPIQAFLLY derived from action of metallo-protease from *L. lactis* on αS-2 casein and cleavage of β-casein by *L. rhamnosus/L. paracasei* have been isolated [8, 36]. Furthermore, a novel peptide FPSHTGMSVPPP from *de novo* synthesis of *L. plantarum* IS10 active against *Aspergillus flavus, Eurotium rubrum,* and *Penicillium roqueforti* have also been reported [38].

4.2.3 CYCLIC DIPETIDES

Cyclic dipeptides (CDPs), often referred as 2,5-dioxopiperazines, are another class of low molecular weight (MW) peptide derived compounds associated with antimicrobial, antiviral, and antitumoral activities [47]. Their MIC is comparatively higher as compared with other LAB metabolites suggesting their mode of inhibitory action as an additive effect with the other metabolites. Bioactive CDPs are chiral molecules with diverse side chains thereby making them suitable candidates for drug delivering agents. They can be used as signal molecules with antiviral and antimicrobial potential [7, 55]. For instance, cyclo(L-Phe-L-Pro), cyclo(L-Ala-L-Val), and cyclo(L-Pro-L-Tyr) activate intracellular signal molecules based on LuxR-quorum sensing systems [26]. Several studies have elaborated their antifungal potential.

Cyclo(glycyl-L-leucyl), cyclo(Phe-Pro), and cyclo(L-Phe-trans-4-OH-L-Pro) were isolated using culture filtrate of *L. plantarum* strains possessing antifungal activities against molds *F. avenaceum, A. fumigatus* and *P. roqueforti* [41, 61]. Recently, three CDPs containing proline produced by *Leuconostoc mesenteroides* LBP-K06 have been recovered from Korean fermented product 'Kimchi' against MDR *S. aureus and S. typhimurium* [33]. In another study, cis-cyclo (Phe-Pro) and cyclo(Val-Pro) produced by strain *L. plantarum* LBP-K10 was found effective against *Candida albicans* [28]. Also, the production of cyclo(Phe-Pro) and cyclo(Phe-OH-Pro) by strains such as *L. sakei* and *L. coryniformis* was suggestive of common LAB metabolites [34].

4.2.4 PHENYLLACTIC ACID (PLA)

Phenyllactic acid (PLA) is an antimicrobial agent with broad specificity possessing both antibacterial and antifungal properties. The compound shows antagonistic activity against many pathogens including *Listeria monocytogenes* [19], *Escherichia coli* serotype O157:H7, *Yersinia enterocolitica* and methicillin-resistant *Staphylococcus aureus* [43] as well as food-spoiling fungi, including *Candida* and *Rhodotorula* strains of yeasts [57] and various molds, such as *Penicillium verrucosum*, *Aspergillus flavus*, and *Penicillium commune* [30]. PLA, a naturally occurring phytochemical in honey, have gained popularity due to: its production in various LAB fermented products, no cytotoxicity to animals and human cell lines, and absence of any objectionable odor. It acts in synergy with other antifungal substances for inhibition of spoilage microflora.

The PLA biosynthesis pathway in LAB is well characterized wherein precursor phenylalanine is metabolized firstly to phenylpyruvic acid (PPA) by transamination reaction of aminotransferase enzyme and subsequent reduction to PLA by a dehydrogenase (Figure 4.2) [37]. Several studies have reported that enzyme lactate dehydrogenase is accounted for transformation of PPA to PLA. The bioconversion of phenylalanine to PLA is a rate-determining step and creation of later can be enhanced 14X by adding precursor PPA to the growth medium [32]. Presence of tyrosine in medium suppresses the production of PLA as aromatic aminotransferase is also involved in catalyzing tyrosine to 4-hydroxy PPA. Replacement of peptone with corn steep liquor in MRS agar further improves PLA production. The PLA-producing strains had been found to restrict the "bread rope" defect mediated by aerobic spore-formers (*Bacillus* spp.) and the spoilage molds in yeast-leavened doughs [18].

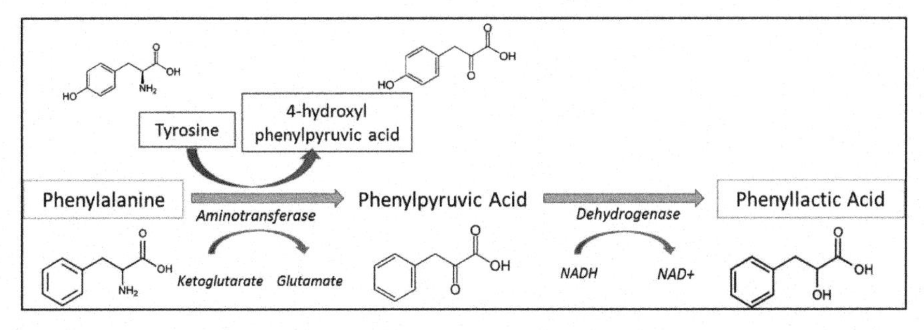

FIGURE 4.2 Possible pathways of phenyllactic acid synthesis. (Source: Modified significantly from Yu et al. [64].)

4.2.5 VOLATILE COMPOUNDS

Volatile organic compounds are produced during mixed acid or hetero-fermentation by many LAB, which include ethanol, diacetyl, and carbon dioxide. Volatiles are involved in antifungal activity in synergism with other metabolites. However, culture filtrate soon gets deprived of these volatiles during cell removal stages of filtration, clarification, and heating of supernatant before mixing with agar. Hence, their antifungal activity is often overlooked.

Diacetyl (2,3-butanodione) is a routinely occurring volatile flavoring component in fermented milk products produced usually by citrate fermenting LAB strains with concentration ranging from 0.045 and 27 μg/g and exhibits antimicrobial activities at low pH. The antimicrobial activity of diacetyl can be attributed to inactivation of arginine-binding proteins thereby limiting arginine utilization [27]. Other volatile compounds exhibiting antifungal activity associated with certain LAB includes carbon dioxide and ethanol. The CO_2 and ethanol are mainly produced by strains of heterofermentative LAB. Antimicrobial effect of CO_2 is due to development of anaerobiosis there by inhibiting biocatalytic decarboxylation causing subsequent buildup of CO_2 in the membrane lipophilic bilayer causing malfunctioning in permeability. Ethanol, on the other hand, helps in denaturing proteins and dissolving lipids against many fungi.

4.2.6 FATTY ACIDS

Selected strains of lactobacilli and lactococci possess lipid hydrolyzing spectrum that can lead to the production of vast amounts of free fatty acids in certain products, e.g., fermented sausages. Fatty acids and their hydroxyl derivatives exhibit both antibacterial and antimycotic properties with longer chain length acids contributing more towards antagonism than short-chain fatty acids. For instance, Lauric (C_{12}) and capric (C_{10}) fatty acids were remarkably effective against *C. albicans* [5].

The structure of fatty acid containing at least one hydroxyl group and one degree of unsaturation in their carbon moiety plays a vital role in antifungal activity against many yeasts and molds. However, yeasts are substantially more responsive to the fatty acids than molds. Antifungal fatty acids disrupt the membrane integrity by partitioning the lipid bilayers of fungal species. Cell wall instability and permeability causes uncontrolled release of intracellular components, ultimately resulting in cytoplasmic collapse of fungal cells [3]. Racemic mixtures of four antifungal hydroxylated fatty acids (derivatives of decanoic, dodecenoic, dodecnoic, and tetradecanoic acids) isolated from *L. plantarum* MiLAB 14 exhibited strong inhibition against yeasts than molds with MICs 10–100 mg/ml [59]. In another study of sourdough, supplemented with linoleic acid, *L. hammesii* DSM 16381 showed conversion of linoleic acid to a hydroxylated $C_{18:1}$ fatty acid with anti-yeast activity against *A. niger* [6].

4.2.7 REUTERIN

Reuterin is a comprehensive antimicrobial metabolite predominantly produced by food grade *L. reuterivia* conversion of glycerol under anaerobic conditions. Other reuterin producing LAB encompasses *L. brevis*, *L. coryniformis*, *L. collinoides*, and *L. buchneri*. The compound is effective against an array of Gram-positive and Gram-negative bacteria including spores, fungi, and protozoa [4]. Metabolism of reuterin from glycerol in *L. reuteri* has been attributed to presence of Glycerol/diol dehydrase regulating biotransformation of glycerol to 3-Hydroxy propionaldehyde (HPA) and 1,3-propanediol. This bioconversion is regulated by *pdu* (propanediol utilization) operon encoding genes for glycerol/diol dehydratase in *L. coryniformis*. The mechanistic action of reuterin against *E. coli* has been correlated to oxidative stress to cell as evidenced by up-regulation of genes controlled by OxyR, a transcriptional regulator, responsive to oxidative stress [56].

Till 2016, reuterin was thought to contain 3-HPA, its hydrate, and dimer [62]. Now, one more compound called 'acrolein' was added to compound mixture of reuterin. Acrolein was formed as spontaneous dehydration of 3-HPA in aqueous form. This multi-component system thought to create imbalance in intracellular redox state by alteration of thiol groups in proteins and related molecules thereby exhausting glutathione and modifying functional enzymes [21]. Researchers have proposed suppression of ribonuclease activity by reuterin thereby interfering with DNA biosynthesis [20]. Often, it is miscorrelated with reutericyclin, which is a tetramic acid generated by strains of *L. reuteri* active against some strains of Gram-positive bacteria.

4.2.8 OTHER ANTIFUNGAL COMPOUNDS FROM LACTIC ACID BACTERIA (LAB)

Certain LAB also produce hydrogen peroxide (H_2O_2) utilizing atmospheric oxygen as they possess flavoprotein oxidases and are catalase (CAT) negative. The oxidizing power of H_2O_2 on the bacterial cell provides antimicrobial effect by generating free radicals responsible for damaging structures of cellular proteins. This effect is further potentiated by lacto-peroxidase-thiocyanate-peroxide system involving the reaction between H_2O_2 and thiocyanate to form hypothiocyanate and other intermediary products through catalysis by lacto-peroxidase. This system is effective against yeast *Candida albicans*. MRS agar (*De Man*, Rogosa, and Sharpe agar) is not a preferred substrate when screening cultures for H_2O_2 production because H_2O_2 is readily decomposed in this medium; presumably because of CAT activity in the yeast extract [49].

Some other types of miscellaneous antifungal compounds reported from LAB include nucleosides having nitrogenous base attached to a sugar molecule. Two nucleosides having cytosine base (Cytidine and 2-deoxycytidine) have been reported from the culture filtrate of *L. amylovorus* DSM 19280 with MIC values >200 mg/ml against *A. fumigatus* J9 [51]. Some more antifungal compounds from LAB such as Lactones have been reported from two *L. plantarum* isolates from beer and kimchi, which has shown to elicit antimicrobial and antiviral responses. Antifungal lactones-mevanolactone-produced by *L. plantarum* VTT was first reported by Niku-Paavola [41]; and δ-dodecalactone responsible for fruity flavor was purified from cell extract of *L. plantarum* AF1 and exhibited antifungal activity against *Aspergillus* genus and *P. roqueforti* [63]. The antifungal effect of lactones attributed to adsorption on cell membrane leading to increase in membrane fluidity.

4.3 POTENTIAL USE OF ANTIFUNGAL LAB IN FOOD SYSTEMS

With the advent of rapid expansion in global food industry and emergence of diversified segments of foods, maintenance of the highest quality and freshness of food to the end consumer will always be a priority area for industries and regulatory bodies. According to Food and Agriculture Organization's (FAO) SAVE FOOD initiative, nearly 1.3 billion tons of food produced each year get lost or wasted with specific commodity-wise waste corresponding to 45%, 45%, 35%, 30%, 20% and 20% in sector of fruits and vegetables, roots, and tubers, fish, and seafood, cereals, dairy, and meat products, respectively. Further, loss in developing countries comprises of 40% at postharvest and processing levels. On the other hand, losses in industrialized countries account for 40% at retail and consumer levels thereby warranty preservation of food products throughout the supply chain. The application of food-grade LAB to counter fungal spoilage and toxicity appears to be a promising approach to increase shelf-life and minimize the use of physical and chemical preservation treatments. Potential of antifungal LAB to evade fungal spoilage can be employed in many foods and feed formulations as live-cell and their metabolites in fermented products, use of microencapsulated purified biocidal components in biodegradable matrix in food formulations or as antimicrobial agents on packaging films, etc. Hence, these active bioprotective cultures and their metabolites can provide a shielding mechanism against fungal spoilers in post harvested fruits and vegetables, processed, and raw food items including bakery and dairy products and animal feed *viz.* silages (Table 4.2).

TABLE 4.2 Potential Use of Antifungal Compounds as Bioprotectants in Foods

Food Class	Food Type	Antifungal LAB	Activity Spectrum	References
Fruits and vegetables	Cucumber	*L. plantarum* CUK501	*A. flavus* and *F. graminaerum*	[54]
	Apple	Lactic acid bacteria strains LAB 13/43	*P. expansum*	[35]
	Tomato puree	*L. fermentum* YML014	*P. expansum*, *A. Niger* and *A. flavus*	[1]
Dairy products	Yogurt	*L. paracasei* and *L. rhamnosus*	*Debaryomyces hansenii* and *R. mucilaginosa*	[17]
	Cheddar cheese	*L. plantarum*	*P. roqueforti*	[66]
	Cultured cream and Semi-hard cheese	*L. plantarum*, *L. harbinensis* and *L. rhamnosus*	*P. commune*, *M. racemosus* and *R. mucilaginosa*	[53]
Bakery	Sourdough	*L. plantarum 21B*	*A. niger* FTDC 3227	[29]
	Bread	*L. amylovorus*	*A. niger*, *P. roqueforti*, *F. culmorum* and *P. expansum*	[51]
	Pound cake and milk bread roll			[31]
Animal feed	Corn silage	*L. buchneri*	Yeasts	[42]
		L. buchneri, *L. plantarum* and *P. acidilaciti*	Yeasts	[46]
		LAB	*Fusarium*	[40]

TABLE 4.2 *(Continued)*

Food Class	Food Type	Antifungal LAB	Activity Spectrum	References
Miscellaneous	Barley malt extract fermentation	*L. brevis* R2D and *L. plantarum* FST1.7 (2/2)	*Fusarium culmorum*	[45]
	Fruit juice	*L. plantarum*	*R. mucilaginosa*	[14]
	Raw smoked sausages	*Lc. lactis* ssp. *lactis* strains	*E. repens*	[60]
	Unprocessed poultry meat	*L. acidophilus*	*A. alternate*	[22]

4.3.1 FRUITS AND VEGETABLES

Raw and processed fruits as well as vegetables are vulnerable to fungal spoilage due to their perishable nature having high water activity and acidity. During their long-term storage after postharvest, they are easily attacked by molds such as *Fusarium*, *Mucor*, *Penicillium*, *Alternaria*, and *Botrytis* species giving rise to postharvest diseases. Countless studies had been suggested for enhancing keeping quality of fruits and vegetables post-harvest during their transportation and storage by LAB or their cell-free supernatants. The benign role of *L. plantarum* in restricting *A. flavus* and *F. graminaerum* growth for shelf-life extension of cucumber had been studied [54]. In another study, *L. fermentum* YML014 provided a strong antimycotic environment for *P. expansum*, *A. niger*, and *A. flavus* in tomato puree model [1]. The LAB strains 13/43 found highly potent against blue mold *P. expansum* when sprayed over the surface of apple [35].

4.3.2 DAIRY PRODUCTS

Fermented milk products especially cheese and yogurt despite possessing a competitive environment for the growth of contaminants are susceptible to fungal attack including psychrotrophicmolds capable to persevere refrigeration temperatures and low oxygen environment. Use of antifungal strains alone or in combination with specific adjunct cultures can promote the shelf-life of fermented products. For instance, selected strains of adjunct cultures *L. harbinensis* and *L. rhamnosus* exhibited bioprotective properties counter to *Rhizopus mucilaginosa* and *Debaryomyces hansenii* in yogurt; antifungal *L. plantarum* isolates inhibited growth of *P. roqueforti* during cheddar cheese production [17, 66]. *In situ* assays utilizing adjunct LAB strains combinations of *L. plantarum* with *L. harbinensis* or *L. rhamnosus* delayed growth of fungal targets, e.g., *Penicillum, Mucor,* and *Rhodotorula* in cultured cream and semi-hard cheeses without any significant effect on sensory attributes [53].

4.3.3 BAKERY

Fungal spoilage in bakery industry largely contributes to huge economic and health costs. LAB as natural protectants offers significant potential as an alternative to chemical preservatives like benzoates and sorbates. Use of antifungal strains of *L. plantarum* and *L. amylovorus* in co-fermentation with *S. cerevisiae* during bread/sourdough making can improve upon shelf-life up to 7 days and quality of product with inhibitory action against *Aspergillus, Candida, Fusarium,* and *Penicillium* molds [29, 51]. Furthermore, a mix of *Lactobacillus* strains as whole cells in sourdough making inhibited *Cladosporium sphaerospermum, Eurotium repens,* and *Aspergillus niger* in pound cake and milk bread roll [31].

4.3.4 ANIMAL FEED

Addressing fungal decay in animal feed during storage is another potential application area as LAB plays a vital role in silage. Silage is produced by fermentation of water-soluble carbohydrates present in green fodder to organic acids under anaerobic conditions. Incorporation of oxygen during preparation and storage of silage pave the way for growth of aerobic fungi leading to spoilage and decreased nutritive value. Plethora of reports are available wherein hetero-fermentative *L. buchneri* is used as silage additive which results in high quality feeds by production of antimicrobial metabolites during anaerobic degradation of lactic acid as antifungal agents. In combination with rapid lactic acid-producing homo-fermentative LAB strains, forage preservation is further enhanced [42, 46]. Other antifungal compounds, e.g., azealic acid have been found in silage inoculants thus prospecting novelty of such strains in increasing keeping quality of silage. Finally, LAB supplementation in silage were found effective in controlling mycotoxigenic *Fusarium* strains and reduce toxin availability [40]. Feedtech Silage F3000, a commercial bioprotectant as silage inoculant contains strain of *L. plantarum* MiLAB 393.

4.3.5 MISCELLANEOUS APPLICATIONS

The addition of starters *L. brevis* R2D and *L. plantarum* FST1.7 during barley malt extract fermentation have been found to suppress the growth of *Fusarium culmorum* [45]. Beverages, when added with LAB especially *L. plantarum,* can reduce yeast count *viz. R. mucilaginosa.* Shelf-life of rice cake starters, raw poultry meat, and raw smoked sausages has also been extended with the reduction in the count of spoilage fungi using antifungal LAB and its metabolites [14, 22, 52, 60]. Commercial antifungal formulations based on bioprotectant LAB and its metabolites are also available in the market for application in different food systems (Table 4.3).

TABLE 4.3 Commercial Antifungal Formulations based on LAB and its Metabolites

Product Name	Application Scope	Protective Cultures	Company
BEfresh	Fresh fermented milk products	*L. paracasei* and *P. freudenreichii* subsp. *shermanii*	Handary
BIOPROX® RP 80	Stirred/set yogurt, fermented milk, and cheeses	*L. rhamnosus* and *L. plantarum*	Bioprox
BIOPROX® RP 83	Stirred/set yogurt, fermented milk, and cheeses	*L. plantarum*	
BIOPROX® RP 94	Fresh/continental cheeses and cheddar cheese		
Dairy Safe™	Cheeses	Defined Mesophilic D-type; Lactic acid bacteria	CSK Food Enrichment B.V.
Delvo Guard	Yogurt, sour cream and fresh cheese	*L. rhamnosus* and *L. sakei*	DSM
FreshQ culture series	Cottage cheese, yogurt, sour cream	*L. rhamnosus* and *L. paracasei*	CHR Hansen
Hi Shield P	Bakery and Salad dressings	Corn-Fermentate of LAB and yeasts	HI-FOOD S.p.A.
Holdbac culture series	Fermented dairy products including cheese and yogurt	*L. rhamnosus*, *L. plantarum* and *P. freudenreichii* subsp. *shermanii*	DuPont Danisco

TABLE 4.3 *(Continued)*

Product Name	Application Scope	Protective Cultures	Company
Inhibit culture series	Cheese dips and spreads, Bakery products, Meats	Fermentate (wheat, whey, brown rice) of *P. freudenreichii*	Mezzoni foods
M-CULTURE Safe 3100 SSL	Raw sausages	*L. plantarum* and *L. curvatus*	Meat cracks Technology GmbH
MicroGard	Fermented sausages, bakery/ dairy products and cured meat	Skim milk- Fermentate of *P. freudenreichii* subsp. *shermani*	DuPont Danisco

During the incorporation of different antifungal agents from LAB in the application areas (Table 4.3), optimization for enhancement of antifungal activity of either LAB or its bioactive components are other areas to be taken into consideration. This can be achieved by availing cooperative behavior of microorganisms as co-cultures/adjunct cultures, addition of specific precursors to up-regulate desired biosynthetic pathways, e.g., glycerol for reuterin production, induction of stress conditions to enable LAB to produce diverse metabolites *via* different pathways and ascertaining association between bioprotective cultures with active molecules such as chitosan.

4.4 SUMMARY

LAB have immense potential to control spoilage and pathogenic fungal microflora in food products. They are involved in production of multitude of antimicrobial metabolites ranging from primary organic acids to bacteriocins and others including reuterin, PLA, fatty acids, etc. This chapter delineates the role of these antifungal metabolites produced by LAB along with their mechanistic role in controlling spoilage contaminants. The biocidal activities of these metabolites arise due to interconnected inhibition mechanisms of membrane destabilization, proton gradient interference, inhibition of enzymes involved in cellular metabolism or induction of oxidative stress. This chapter further summarizes applications of antifungal LAB and their metabolites in biopreservation of wide range of food commodities. Some of the successful commercial LAB strains as bioprotectants used by selected companies have also been documented. The 'clean label' approach of biopreservation availing benign role of LAB and their metabolites along with interventions of modern biotechnology can hold great promise of safer and quality products acceptable to both consumers and regulatory bodies.

KEYWORDS

- **antifungal compounds**
- **antimicrobials**
- **biopreservation**
- **cyclic peptides**
- **lactic acid bacteria (LAB)**
- **reuterin**

REFERENCES

1. Adedokun, E. O., Rather, I. A., Bajpai, V. K., & Park, Y. H., (2016). Biocontrol efficacy of *Lactobacillus fermentum* YML014 against food spoilage molds using the tomato puree model. *Frontiers in Life Science*, *9*, 64–68.
2. Aunsbjerg, S. D., Honore, A. H., Marcussen, J., Ebrahimi, P., Vogensen, F. K., Benfeldt, C., Skov, T., & Knøchel, S., (2015). Contribution of volatiles to the antifungal effect of *Lactobacillus paracasei* in defined medium and yogurt. *International Journal of Food Microbiology*, *194*, 46–53.
3. Avis, T. J., & Belanger, R. R., (2001). Specificity and mode of action of the antifungal fatty acid cis-9-heptadecenoic acid produced by *Pseudozyma flocculosa*. *Applied and Environmental Microbiology*, *67*, 956–960.
4. Axelsson, L. T., Chung, T. C., Dobrogosz, W. J., & Lindgren, S. E., (1989). Production of a broad spectrum antimicrobial substance by *Lactobacillus reuteri*. *Microbial Ecology in Health and Disease*, *2*, 131–136.
5. Bergsson, G., Arnfinnsson, J., Steingrimsson, O., & Thormar, H., (2001). *In vitro* killing of *Candida albicans* by fatty acids and monoglycerides. *Antimicrobial Agents and Chemotherapy*, *45*, 3209–3212.
6. Black, B. A., Zannini, E., Curtis, J. M., & Ganzle, M. G., (2013). Antifungal hydroxyl fatty acids produced during sourdough fermentation: Microbial and enzymatic pathways, and antifungal activity in bread. *Applied and Environmental Microbiology*, *79*, 1866–1873.
7. Borthwick, A. D., (2012). 2,5-Diketopiperazines: Synthesis, reactions, medicinal chemistry, and bioactive natural products. *Chemical Reviews*, *112*, 3641–3716.
8. Bougherra, F., Dilmi-Bouras, A., Balti, R., Przybylski, R., Adoui, F., Elhameur, H., Chevalier, M., et al., (2017). Antibacterial activity of new peptide from bovine casein hydrolyzed by a serine metalloprotease of *Lactococcus lactis* subsp. *lactis* BR16. *Journal of Functional Foods*, *32*, 112–122.

9. Broberg, A., Jacobsson, K., Strom, K., & Schnurer, J., (2007). Metabolite profiles of lactic acid bacteria in grass silage. *Applied and Environmental Microbiology*, *73*, 5547–5552.
10. Brosnan, B., Coffey, A., Arendt, E. K., & Furey, A., (2012). Rapid identification, by use of the LTQ Orbitrap hybrid FT mass spectrometer, of antifungal compounds produced by lactic acid bacteria. *Analytical and Bioanalytical Chemistry*, *403*, 2983–2995.
11. Chen, C., Chen, X., Jiang, M., Rui, X., Li, W., & Dong, M., (2014). A newly discovered bacteriocin from *Weissella hellenica* D1501 associated with Chinese Dong fermented meat (NanxWudl). *Food Control*, *42*, 116–124.
12. Corsetti, A., Gobbetti, M., Rossi, J., & Damiani, P., (1998). Antimould activity of sourdough lactic acid bacteria: Identification of a mixture of organic acids produced by *Lactobacillus sanfrancisco* CB1. *Applied Microbiology and Biotechnology*, *50*, 253–256.
13. Cray, J. A., Stevenson, A., Ball, P., Bankar, S. B., Eleutherio, E. C., Ezeji, T. C., Singhal, R. S., et al., (2015). Chaotropicity: A key factor in product tolerance of biofuel-producing microorganisms. *Current Opinion in Biotechnology*, *33*, 228–259.
14. Crowley, S., Mahony, J., & Van, S. D., (2012). Comparative analysis of two antifungal *Lactobacillus plantarum* isolates and their application as bioprotectants in refrigerated foods. *Journal of Applied Microbiology*, *113*, 1417–1427.
15. Crowley, S., Mahony, J., & Van, S. D., (2013). Current perspectives on antifungal lactic acid bacteria as natural bio-preservatives. *Trends in Food Science and Technology*, *33*, 93–109.
16. Dang, T. D. T., Vermeulen, A., Ragaert, P., & Devlieghere, F., (2009). A peculiar stimulatory effect of acetic and lactic acid on growth and fermentative metabolism of *Zygosaccharomyces bailii*. *Food Microbiology*, *26*, 320–327.
17. Delavenne, E., Ismail, R., Pawtowski, A., Mounier, J., & Barbier, B., (2012). Assessment of *lactobacilli* strains as yogurt bioprotective cultures. *Food Control*, *30*, 206–213.
18. Di Biase, M., Lavermicocca, P., Lonigro, S. L., & Valerio, F., (2014). *Lactobacillus brevis*-based bioingredient inhibits *Aspergillus niger* growth on pan bread. *Italian Journal of Agronomy*, *9*, 146–151.
19. Dieuleveux, V., Lemarinier, S., & Gueguen, M., (1998). Antimicrobial spectrum and target site of D-3-phenyllactic acid. *International Journal of Food Microbiology*, *40*, 177–183.
20. Dobrogosz, W. J., Casas, I. A., Pagano, G. A., Sjöberg, B. M., Talarico, T. L., & Karlsson, M., (1989). *Lactobacillus reuteri* and the enteric microbiota. In: Grubb, R., Midtvedt, T., & Norin, E., (eds.), *The Regulatory and Protective Role of the Normal Microflora* (Vol. 28, pp. 283–292). Macmillan, London - England.
21. Engels, C., Schwab, C., Zhang, J., Stevens, M. J. A., Bieri, C., Ebert, M. O., McNeill, K., et al., (2016). Acrolein contributes strongly to antimicrobial and heterocyclic amine transformation activities of reuterin. *Scientific Reports*, *6*, 36246.
22. Garcha, S., & Natt, N. K., (2011). *In situ* control of food spoilage fungus using *Lactobacillus acidophilus* NCDC 291. *Journal of Food Science and Technology*, *49*, 643–648.
23. Garmiene, G., Salomskiene, J., Jasutiene, I., Macioniene, I., & Miliauskiene, I., (2010). Production of benzoic acid by lactic acid bacteria from *Lactobacillus*, *Lactococcus* and *Streptococcus* genera in milk. *Milchwissenschaft*, *65*, 295–298.
24. Guimarães, A., Venancio, A., & Abrunhosa, L., (2018). Antifungal effect of organic acids from lactic acid bacteria on *Penicillium nordicum*. *Food Additives and Contaminants*, *35*(9), 1803–1818.

25. Guo, J., Brosnan, B., Furey, A., Arendt, E., Murphy, P., & Coffey, A., (2012). Antifungal activity of *Lactobacillus* against *Microsporum canis*, *Microsporum gypseum*, and *Epidermophyton floccosum*. *Bioengineered*, *3*, 104–113.
26. Holden, M. T., Chhabra, S., DeNys, R., & Stead, P., (1999). Quorum-sensing cross talk: Isolation and chemical characterization of cyclic dipeptides from *Pseudomonas aeruginosa* and other gram-negative bacteria. *Molecular Microbiology*, *33*, 1254–1266.
27. Jay, J. M., (2000). *Modern Food Microbiology* (6th edn., p. 790). Van Nostrand Reinhold, New York-NY, USA.
28. Kwak, M. K., Liu, R., Kim, M. K., Moon, D., Kim, A. H., Song, S. H., et al., (2014). Cyclic dipeptides from lactic acid bacteria inhibit the proliferation of pathogenic fungi. *Journal of Microbiology*, *52*, 64–70.
29. Lavermicocca, P., Valerio, F., Evidente, A., Lazzaroni, S., Corsetti, A., & Gobbetti, M., (2000). Purification and characterization of novel antifungal compounds from the sourdough *Lactobacillus plantarum* strain 21B. *Applied and Environmental Microbiology*, *66*, 4084–4090.
30. Lavermicocca, P., Valerio, F., & Visconti, A., (2003). Antifungal activity of phenyllactic acid against molds isolated from bakery products. *Applied and Environmental Microbiology*, *69*, 634–640.
31. Le Lay, C., Mounier, J., Vasseur, V., Weill, A., Le Blay, G., Barbier, G., & Coton, E., (2016). *In vitro* and *in situ* screening of lactic acid bacteria and propionic bacteria antifungal activities against bakery product spoilage molds. *Food Control*, *60*, 247–255.
32. Li, X., Jiang, B., & Pan, B., (2007). Biotransformation of phenylpyruvic acid to phenyllactic acid by growing and resting cells of a *Lactobacillus* sp. *Biotechnology Letters*, *29*, 593–597.
33. Liu, R., Kim, A. H., Kwak, M. K., & Kang, S. O., (2017). Proline-based cyclic dipeptides from Korean fermented vegetable Kimchi and from *Leuconostocmes enteroides* LBP-K06 have activities against multidrug-resistant bacteria. *Frontiers in Microbiology*, *8*, 761–767.
34. Magnusson, J., Strom, K., Roos, S., Sjogren, J., & Schnurer, J., (2003). Broad and complex antifungal activity among environmental isolates of lactic acid bacteria. *FEMS Microbiology Letters*, *219*, 129–135.
35. Matei, G. M., Matei, S., Matei, A., Cornea, C. P., Draghici, E. M., & Jerca, I. O., (2016). Bioprotection of fresh food products against blue mold using lactic acid bacteria with antifungal properties. *Romanian Biotechnological Letters*, *21*, 11201–11208.
36. McNair, L. K. F., Siedler, S., Vinther, J. M., Hansen, A. M., Neves, A. R., Garrigues, C., Jager, A. K., et al., (2018). Identification and characterization of a new antifungal peptide in fermented milk product containing bioprotective *Lactobacillus* cultures. *FEMS Yeast Research*, *18*(8), 231–238.
37. Mu, W., Yu, S., Zhu, L., Zhang, T., & Jiang, B., (2012). Recent research on 3-phenyllactic acid, a broad-spectrum antimicrobial compound. *Applied Microbiology and Biotechnology*, *95*, 1155–1163.
38. Muhialdin, B. J., Hassan, Z., Bakar, F. A., & Saari, N., (2016). Identification of antifungal peptides produced by *Lactobacillus plantarum* IS10 grown in the MRS broth. *Food Control*, *59*, 27–30.
39. Muhialdini, B. J., Hassan, Z., Sadon, S. K., Zulkifli, N. A., & Azfari, A., (2011). A. Effect of pH and heat treatment on antifungal activity of *Lactobacillus fermentum*

T-007, *Lactobacillus pentosus* G004 and *Pediococcus pentosaceus* T-010. *Innovative Romanian Food Biotechnology*, *8*, 41–53.
40. Niderkorn, V., (2007). *Biotransformation and Sequestration Activities of Fusariotoxins in Fermentative Bacteria for the Detoxification of Corn Silage* (p. 208). Doctorate dissertation, Université Blaise Pascal University, Clermont-Ferrand-Theix, France.
41. Niku-Paavola, M. L., Laitila, A., Mattila-Sandholm, T., & Haikara, A., (1999). New types of antimicrobial compounds produced by *Lactobacillus plantarum*. *Journal of Applied Microbiology*, *86*, 29–35.
42. Nishino, N., Wada, H., Yoshida, M., & Shiota, H., (2004). Microbial counts, fermentation products, and aerobic stability of whole crop corn and a total mixed ration ensiled with and without inoculation of *Lactobacillus casei* or *Lactobacillus buchneri*. *Journal of Dairy Science*, *87*, 2563–2570.
43. Ohhira, I., Kuwaki, S., Morita, H., Suzuki, T., Tomita, S., Hisamatsu, S., Sonoki, S., & Shinoda, S., (2004). Identification of 3-phenyllactic acid as a possible antibacterial substance produced by *Enterococcus faecalis* TH10. *Biocontrol Science*, *9*, 77–81.
44. Özcelik, S., Kuley, E., & Özogul, F., (2016). Formation of lactic, acetic, succinic, propionic, formic and butyric acid by lactic acid bacteria. *LWT*, *73*, 536–542.
45. Peyer, L. C., & Zannini, E., (2016). Lactic acid bacteria as sensory biomodulators for fermented cereal-based beverages. *Trends in Food Science and Technology*, *54*, 17–25.
46. Reich, L. J., & Kung, L., (2010). Effects of combining *Lactobacillus buchneri* 40788 with various lactic acid bacteria on the fermentation and aerobic stability of corn silage. *Animal Feed Science and Technology*, *159*, 105–109.
47. Rhee, K. H., (2004). Cyclic dipeptides exhibit synergistic, broad spectrum antimicrobial effects and have anti-mutagenic properties. *International Journal of Antimicrobial Agents*, *24*, 423–427.
48. Rizzello, C. G., Cassone, A., Coda, R., & Gobbetti, M., (2011). Antifungal activity of sourdough fermented wheat germ used as an ingredient for bread making. *Food Chemistry*, *127*, 952–959.
49. Rodrıguez, J. M., Martınez, M. I., Suarez, A. M., & Martınez, J. M., (1997). Research note: Unsuitability of the MRS medium for the screening of hydrogen peroxide-producing lactic acid bacteria. *Letters in Applied Microbiology*, *25*, 73–74.
50. Ross, R. P., Morgan, S., & Hill, C., (2002). Preservation and fermentation: Past, present and future. *International Journal of Food Microbiology*, *79*, 3–16.
51. Ryan, L. A. M., Zannini, E., Dal, B. F., Pawlowska, A., Koehler, P., & Arendt, E. K., (2011). *Lactobacillus amylovorus* DSM 19280 as a novel food-grade antifungal agent for bakery products. *International Journal of Food Microbiology*, *146*, 276–283.
52. Salas, M. L., Mounier, J., Valence, F., Coton, M., & Thierry, A., (2017). Antifungal microbial agents for food biopreservation: A review. *Microorganisms*, *5*(37), 218–227.
53. Salas, M. L., Thierry, A., Lemaître, M., & Garric, G., (2018). Antifungal activity of lactic acid bacteria combinations in dairy mimicking models and their potential as bioprotective cultures in pilot scale applications. *Frontiers in Microbiology*, *9*, 1787–1792.
54. Sathe, S. J., Nawani, N. N., Dhakephalkar, P. K., & Kapadnis, B. P., (2007). Antifungal lactic acid bacteria with potential to prolong shelf-life of fresh vegetables. *Journal of Applied Microbiology*, *103*, 2622–2628.
55. Sauguet, L., Moutiez, M., Li, Y., Belin, P., & Seguin, J., (2011). Cyclodipeptide synthases, a family of class-I aminoacyl-tRNA synthetase-like enzymes involved in non-ribosomal peptide synthesis. *Nucleic Acids Research*, *39*, 4475–4489.

56. Schaefer, L., Auchtung, T. A., Hermans, K. E., Whitehead, D., & Borhan, B., (2010). The antimicrobial compound reuterin (3-hydroxypropionaldehyde) induces oxidative stress via interaction with thiol groups. *Microbiology, 156*, 1589–1599.
57. Schwenninger, S. M., Lacroix, C., Truttmann, S., Jans, C., Sporndli, C., Bigler, L., & Meile, L., (2008). Characterization of low-molecular-weight antiyeast metabolites produced by a food-protective *Lactobacillus-Propionibacterium* coculture. *Journal of Food Protection, 71*, 2481–2487.
58. Siedler, S., Balti, R., & Neves, A. R., (2019). Bioprotective mechanisms of lactic acid bacteria against fungal spoilage of food. *Current Opinion in Biotechnology, 56*, 138–146.
59. Sjogren, J., Magnusson, J., Broberg, A., Schnurer, J., & Kenne, L., (2003). Antifungal 3-hydroxy fatty acids from *Lactobacillus plantarum* MiLAB 14. *Applied and Environmental Microbiology, 69*, 7554–7557.
60. Stoyanova, L. G., Ustyugova, E. A., Sultimova, T. D., & Bilanenko, E. N., (2010). New antifungal bacteriocin-synthesizing strains of *Lactococcus lactis* ssp. *lactis* as the perspective biopreservatives for protection of raw smoked sausages. *American Journal of Agricultural and Biological Sciences, 4*, 477–485.
61. Strom, K., Sjogren, J., Broberg, A., & Schnurer, J., (2002). *Lactobacillus plantarum* MiLAB 393 produces the antifungal cyclic dipeptides cyclo(L-Phe-L-Pro) and cyclo(L-Phe-trans-4-OH-L-Pro) and 3-phenyllactic acid. *Applied and Environmental Microbiology, 68*, 4322–4327.
62. Vollenweider, S., & Lacroix, C., (2004). 3-Hydroxypropionaldehyde: Applications and perspectives of biotechnological production. *Applied Microbiology and Biotechnology, 64*, 16–27.
63. Yang, E. J., Kim, Y. S., & Chang, H. C., (2011). Purification and characterization of antifungal d-dodecalactone from *Lactobacillus plantarum* AF1 isolated from Kimchi. *Journal of Food Protection, 74*, 651–657.
64. Yu, S., Zhou, C., Zhang, T., Jiang, B., & Mu, W., (2015). Short communication: 3-Phenyllactic acid production in milk by *Pediococcus pentosaceus* SK25 during laboratory fermentation process *Journal of Dairy Science, 98*, 813–817.
65. Zhang, C., Brandt, M. J., Schwab, C., & Ganzle, M. G., (2010). Propionic acid production by co-fermentation of *Lactobacillus buchneri* and *Lactobacillus diolivorans* in sourdough. *Food Microbiology, 27*, 390–395.
66. Zhao, D., (2011). Isolation of antifungal lactic acid bacteria from food sources and their use to inhibit mould growth in cheese. *Master's Degree Thesis* (p. 99). California Polytechnic State University, San Luis Obispo, California, USA.

CHAPTER 5

APPLICATIONS OF ANTIMICROBIAL ENZYMES IN FOODS

VEENA NAGARAJAPPA, PRAVIN D. SAWALE,
SURENDRA N. BATTULA, and VENUS BANSAL

ABSTRACT

Antimicrobial enzymes are part of the living organisms that can cause inhibition in the growth or destruction of bacterial and fungal pathogens. These enzymes are considered as safe, natural, and do not exhibit any toxicity, thus they are increasingly employed in biopreservation of foods as a substitute to chemical preservatives and antibiotics. They exhibit antimicrobial activity through various mechanisms of action, either by hydrolyzing the key components of the bacterial and fungal cell walls or by *in situ* generations of cytotoxic products, resulting in lysis of the microbial cell. Presently, antimicrobial enzymes are used in various food applications either added individually or in a combination of two or more enzymes or combined with other antibacterial agents. Recently, the antibacterial spectrum can be enhanced by physical and chemical treatments and genetic modification may broaden their applications in agriculture, food, healthcare, and medical fields. This chapter focuses on the importance of antimicrobial enzymes, different types including hydrolases, oxido-reductases, and bacteriophage lysins, and their applications, and future potential.

5.1 INTRODUCTION

Antimicrobial enzymes, the main components of living organisms, exert an important role in fighting against microbial and fungal infection through their various defense mechanisms. These enzymes have earned interest in food preservation strategies, because they are natural compounds, considered as safe, and do not exhibit toxicity. The mechanism of action of the antimicrobial agent determines the type and extent of microbial damage. For instance, hydrolytic antimicrobial enzymes can hydrolyze the key components of bacterial and fungal cell walls, whereas oxidoreductase enzymes exert their antimicrobial effects by *in situ* generations of cytotoxic products.

Several preservation techniques have been used to prevent the growth of harmful microorganisms, such as drying, thermal processing, acidification, salting, and addition of synthetic preservatives [10]. In recent years, foods preserved with antimicrobial enzymes are gaining much attention by investigators, because of increasing prevalence of antibiotic-resistant bacterial strains and adverse effects of chemical preservatives. These enzymes are used in food formulations; adding directly into the food matrix, as a surface coating on the product or included in the packaging material, for effectively inhibiting the growth of undesirable microbes in the foods [38]. Several studies have reported increased antibacterial spectrum when combinations of enzymes or with other antimicrobial agents are used in the food systems.

In this chapter, importance of antimicrobial enzymes and their mode of action and food applications are discussed.

5.2 ANTIMICROBIAL ENZYMES

5.2.1 LACTOPEROXIDASE (LP)

Lactoperoxidase (LP) is arranged in a single polypeptide chain of about 80 kDa and contains N-glycosylation sites, one calcium, and one iron molecule per mole. This occurs widely in secretions of salivary, mammary, and lachrymal glands of mammals. Antimicrobial activity of lactoperoxidase system (LPS) is well documented and occurs when the LP enzyme is present along with the thiocyanate ion (SCN^-) or halogen (such as Cl^-, Br^- or I^-) and hydrogen peroxide (H_2O_2).

In human saliva and milk, LPS acts as natural antibacterial system and has proven to be both bacteriostatic and bactericidal for a broad range of Gram-positive [31, 41, 62] and Gram-negative [3, 17, 62] microorganisms. LP is an oxidoreductase, catalyzes thiocyanate oxidation in the presence of H_2O_2, results in formation of oxidation products, principally hypothiocyanite ion, which is a major antimicrobial substance either to inhibit or kill the growth of microorganisms. The mode of action of LPS is shown in the following reactions:

- **Step 1 (Compound I Formation):** LP in the presence of H_2O_2 acts as a relatively specific electron acceptor and thus compound I is formed [35]. In this step, two electrons are transferred from the LP to H_2O_2, which is reduced into water.

$$LP + H_2O_2 \rightarrow \text{Compound I} + H_2O$$
$$\text{(Oxidation state: +2)}$$

- **Step 2 (Halogenation Process):** In the presence of a halide (Cl^-, Br^- or I^-) or a thiocyanate, compound I is reduced back to its native enzymatic form through two electron transfer, while the thiocyanate (SCN^-) is oxidized into a hypothiocyanite ($OSCN^-$).

$$SCN^- + \text{Compound I} \rightarrow \text{Native LP} + OSCN^-$$

Hypothiocyanites (which reduce growth, uptake of oxygen and production of lactic acid by bacteria) are powerful oxidants and oxidize the sulphydryl groups of essential proteins and enzymes [49]. It appears that bacterial cytoplasmic membrane is structurally damaged on exposure to LPS resulting in a potassium and amino acids leakage, thus affecting the uptake of solutes (such as glucose, amino acids, purine, and pyrimidines)into the bacterial cell. Subsequently, nucleic acid and protein synthesis is also affected [49].

5.2.2 GLUCOSE OXIDASE (GOX)

GOX is a glycosylated flavoprotein, and contains two similar subunits of 80 kDa. It catalyzes the oxidation of glucose to D-Glucono-δ-lactone with release of H_2O_2 in the presence of oxygen.

$$\beta\text{-D-glucose} + O_2 \xrightarrow{\text{GOX}} \beta\text{-D-Glucono-}\delta\text{-lactone} + H_2O_2$$

$$\downarrow + \text{ water}$$

$$\text{D-Gluconic acid}$$

Antibacterial activity of GOX is attributed through the production of hydrogen peroxide, which catalyzes the peroxidation of membrane lipids [23] or a reaction with cell proteins, destroying the molecular structure. The fungicidal effect is due to decrease in pH through gluconic acid produced from δ-D-gluconolactone [6]. Extraction of GOX has been carried out from several fungal sources, largely from *Aspergillus* and *Penicillium* species, of which *Aspergillus niger* is mostly used to produce food and diagnostic enzyme owing to its bactericidal and fungicidal effects. Glucose oxidase (GOX) has many applications in industry and bears GRAS status [59]. Tiina and Sandholm [56] reported that growth of food-poisoning organisms (such as *C. perfringens, S. aureus, S. infantis* and *B. cereus, etc.*) were inhibited in a sterile filtered meat medium by the GOX-glucose system.

5.2.3 LYSOZYME (LYZ)

Lysozymes (LYZ) (muramidase) are natural antimicrobial enzymes, and found in most tissues and body fluids of animals and humans, are most common in milk and chicken egg white, which can hydrolyze β-(1,4) linkages between N-acetylmuramic acid and N-acetyl-glycosamine in a peptidoglycan layer. It can also degrade β-(1,4) linkages between N-acetyl-D-glycosamine residues in chitodextrins [61]. It exhibits more effective antimicrobial action towards certain pathogens, especially Gram-positive bacteria rather than Gram-negative, because of structural differences in the cell walls. However, a few Gram-positive bacterial species such as *B. subtillus*, *C. perfringens*, *E. coli, S. aureus,* and *S. typhimurium* were not inhibited by the LYZ [25].

Limited action of enzyme on Gram-negative organisms is because of the presence of lipopolysaccahride layer in the external membrane [27] and is also influenced by the composition and the order of N-acetyloamino sugars in bacterial cell walls. It is however possible to extend its antibacterial activity towards Gram-positive bacteria through modification of enzyme surface hydrophobicity [7, 26] by several methods, such as ultrafiltration treatment, heat denaturation, γ-radiation, high pressure application, covalent attachment of the LYZ with several polymers, treatment with proteases and by genetic modification [25, 34, 46]. The range of antimicrobial activities of enzymes can be increased by certain chelating agents (disodium ethylenediamine tetra-acetic acid (EDTA), disodium pyrophosphate, penta-sodium tripolyphosphate), butylparben, and some naturally occurring antimicrobial substances [16].

5.2.4 β-GLUCANASE

β-Glucanase (β-D-1,3-glucan glucanohydrolase) is a hydrolytic enzyme, which hydrolyzes the β-glucan by breaking β-1,3-glucosidic linkage.

β-Glucan is a major component of fungal cell wall polysaccharide [24]. These enzymes are classified as endo- and exo-β-glucanases, commonly found in plants, fungi, and bacteria. β-Glucanase can provide means for degrading yeast cells and fungal mycelia, utilization of yeast cell walls as biomass resources wasted by brewing industries, preparation of protoplast from fungi for gene manipulation studies and control of pathogenic fungi causing food rotting during storage [1]. These enzymes are usually produced by bacteria including *Bacillus clausii, Bacillus subtilis, Bacillus circulans, Bacillus halodurans, Bacillus licheniformis, Pseudomonas cepacia,* and *Arthrobacter* species and could inhibit the growth of pathogenic fungi [13].

5.2.5 CHITINASE

Chitinases are hydrolases catalyzing the hydrolysis of chitin and are a polymer of N-acetyl-D-glucosamine connected through β-(1,4)-glycosidic linkage [48]. Chitin is a major structural element of cell wall of fungi, exoskeleton of insects, and crustacean shells. Chitinase enzymes are usually produced from plants, animals, bacteria, and fungi; however, they represent structural and mechanical differences among the enzymes produced by these sources. One of the widely studied microorganisms to produce chitinase is *Trichoderma harzianum*. These enzymes are being used as potential biofungicidal agents to prevent many diseases [48].

5.2.6 BACTERIOPHAGE LYSINS

Bacteriophage lysins (endolysins) are potential alternative to antibiotics for fighting against diseases caused by various pathogenic microorganisms. Endolysins are bacteriophage-encoded lytic enzymes, which specifically hydrolyzes peptidoglycan layer of cell wall of Gram-positive strains, when applied exogenously to the bacterial host [4]. These enzymes are more effective towards Gram-positive than Gram-negative microorganisms owing to the outer membrane compositional difference.

These lysins have unique characteristics that make them attractive antibacterial system including greater specificity to pathogens without affecting the natural microflora, low chance of developing bacterial resistance and eliminate colonizing pathogens from mucosal surfaces and biofilms [20]. Many studies have demonstrated that endolysins isolated from various phages have potential application as antibacterial agents [28, 36, 44] and considered as promising anti-infective agents in the field of medicine and biotechnology [21]. Nowadays, bioengineered novel or modified lysins or modular endolysins can be used as effective tool to enhance killing activity, solubility, and with broad lytic spectrum.

5.3 POTENTIAL USE OF ANTIMICROBIAL ENZYMES IN FOOD SYSTEMS

Antimicrobial enzymes are employed in several food applications to control growth of food borne pathogens and spoilage causing organisms. Table 5.1 represents the food applications of hydrolases, oxido-reductase, and phage lysin enzymes used in combinations with other antimicrobial agents.

TABLE 5.1 Food Applications of Important Antimicrobial Enzymes

Food Product	Antimicrobial Enzyme/Agent	Packaging and Storage Condition	Main Results	References
Burrata cheese	LYZ (150, 250 and 500 mg/Kg) and Na_2-EDTA	Air and MAP (5:95, Nitrogen: Carbon dioxide); stored at 8°C	Combined antimicrobial treatment prolongs the shelf-life of cheese especially at high LYZ concentration against *Pseudomonas* spp.	[8]
Burrata cheese	LYZ (500 mg/Kg) and disodium-EDTA (50 mM)	Air and MAP (35:65, Nitrogen: Carbon dioxide)	Combinations of antimicrobial substances and MAP enhanced product shelf-life	[9]
Cuajada (curdled milk)	Individually or in combination of nisin, reuterin, and LPS	Storage for 12 days at 10°C	Combined antimicrobial (nisin, reuterin, and LPS) treatment enhanced the bactericidal activity as observed by pronounced decrease in *L. monocytogenes* and *Staph. aureus* counts during 12 days of storage when compared with control	[2]

TABLE 5.1 *(Continued)*

Food Product	Antimicrobial Enzyme/Agent	Packaging and Storage Condition	Main Results	References
Liquid whole egg	Combination of 500 U GOX and glucose (0.5 g/100 mL)	Stored at refrigeration temperature (7°C).	Bactericidal action against *S. enteritidis, M. luteus* and *B. cereus*, whereas bacteriostatic activity on *P. fluorescens* was observed after a storage period of 5 days	[14]
Mozzarella cheese	LYZ (0.25 mg/mL) & disodium-EDTA (50, 20 & 10 mM/L)	Brine solution containing LYZ and EDTA and storage period of eight days (4°C)	Pseudomonadaceae and coliforms growth inhibited for the first seven days, however, no such effect on the growth of Lactic acid bacteria	[55]
Ostrich meat patties	Lysozyme (LYZ) (250 ppm), nisin (250 ppm), EDTA (20 mM)	vacuum and air	*L. monocytogenes* population decreased to less than 2 log cfu/g by the antimicrobial treatment. However, no such effect on *Enterobacteriaceae* and *Pseudomonas* spp.	[39]
Processed poultry meat	GOX with or without glucose @ 4%		*Salmonella typhimurium* was inhibited during storage period, however, no effect on *Pseudomonas* spp.	[29]
Semi cooked coated chicken	Alone or combination of LYZ (1.5%), EDTA (1.5%), rosemary (0.2%) and oregano oil (0.2%)	Aerobic packaging; stored for 16 days @ 4 & 8°C	Growth of Pseudomonas, yeast, and mold growth was inhibited by using combination of EDTA, LYZ, and rosemary/ oregano oil and shelf-life was extended up to five days when compared to control	[43]

TABLE 5.1 *(Continued)*

Food Product	Antimicrobial Enzyme/Agent	Packaging and Storage Condition	Main Results	References
Shrimp	GOX (1 unit/mL) & catalase addition in glucose (4% w/v)	Stored at 1°C	Inhibition of *P. fluorescens* growth on shrimp	[32]
Skim milk	Individually or in combination of nisin & LPS	Storage for 15 days at 25°C	Bactericidal activity against *L. monocytogenes* was more pronounced when LPS was used after 4 hours of nisin action	[5]
Soft cheese	LYZ (activity equal to 39 unit/mL) (250 and 300 mg/Kg curd)	6 ± 1°C for 15 days	Prolonged shelf-life of soft cheese especially at 300 mg/Kg concentration of LYZ	[15]

5.3.1 LACTOPEROXIDASE (LP)

LPS has potential bactericidal activity especially in raw milk preservation where refrigeration system is not available and is suggested to be one of the most common industrial applications [30]. Addition of sodium thiocyanate and sodium percarbonate or GOX in raw milk could further augment LP activity, which results in reduced milk spoilage by inhibiting the growth of several microorganisms namely streptococci, yeasts, and fungi [53]; and moreover, LPS as such has no negative impact on the organoleptic and physicochemical characteristics of milk and milk products. Antibacterial activity of LPS either individually or combined with other antimicrobial compounds for inhibiting the pathogenic bacteria in various types of foods has been reported by many researchers [2, 5, 18, 63].

5.3.2 GLUCOSE OXIDASE (GOX)

GOX is most commonly utilized in industries for removal of glucose from egg albumin, egg yolk, whole egg, dried meat, and potatoes, before drying, to prevent browning and off-odors caused by Maillard reactions [37] and oxygen removal from orange juice concentrates, syrups, and carbonated soft drink to maintain freshness, color, and extend shelf-life [51]. Antibacterial effect of this enzyme is used for food packaging and storage, thereby so enhancing the keeping quality of foods [57]. Many researchers have reported the antibacterial activity of GOX for preservation of food by inhibiting the growth of harmful microorganisms [14, 56, 60]. Combination of GOX and catalase (CAT) has also been used as an on-board preservation system for seafood (fillets and shrimp) [19, 32, 33, 54].

5.3.3 LYSOZYME (LYZ)

LYZ is a lytic enzyme commonly utilized both in food and pharmaceutical industries as a natural biopreservative because of its inhibitory activity against undesirable microorganisms and non-toxic to humans. It is commonly used as preservative in meat, fish, and dairy products. One of the important industrial applications of LYZ is to hinder late blowing of cheeses during ripening period caused by the *Clostridium tyrobutyricum*, which negatively influences the quality of cheese [7]. The enzyme has GRAS status by US-FDA and can be used as natural additive to inhibit the lactic acid bacterial (LAB) growth in wine and beer production.

5.3.4 β-GLUCANASE

β-Glucanase is used as biological control of pathogenic fungi (*Curvularia affinis, Colletotrichum gloeosporioides, etc.*) via degrading the β-glucan, a polysaccharide present in fungal cell wall [13, 45]. To avoid the accumulation of barley beta-glucan, these enzymes are used in the brewing process for degradation. The enzyme can be utilized to produce soluble β-glucan, which can act as potential immunostimulator [42], and for the preparation of fungal protoplast [47].

5.3.5 CHITINASE

Chitinase enzymes have been used in waste management, as biocontrol agent and in health care [48]. Chitanases can be used for conversion of chitinous waste to chitooligomers, which have various applications especially in food, biochemical, and other industries. The depolymerized compounds obtained by the chitinous waste can be used for production of single-cell proteins [50], biofertilizers [52], and protoplast from fungal sources. The enzyme can also act as biopesticides that can be used as substitute to chemical pesticides [40]. These enzymes have also been used as component of packaging material intended for foods and released slowly from such packaging during the storage of food [47].

5.3.6 BACTERIOPHAGE LYSINS

Endolysin genes (ply118/ply511) from *L. monocytogenes* phage are cloned and expressed in dairy starter culture *L. lactis*, which has potential as biopreservative against pathogenic Listeria in food [22]. Truncated lysin (pSL-PL511ΔC) showed increased catalytic activity than native enzyme and can be used for secretion of enzymes and as starter culture for fermented milk [22]. *Clostridium perfringens* strain causes food-borne illness by producing enterotoxin and is also linked with food spoilage. Zimmer et al. [64] proposed that phage Ply3626 has a potential biopreservative action towards *Clostridium perfringens* in food, feed, raw poultry products, etc. *L. lactis* phage endolysins are used in dairy industry to control cheese ripening process [12, 58]. De Ruyter et al. [11] studied the holin-endolysin protein from lactococcal phage phi US3 and induced the death of *L. lactis* to release the enzymes, responsible for accelerating the cheese ripening process and thus involved in flavor production.

5.4 FUTURE POTENTIAL

Antimicrobial enzymes are already being used in food applications, but to a lesser extent owing to high production cost for extracting and purifying them from native sources. However, the advances in genetic engineering and synthetic biology approaches have paved the way for cloning the enzymes and their sequencing in host cells thereby making commercial production of enzyme a practical viability. Other than production cost, the public and legal health concern, adverse effects of enzymes, and their stability in the range of food manufacturing systems need to be addressed [47].

Bacteriophage lysins can be potential alternatives to conventional antibiotics for effectively treating food-producing animals and humans and are one of the viable options for addressing the problem of bacterial drug resistance. More number of clinical trials need to be conducted to evaluate the pharmacokinetic and safety aspects of the phage endolysins in humans to treat infection caused by the pathogenic bacteria.

5.5 SUMMARY

The chapter focuses on the potential antimicrobial enzymes and their application in food systems in a comprehensive manner. The chapter also documents the importance of antimicrobial enzymes and their sources and mechanism of actions, different types of antimicrobial enzymes including hydrolases, oxidoreductases, and bacteriophage endolysins, their food applications, and future potential.

KEYWORDS

- **antimicrobial enzyme**
- **bacteriophage lysin**
- **chitinase**
- **glucose oxidase**
- **lactoperoxidase**
- **lysozyme**
- **β-glucanase**

REFERENCES

1. Aono, R., Hammura, M., Yamamoto, M., & Asano, T., (1995). Isolation of extracellular 28- and 42-kilodalton β-1,3-glucanases and comparison of three β-1,3-glucanases produced by *Bacillus circulans* IAM1165. *Applied and Environmental Microbiology*, *61*(1), 122–129.
2. Arqués, J. L., Rodríguez, E., Nuñez, M., & Medina, M., (2008). Antimicrobial activity of nisin, reuterin, and lactoperoxidase system on *Listeria monocytogenes* and *Staphylococcus aureus* in Cuajada, a semisolid dairy product manufactured in Spain. *Journal of Dairy Science*, *91*(1), 70–75.
3. Björck, L., Rosén, C., Marshall, V., & Reiter, B., (1975). Antibacterial activity of the lactoperoxidase system in milk against pseudomonas and other gram negative bacteria. *Applied Microbiology*, *30*(2), 199–204.
4. Borysowski, J., Weber-Dabrowska, B., & Gorski, A., (2006). Bacteriophage endolysins as a novel class of antibacterial agent. *Experimel. Biology and Medicine, 231*(4), 366–377.

5. Boussouel, N., Mathieu, F., Revol-Junelles, A. M., & Millière, J. B., (2000). Effects of combinations of lactoperoxidase system and nisin on the behavior of *Listeria monocytogenes* ATCC 15313 in skim milk. *International Journal of Food Microbiology, 61*(2–3), 169–175.
6. Bradshaw, C. E., (2011). *In vitro* comparison of the antimicrobial activity of honey, iodine, and silver wound dressings. *Bioscience Horizons: The International Journal of Student Research, 4*(1), 61–70.
7. Cegielska-Radziejewska, R., Lesnierowski, G., & Kijowski, J., (2003). Antibacterial activity of lysozyme modified by the membrane technique. *Electronic Journal of Polish Agricultural Universities, 6*(2), 2–10.
8. Conte, A., Brescia, I., & Del, N. M. A., (2011). Lysozyme/EDTA disodium salt and modified- atmosphere packaging to prolong the shelf-life of burrata cheese. *Journal of Dairy Science, 94*(11), 5289–5297.
9. Costa, C., Lucera, A., Conte, A., Zambrini, A. V., & Del, N. M. A., (2017). Technological strategies to preserve burrata cheese quality. *Coatings, 7*, 97–105.
10. Davidson, P. M., & Taylor, T. M., (2007). Chemical preservatives and natural antimicrobial compounds: Chapter 33. In: Doyle, P., & Beuchat, L., (eds.), *Food Microbiology: Fundamentals and Frontiers* (pp. 713–734). American Society for Microbiology Press, Washington, DC.
11. De Ruyter, P. G., Kuipers, O. P., Meijer, W. C., & De Vos, V. M., (1997). Food-grade controlled lysis of *Lactococcus lactis* for accelerated cheese ripening. *Nature Biotechnology, 15*(10), 976–979.
12. Deutsch, S. M., Guezenec, S., Piot, M., Foster, S., & Lortal, S., (2004). Mur-LH, the broad-spectrum endolysin of *Lactobacillus helveticus* temperate bacteriophage phi-0303. *Applied and Environmental Microbiology, 70*(1), 96–103.
13. Dewi, R. T. K., Mubarik, N. R., & Suhartono, M. T., (2016). Medium optimization of β-glucanase production by *Bacillus subtilis* SAHA 32.6 used as biological control of oil palm pathogen. *Emirates Journal of Food and Agriculture, 28*(2), 116–125.
14. Dobbenie, D., Uyttendaele, M., & Debevere, J., (1995). Antibacterial activity of the glucose oxidase/glucose system in liquid whole egg. *Journal of Food Protection, 58*(3), 273–279.
15. Doosh, K. S., & Abdul-Rahman, S. M., (2014). Effect of lysozyme isolated from hen egg white in elongation the shelf-life of Iraqi soft cheese made from buffalo milk. *Pakistan Journal of Nutrition, 13*(11), 635–641.
16. Durance, T. D., (1994). Separation, purification, and thermal stability of lysozyme and avidin from chicken egg white. In: Sim, J. S., & Nakai, S., (eds.), *Egg Uses and Processing Technologies, New Developments* (pp. 77–93). International CAB, Wallingford-UK.
17. Earnshaw, R. G., Banks, J. G., Francotte, C., & Defrise, D., (1990). Inhibition of *Salmonella typhimurium* and *Escherichia coli* in an infant milk formula by activated lactoperoxidase system. *Journal of Food Protection, 53*(2), 170–172.
18. Elotmani, F., & Assobhei, O., (2003). In vitro inhibition of microbial flora of fish by nisin and lactoperoxidase system. *Letters in Applied Microbiology, 38*(1), 60–65.
19. Field, C. E., Pivarnik, L. F., Barnett, S. M., & Rand, Jr. A. G., (1986). Utilization of glucose oxidase for extending the shelf-life of fish. *Journal of Food Science, 51*(1), 66–70.
20. Fischetti, V. A., (2010). Bacteriophage endolysins: A novel anti-infective to control gram-positive pathogens. *International Journal of Medical Microbiology, 300*(3), 357–362.

21. Fischetti, V. A., Nelson, D., & Schuch, R., (2006). Reinventing phage therapy: Are the parts greater than the sum? *Nature Biotechnology, 24*(12), 1508–1511.
22. Gaeng, S., Scherer, S., Neve, H., & Loessner, M. J., (2000). Gene cloning and expression and secretion of *Listeria monocytogenes* bacteriophage-lytic enzymes in *Lactococcus lactis*. *Applied and Environmental Microbiology*, *66*(7), 2951–2958.
23. Harel, S., & Kanner, J., (1985). Hydrogen peroxide generation in ground muscle tissues. *Journal of Agricultural and Food Chemistry*, *33*(6), 1186–1188.
24. Hong, T. Y., Cheng, C. W., Huang, J. W., & Meng, M., (2002). Isolation and biochemical characterization of an endo-1,3-β-glucanase from *Streptomyces sioyaensis* containing a C-terminal family 6 carbohydrate-binding module that binds to 1,3-β-glucan. *Microbiology*, *148*, 1151–1159.
25. Hughey, V. L., & Johnson, E. A., (1987). Antimicrobial activity of lysozyme against bacteria involved in food spoilage and food-borne disease. *Applied and Environmental Microbiology*, *53*(9) 2165–2170.
26. Ibrahim, H. R., Higashiguchi, S., Juneja, L. R., Kim, M., & Yamamoto, T., (1996). A structural phase of heat-denatured lysozyme with novel antimicrobial action. *Journal of Agricultural and Food Chemistry*, *44*(6), 1416–1423.
27. Ibrahim, H. R., Kato, A., & Kobayashi, K., (1991). Antimicrobial effects of lysozyme against Gram-negative bacteria due to covalent binding of palmitic acid. *Journal of Agricultural and Food Chemistry, 39*(11), 2077–2082.
28. Jado, I., Lopez, R., Garcia, E., Fenoll, A., Casal, J., & Garcia, P., (2003). Phage lytic enzymes as therapy for antibiotic-resistant *Streptococcus pneumoniae* infection in a murine sepsis model. *The Journal of Antimicrobial Chemotherapy, 52*(6), 967–973.
29. Jeong, D. K., Harrison, M. A., Frank, J. F., & Wicker, L., (1992). Trials on the antimicrobial effect of glucose oxidase on chicken breast and muscle. *Journal Food Safety*, *13*(1), 43–49.
30. Jooyandeh, H., Aberoumand, A., & Nasehi, B., (2011). Application of lactoperoxidase system in fish and food products: A review. *American-Eurasian Journal of Agricultural and Environmental Sciences*, *10*(1), 89–96.
31. Kamau, D. N., Doores, S., & Pruitt, K. M., (1990). Antibacterial activity of lactoperoxidase system against *Listeria monocytogenes* and *Staphylococcus aureus* in milk. *Journal of Food Protection*, *53*(12), 1010–1014.
32. Kantt, C. A., Bouzas, J., Dondero, M., & Torres, J. A., (1993). Glucose oxidase/catalase solution for on-board control of shrimp microbial spoilage: Model studies. *Journal of Food Science*, *58*(1), 104–107.
33. Kantt, C. A., & Torres, J. A., (1993). Growth inhibition by glucose oxidase of selected organisms associated with the microbial spoilage of shrimp (*Pandalus jordani*): *In vitro* model studies. *Journal of Food Protection*, *56*(2), 147–152.
34. Kijowski, J., Marciszewska, C., & Cegielska-Radziejewska, R., (2002). Quality and microbiological stability of chilled chicken breast muscles treated with a lysozyme solution. *Polish Journal of Food and Nutrition Sciences*, *11/52*(2), 47–54.
35. Kohler, H., & Jenzer, H., (1989). Interaction of lactoperoxidase with hydrogen peroxide: Formation of enzyme intermediates and generation of free radicals. *Free Radical Biology and Medicine*, *6*(3), 323–339.
36. Loeffler, J. M., Djurkovic, S., & Fischetti, V. A., (2003). Phage lytic enzyme Cpl-1 as a novel antimicrobial for pneumococcal bacteremia. *Infection and Immunity*, *71*(11), 6199–6204.

37. Low, N., Jiang, Z., Ooraikul, B., Dokhani, S., & Palcic, M. M., (1989). Reduction of glucose content in potatoes with glucose oxidase. *Journal of Food Science*, *54*(1), 118–121.
38. Lucera, A., Costa, C., Conte, A., & Del, N. M. A., (2012). Food applications of natural antimicrobial compounds. *Frontiers in Microbiology*, *3*, 287–295.
39. Mastromatteo, M., Lucera, A., Sinigaglia, M., & Corbo, M. R., (2010). Synergic antimicrobial activity of lysozyme, nisin, and EDTA against *Listeria Monocytogenes* in ostrich meat patties. *Journal of Food Science, 75*(7), M422–M429.
40. Melchers, L. S., & Stuiver, M. H., (2000). Novel genes for disease-resistance breeding. *Current Opinion in Plant Biology*, *3*(2), 147–152.
41. Mickelson, M. N., (1979). Antibacterial action of lactoperoxidase-thiocyanate-hydrogen peroxide on *Streptococcus agalactiae*. *Applied and Environmental Microbiology*, *38*(5), 821–826.
42. Mohagheghpour, N., Dawson, M., & Hobbs, P., (1995). Glucans as immunological adjuvants. *Advances in Experimental Medicine and Biology, 383*, 13–22.
43. Ntzimani, A. G., Giatrakou, V. I., & Savvaidis, I. N., (2010). Combined natural antimicrobial treatments (EDTA, lysozyme, and rosemary and oregano oil) on semi cooked coated chicken meat stored in vacuum packages at 4°C: Microbiological and sensory evaluation. *Innovative Food Science and Emerging Technologies, 11*(1), 187–196.
44. O'Flaherty, S., Coffey, A., Meaney, W., Fitzgerald, G. F., & Ross, R. P., (2005). The recombinant phage lysin LysK has a broad spectrum of lytic activity against clinically relevant staphylococci, including methicillin-resistant *Staphylococcus aureus*. *Journal of Bacteriology, 187*(20), 7161–7164.
45. Pitson, S. M., Seviour, R. J., & McDougall, B. M., (1993). Non-cellulolytic fungal β-glucanases: Their physiology and regulation. *Enzyme and Microbial Technology*, *5*(3), 178–192.
46. Proctor, V. A., & Cunningham, F. E., (1988). The chemistry of lysozyme and its use as a food preservative and a pharmaceutical. *Critical Reviews in Food Science and Nutrition*, *26*(4), 359–395.
47. Ramos, O. S., & Malcata, F. X., (2011). Food grade enzymes: Chapter 3.48, In: Moo-Young, M., (ed.), *Comprehensive Biotechnology* (2nd edn., Vol. 3, pp. 555–569). Elsevier, USA.
48. Rathore, A. S., & Gupta, R. D., (2015). Chitinases from bacteria to human: Properties, applications, and future perspectives. *Enzyme Research*, p. 8. Article ID-791907.
49. Reiter, B., & Härnulv, G., (1984). Lactoperoxidase antibacterial system: Natural occurrence, biological functions, and practical applications. *Journal of Food Protection*, *47*(9), 724–732.
50. Revah-Moiseev, S., & Carroad, P. A., (1981). Conversion of the enzymatic hydrolysate of shellfish waste chitin to single-cell protein. *Biotechnology and Bioengineering*, *23*(5), 1067–1078.
51. Sagi, I., & Mannheim, C. H., (1990). The effect of enzymatic oxygen removal on quality of unpasteurized and pasteurized orange juice. *Journal of Food Processing and Preservation*, *14*(4), 253–266.
52. Sakai, K., Yokota, A., Kurokawa, H., Wakayama, M., & Moriguchi, M., (1998). Purification and characterization of three thermostable endochitinases of a noble *Bacillus* strain, MH-1, isolated from chitin-containing compost. *Applied and Environmental Microbiology*, *64*(9), 3397–3402.

53. Seifu, E., Buys, E. M., & Donkin, E. F., (2005). Significance of the lactoperoxidase system in the dairy industry and its potential applications: A review. *Trends in Food Science and Technology*, *16*(4), 137–154.
54. Shaw, S. J., Bligh, E. G., & Woyewoda, A. D., (1986). Spoilage pattern of Atlantic cod fillets treated with glucose oxidase/gluconic acid. *Canadian Institute of Food Science and Technology Journal, 19*(1), 3–6.
55. Sinigaglia, M., Bevilacqua, A., Corbo, M. R., Pati, S., & Del, N. M. A., (2008). Use of active compounds for prolonging the shelf-life of mozzarella cheese. *International Dairy Journal*, *18*(6), 624–630.
56. Tiina, M., & Sandholm, M., (1989). Antimicrobial effect of the glucose oxidase-glucose system on food-poisoning organism. *International Journal of Food Microbiology, 8*(2), 165–174.
57. Vartiainen, B. J., Rättö, M., & Paulussen, S., (2005). Antimicrobial activity of glucose oxidase-immobilized plasma activated polypropylene films. *Packaging Technology and Science*, *18*(5), 243–251.
58. Vasala, A., Valkkila, M., Caldentey, J., & Alatossava, T., (1995). Genetic and biochemical characterization of the *Lactobacillus delbrueckii* subsp. *Lactis* bacteriophage LL-H lysin. *Applied and Environmental Microbiology*, *61*(11), 4004–4011.
59. Wong, C. M., Wong, K. H., & Chem, X. D., (2008). Glucose oxidase: Natural occurrence, function, properties, and industrial applications. *Applied Microbiology and Biotechnology, 78*(6), 927–938.
60. Yoo, W., & Rand, A. G., (2006). Antibacterial effect of glucose oxidase on growth of *Pseudomonas fragi* as related to pH. *Journal of Food Science*, *60*(4), 868–871.
61. Yoshimura, K., Toibana, A., & Nakahama, K., (1988). Human lysozyme: Sequencing of a cDNA, and expression and secretion by *Saccharomyces cerevisiae. Biochemical and Biophysical Research Communications, 150*(2), 794–801.
62. Zapico, P., Gaya, P., Nuñez, M., & Medina, M., (1995). Activity of Goat's milk lactoperoxidase system on *Pseudomonas fluorescens* and *Escherichia coli* at refrigeration temperatures. *Journal of Food Protection*, *58*(10), 1136–1138.
63. Zapico, P., Medina, M., Gaya, P., & Nuñez, M., (1998). Synergistic effect of nisin and lactoperoxidase system on *Listeria monocytogenes* in skim milk. *International Journal of Food Micirobiology*, *40*(1–2), 35–42.
64. Zimmer, M., Vukov, N., Scherer, S., & Loessner, M. J., (2002). The murien hydrolase of the bacteriophage phi3626 dual lysis system is active against all tested *Clostridium perfringens* strains. *Applied and Environmental Microbiology*, *68*(11), 5311–5317.

CHAPTER 6

HEALTH BENEFITS OF ANTIMICROBIAL PEPTIDES

SUNITA MEENA, KAPIL SINGH NARAYAN, and SANDEEP KUMAR

ABSTRACT

Antimicrobial peptides (AMPs) are small peptides (12–50 amino acids), which are active against microbes including bacteria, yeast, molds, fungi, and other parasitic infections. These are ribosomally synthesized peptides secreted by most organisms and play an important role in host defense. They are specific and cause no damage to the host cell/producer cell. AMPs are mostly active on the target membrane by direct binding with the membrane phospholipids and create the pores. These pores disturb proton motive force, releases intracellular molecules causing target cell death. AMPs are structurally classified into α-helical, β-sheet, and extended peptides based on their amino acid sequence. Barrel-stave, toroidal, carpet, and detergent-like models have been proposed for their action mechanism which differs from species to species and within species. AMPs produced by mammals function as immune-modulators and have a crucial role in innate immunity. AMPs also have other bioactivities such as antioxidant, antitumor, antimicrobial, and many more which makes them popular for wide applications.

6.1 INTRODUCTION

Antimicrobial peptides (AMPs) are peptides, which act against microbes, such as bacteria, yeast, molds, parasites, and fungi. AMPs are commonly recognized as defense peptides of the host. Till today, 3042 AMPs are reported in antimicrobial peptide database (APD) (Figure 6.1). AMPs are produced by bacteria, fungi, insects, fishes, animals, and plants and intensively involved in their innate immune system and function as immunomodulators. These are mostly active on membrane of the target to create poration complexes leading to release of intracellular substances, such as ions, depletion of proton motive force and thereby causing death of the target cell. These are highly specific and non-damaging to the producer cells. The mechanism of action differs from species to species and within species.

AMPs may target the plasma membrane and interfere with DNA synthesis, protein synthesis, and protein folding. Generally, cationic AMPs act in a three-step process: beginning with encounter with the target where interaction is electrostatic in nature; followed by binding of peptides on the negatively charged target cell membrane resulting in insertion of peptide oligomers; and involving hydrophobic surface of poration complex to interact with nonpolar core of lipid membrane with formation of hydrophilic pore leading to leakage of intracellular metabolites leading to cell death. Due to their use in food as a safe additive and therapeutic potentials, AMPs have attracted scientific attention in recent years.

The aim of the chapter is to explore the role of AMP, their sources, their mechanism of action and their emerging therapeutic potential.

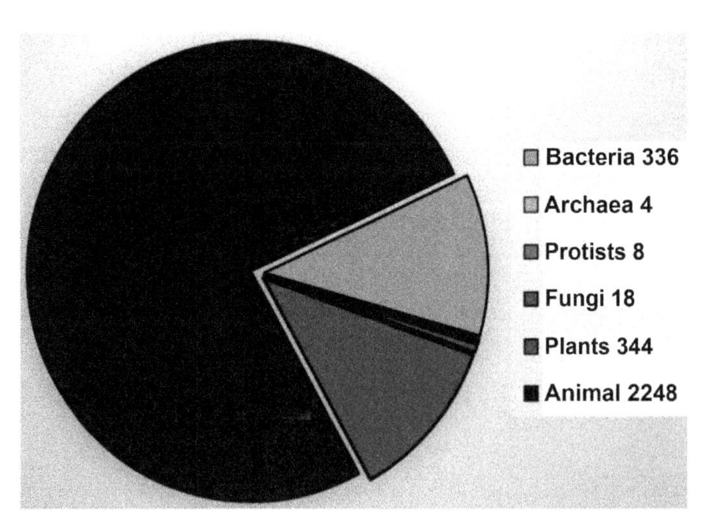

FIGURE 6.1 Total number (3042) of antimicrobial peptides reported as of December 2018 in antimicrobial peptide database (APD).

6.2 SOURCES OF ANTIMICROBIAL PEPTIDES (AMPS)

AMPs are synthesized by most of the organisms but some of well effective AMPs are also being produced synthetically [96] (Table 6.1).

TABLE 6.1 List of Antimicrobial Peptides (AMPs)

Source of Origin	Peptide	References
Bacteria (Gram Positive)		
Erwinia tasmaniensis	Tasmancin	[107]
Escherichia coli	Colicin and Microcin	[18]
Halophilic archaea	Halocin	[88]
Pseudomonas aeruginosa	Pyocin	[120]
Pseudomonas chlororaphis	Tailocin	[35]
Vibrio spp.	Vibriocin	[123]
Bacteria (Gram Negative)		
Actinoplanes liguriae	Gardimycin	[115]
Bacillus cereus	Cerein7	[112]
Bacillus subtilis	Subtilosin A	[152]
Carnobacterium maltaromaticum	Carnocyclin A	[86]
Enterococcus faecalis	Enterolysin A	[108]
Lactobacillus plantarum	Plantaricin	[163]
Lactobacillus sakei	Sakacins	[106]
Lactococcus lactis	Nisin/Bisin	[27]
Lactococcus lactis	Lacticin	[99]
Lactococcus lactis	Lactococcin A	[67]
Lactococcus lactis, Listeria monocytogenes, and Bacillus subtilis	Lacticin	[137]
Microbispora corallina	Microbisporicin	[45]
Micrococcus varians	Variacin	[122]
Paenibacillus polymyxa	Paenibacillin	[69]
Pediococcus pentosaceus/ acidilacticii	Pediocin PA-1	[28]
Staphylococcal lantibiotics	Epidermin and gallidermin	[57]
Staphylococcus aureus	Aureocin A70	[23]
Staphylococcus simulans	Lysostaphin	[155]
Staphylococcus warneri	Warnericin	[102]

TABLE 6.1 *(Continued)*

Source of Origin	Peptide	References
Streptococci/Streptomyces	Lantibiotic	[37]
Streptococcus mutans	Mutacin	[58]
Sulfolobus acidocaldarius/tokodaii	Sulfolobicin	[40]
Fungi		
Coprinopsis cinerea	Copsin	[42]
Mycogone cervina	Cervinins	[171]
*Paecilomyceslilacinus*A-26	Lipoamino peptides	[113]
Trichoderma viride	Alamethicin	[5]
Fish		
Arctic char (*Salvelinus alpines*), Brown trout (*Salmo truttafario*), Atlantic cod (*Gadus morhua*), and Grayling (*Thymallus thymallus*)	Cathelicidin	[15, 20, 93, 141]
Atlantic cod (*Gadusmorhua*), Rainbow trout (*Oncorhynchus mykiss*)	H2B	[12, 110]
Atlantic halibut (*Hippoglossus hippoglossus*)	Hipposin	[14]
Atlantic salmon (*Salmo salar*)	AsCath 1, 2	[24]
Carp (*Cyprinus carpio*), Rainbow trout (*Oncorhynchus mykiss*)	LEAP2, LEAP2-A, B	[89, 182]
Channel catfish (*Ictalurus punctatus*)	NK-lysin	[169]
Channel catfish (*Ictalurus punctatus*)	HbβP-1,2,3	[159]
European bass (*Dicentrarchus labrax*)	Dicentracin	[136]
Golden-striped grouper (*Grammistes sexlineatus*), Soapfish (*Pogonoperca punctata*)	Gramminstin Gs1, Pp1	[79]
Grouper (*Epinephelus coioides*)	Epinecidin-1	[15]
Hagfish (*Myxine glutinous*)	HFIAP-1, 3	[161]
Japanese flounder (*Paralichthy solivaceus*)	JF-1. 2, L4, L6	[36, 65]
Mudfish (*Misgurnus anguillicaudatus*)	Misgurin	[116]
Red sea bream (*Chrysophrys major*)	Chrysophsin 1, 2, 3	[70]
Teleost fish (*Cyprinus carpio*)	apoA-I	[29]

TABLE 6.1 *(Continued)*

Source of Origin	Peptide	References
Turbot (*Scophthalmus maximus*), Red sea bream (*Chrysophrys major*), Zebrafish (*Danio rerio*)	Hepcidin	[25, 26, 36, 147]
Insects and Animals		
Amphibians	Magainin, dermaseptin, Buforins	[3, 118, 181]
Cattle	Indolicidin	[3, 181]
Dipterans	Diptericins, attacins	[19]
Frog	Brevinins	[3, 181]
Fruit fly	Drosocin, metchnikowins, pyrrhocoricin,	[1, 19, 34]
Hemipteran	Metalnikowin, Thanatin	[19, 181]
Horseshoe crab	Tachyplesin II, Tachycitin	[3, 19, 181]
Insects	Defensins	[3, 181]
Insects, pig	Cecropins, Protegrin I	[34]
Lepidopteran	Heliomicin	[84]
Mosquito	Gambicin	[164]
Scorpion	Androctonin	[19, 181]
Shrimp	Penaeidins	[34]
Tunicates	Clavanin, styelin	[3]
Human		
	Amylin	[167]
	Cathelicidin (LL-37)	[6]
	Histatins (1,3,5)	[64]
	Lysozyme	[44]
	Substance P	[39]
	Ubiquicidin	[17]
	α-defensins (HNP1, 4, HD5, 6)	[77]
	β-amyloid peptide 1–42	[151]
	β-defensins	[76]
Plants		
Benincasa hispida	Hispidulin	[145]
Brassica napus	Peptides	[21]
Capsella bursa-pastoris	Shepherins	[117]

TABLE 6.1 *(Continued)*

Source of Origin	Peptide	References
Carica candamarcensis, Heligmosomoides polygyrus, Ananas comosus, Carica papaya and Cryptostegia grandiflra	Proteinases	[154]
Cicer arietinum	Cicerin/Arietin	[80]
Hevea brasiliensis	Heveins	[13]
Impatiens balsamina	Impatiens	[168]
Lens culinaris	Lc-def	[146]
Macadamia integrifolia	Vicilin-like	[95]
Malvaparviflra, Raphanus sativus	2S albumin	[175]
Oldenlandia affis	Cyclotides: kalata B1 and B2	[74]
Phaseolus lunatus	Lunatusin	[173]
Phaseolus vulgaris	Peptide PvD1/Defensin-like	[31]
Phaseolus vulgaris	Vulgarinin	[174]
Phytolacca americana	Knottin	[50]
Solanum tuberosum	Snakins	[1]
Spinacia oleracea	Peptide So-D1	[143]
Triticum aestivum	Thonein: alpha-1-purothionin	[131]
Triticum aestivum	Puroindolines: PIN*A* and PIN*B*	[66]
Triticum aestivum and Hurdeum vulgare	Defensins	[135]
Vitis vinifera	PR1, PR2 Chitinases	[41]
Food Products		
Cheese	αS2-CN, β-CN fragments	[132]
Collagen	Peptide fragments (AKGANGAPGIAGAPGFPGA–RGPSGPQGPSGPP, GPRGF, LQGM, PAGNPGADGQPGAKGANGAP, GAXGLXGP, QGAR, VGPV, LQGMH, LC)	[7, 47, 87, 133]
Crescenza	β-CN fragments	[150]
Emmental	αs 1-CN and β-CN fragments	[48]
Fermented products	Peptide fragments (GG, DM)	[16]
Gouda	αS1-CN, β-CN fragments	[34]
Manchego	Ovineβ-CN, αs1-CN fragments	[55]

TABLE 6.1 *(Continued)*

Source of Origin	Peptide	References
Meat	Peptide fragments (KRQKYD, KAPVA, EKERERQ, PTPVT, GLSDGEWQ, GFHI, RPR, DFHING, APPPPAEVPEVHEEVH, FHG, APPPPAEVPEVHEEVH, LPLGG, PPPAEVPEVHEEVH, IPITAAKASRNIA, APPPPAEVP, FAGGRGG)	[4, 10, 46, 72]
Milk	Ovine as 2-CN fragments, as1-casein f (1–23)	[91, 105]
	Bovine αS1-casein	[103]
	Peptide fragments (VLPVPQK), (PYVRYL, LVYPFTGPIPN)	[124, 130, 178]
Synthetic Peptides		
Anti-fungal and Anti-bacterial	Mir-KGK-NH_2, ICHHCI-OH, CAMEL, Laur-CKK-NH_2 dimer, Citropin	[73]

6.3 ANTIMICROBIAL PEPTIDE DATABASES (APDS)

The AMPs are most potent defensive peptides ranging from prokaryotes to eukaryotes from millions of years. In 1922, Alexander Fleming discovered the first antimicrobial peptide, called lysozyme (LYZ) [56]. Since then, hundreds of AMPs are being identified every year. Earlier, the number of known AMPs was limited therefore, these AMPs were managed manually. However, with the increasing number of AMPs, it became impractical to handle these peptides manually therefore; databases were established to handle these. Antimicrobial sequence database (AMSdb) was the first database to manage the reported AMPs [51]. In 1988, this database was made available online in the format identical to SWISS-Prot [8]. Later it was not updated; therefore, with time new updated databases were established to handle newly discovered AMPs (Table 6.2).

TABLE 6.2 AMPs:Database List

Database	Description	References
AntiBP2	Antibacterial peptide prediction server	[85]
Anti-TbPdb	Anti-TB Peptides database	[160]

TABLE 6.2 *(Continued)*

Database	Description	References
APD2	AMPs Database	[166]
AVPdb	Database for antiviral peptides	[126]
BaAMPs	Database for biofilm targeting AMPs	[33]
BACTIBASE	Bacteriocin data repository	[61]
BAGEL4	Bacteriocins database	[162]
BIOPEP	Database for bioactive peptide sequence processing	[104]
CAMP	Sequences and structural database of AMPs	[157]
Class AMP	Database for classification of AMPs	[78]
CyBase	Cyclotides database having anti-HIV, anti-bacterial, and insecticidal activity	[165]
DADP	Database for amphibian AMPs	[111]
DBAASP	Antimicrobial Activity and Structural database	[53]
DEFENSINS Knowledge base	Database for manually curated AMPs of defensin family	[142]
EnzyBase	Database of enzybiotics (lysins, autolysins, lysozymes, and large bacteriocins)	[176]
HIPdb	HIV Inhibiting Peptides database	[125]
InverPep	Invertebrate AMPs database	[54]
LAMP	A database linking AMPs	[183]
MilkAMP	A comprehensive database of AMPs of dairy origin	[156]
Peptaibol	Database for antibiotic peptides having antimicrobial activity	[170]
PhytAMP	Plant AMPs database	[60]
YADAMP	AMPs Database	[121]

Adapted from the APD and SATPDB databases.

6.4 CLASSIFICATION OF ANTIMICROBIAL PEPTIDES (AMPS)

6.4.1 STRUCTURAL CLASSIFICATION OF AMPS

Structural and physicochemical attributes of any molecule play a vital role with its activity and specificity for the target. The structure of a protein depends on amino acid sequence, which it contains because some AMPs also contain modified amino acids like lantibiotics. Depending on their structure and functional relationship, AMPs are divided in following three main groups:

1. **α-Helical AMPs:** These are short peptides (<40 amino acids) and amphipathic in nature. They target a broad range of pathogens and their activity depends on seven parameters: size, charge, sequence, percent helical content, hydrophobicity, amphipathicity, and content of hydrophilic and hydrophobic amino acids. When present in aqueous solutions, these AMPs remain unstructured and retain the defined amphipathical α-helix structure when they interact with cell membranes [180]. Most of these AMPs are membrane active antimicrobials and form pores in the microbial membrane; are cationic in nature that facilitates their interaction with anionic surface of microbes and causes its insertion into the cell membrane [3]. The prevalence of these peptides has been reported in hemolymph of drosophila [100], skin secretions of amphibians [9], vacuoles, wounds, blisters, and epithelia of mammals [49, 109]. The well-studied examples of α-helical AMPs are alamethicin, gramicidin, pardaxin, and magainin (Figure 6.2a).
2. **β-Sheet AMPs:** The β-sheet is considered a more stable structure because at least two β-strands are joined by disulfide bonds and hydrophobic interactions form β-sheet peptides [154]. The disulfide bond is formed by the conserved cysteine residues and therefore reduces the proteolytic degradation of AMPs [32]. In addition, removing the disulfide bonds from lactoferricin (Lf) and tachyplesin retains their antimicrobial activity. However, these peptides lose their hemolytic activity [52, 127]. In 2011, Schroeder reported that reduction of disulfide bonds distort the structure of β-defensin-1 (human) that improves the antimicrobial activity for *Lactobacillus* and *Candida* [140]. These peptides are rigid enough therefore they do not lose their β-sheet structure in aqueous solution unlike α-helical AMPs [180]. The examples of β-sheet peptides are: defensins, tachyplesins, polyphemusin, and cathelidins (Figure 6.2b). Defensin peptides form a large group of β-sheet peptides. These peptides are produced as inactive precursors in macrophages, neutrophils, and epithelial cells [84, 119]. Most of these β-sheet peptides are membrane-active but some of them also bind to minor groove of nucleic acid and interfere in protein-DNA interactions such as in case of tachyplesin produced from horseshoe crabs [84]. The other β-sheet peptides, such as bovine

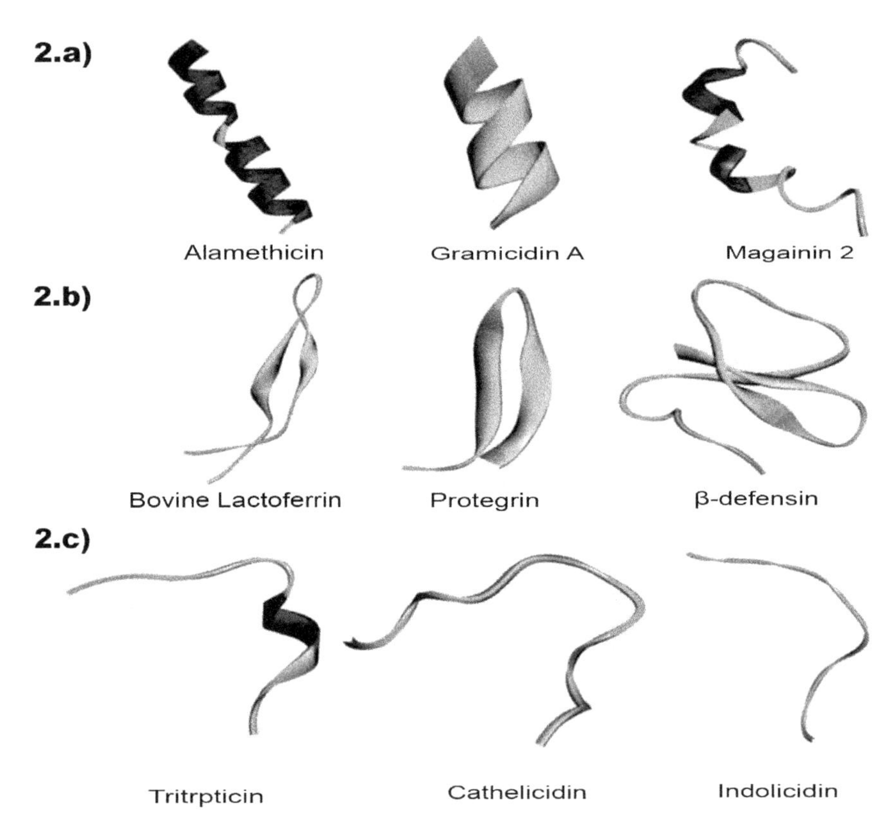

FIGURE 6.2 Structure-based classification: α helical (2a); β sheet (2b; and extended (2c). *Note:* Structures are from Protein Data Bank (PDB ID codes Alamethicin, 1AMT; Gramicidin A, 1NRM; Magainin-2, 2LSA; Bovine Lactoferrin, 1LFC; Protegrin, 1PG1; β-defensin, 1KJ5; Tritrpticin, 1D6X; Cathelicidin, 5XRX; Indolicidin, 1G89).

Lf, inhibit the ATP dependent multi-drug efflux pump by disturbing proton motive force and potential of the membrane of target cell. Therefore, Lf is also used synergistically with other antimicrobials to kill the resistant target cells [97].

3. **Extended AMPs:** These have unique extended coil structure. These peptides do not fold in a regular secondary structure (Figure 6.2c). These have high proportion of arginine, tryptophan, and proline. The shortest AMPs, RRWQWR from bovine Lf and RAWVAWR from human, have been identified [90]. Indolicidin is best characterized extended AMP isolated from bovine leukocytes with high tryptophan content. Indolicidin has also been utilized as a guide to design better

AMPs having enhanced activity such as omiganan developed by Migenix [101].

6.4.2 BONDING BASED CLASSIFICATION OF AMPS

This classification depends on the interaction of polypeptides and independent of 3D structure, source, and activity as all peptides do not have 3D structure. According to this classification, AMPs are divided into four classes:

1. **Class I:** AMPs are linear and not connected via covalent bonding (Figure 6.3a). Based on number of polypeptide chains, class I AMPs are divided into two subclasses: (a) linear single-chain peptides (magainins and LL-37); and (b) two linear peptides connected by non-covalent interactions (enterocin L50).
2. **Class II:** AMPs have chemical bonding between side chains of different peptides (Figure 6.3b). Class IIAMPs are further categorized into two subclasses based on the connections between side chains: (a) a side chain-side chain connection within a single peptide chain, e.g., defensin-like AMPs; and (b) lantibiotics or between two different peptide chains, e.g., distinctin, halocidin, centrocin, and lacticin-3147.
3. **Class III:** AMPs have chemical interactions between side-chain and backbone of peptides (Figure 6.3c). These are further classified based on chemical bonds: (a) peptides connecting via amide bond (Microcin J25 and Lariatins), ester bond (Fusaricidin A); and (b) thioester bond (Thuricin CD).
4. **Class IV:** AMPs include the circular AMPs that are connected via the backbone of one peptide to the backbone of other peptides, e.g., AS-48, subtilosin A, and cyclotides (Figure 6.3d).

6.5 MECHANISMS OF ACTION OF AMPS

Generally, the accepted mechanism of AMPs is the disintegration of cell membranes that leads to ions and metabolites leakage, resulting in cell death. However, there are evidences to report that AMPs also exert their effects through intracellular targets. Even if AMPs act on intracellular targets, their contact with membrane of target cell is necessary for their activities [62]. Based on action mechanisms, AMPs are divided in the following two groups.

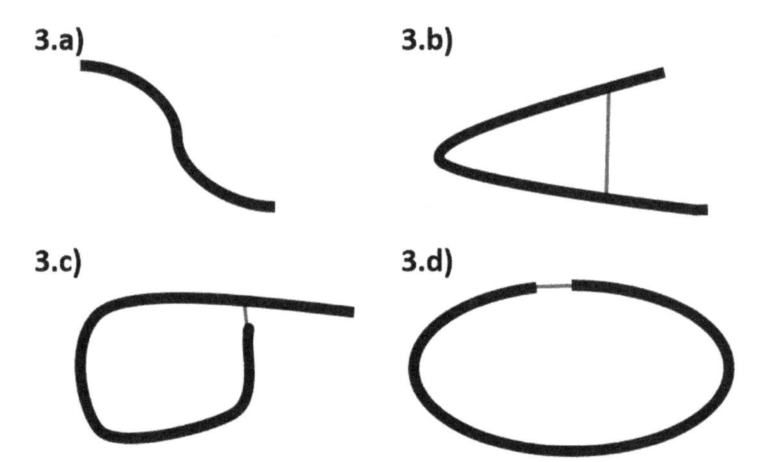

FIGURE 6.3 Bonding based classification: (a) Class I linear peptides, e.g., magainins; (b) Class II side chain-side chain bonded peptides, e.g., defensins; (c) Class III side chain-back bone bonded peptides, e.g., lassos; and (d) Class IV back bone-back bone bonded peptides, e.g., cyclotides.

6.5.1 MEMBRANE ACTIVE AMPS

AMPs have the inherent ability to target and interact with bio-membrane. Initially, AMPs interact with membrane via electrostatic interaction. Most membrane-active AMPs are cationic and hydrophobic, i.e., amphipathicity is the characteristic of these AMPs. The positively charged peptide interacts with anionic membrane lipids via electrostatic interactions and hydrophobic part of peptide helps to insert into cell membranes [92, 98]. Lysine and arginine residues of AMPs involve strong interaction with phospholipids of membrane. After interaction, the critical concentration of AMPs is essential for antimicrobial activity. After reaching the threshold concentration, the peptides start to self-associate on membrane and penetrate deeper causing pore formation to disrupt the membrane. The lipid composition and membrane fluidity also affect the activity of AMPs. The following four models have been reported in the literature to portray the action mechanism:

1. **Barrel-Stave Model:** It explains the pore formation by AMPs inside the membrane. The first barrel stave model was proposed for alamethicin to form single channel in membrane [11]. According to this model, AMPs reach the membrane and interact via electrostatic

force. After reaching the threshold concentration, the peptides go through some conformational changes and self-aggregate. These peptides adopt their amphipathic form, enter, and laterally diffuse deeply inside the lipid bilayer to form pore-like "barrel." "Stave" denotes individual spokes of barrel (Figure 6.4a). Initially, these staves align parallel to the membrane and thereafter, barrels are created. The barrel of AMPs is perpendicularly inserted into cell membrane [38, 179].

2. **Toroidal Model:** The toroidal pores are like barrel-stave pores. In this model, AMPs align themselves either in perpendicularly or tilted plane to the lipid bilayer structure (Figure 6.4b) [59]. The hydrophilic part of peptide faces towards the pore and the hydrophobic part associates with lipid part of membrane bilayer. These peptides induce the bending of membrane lipids. Magainins, protegrins, and melittin peptides distort the normal segregation of polar and non-polar part of membrane [172]. In these types of pores, the polar side of peptides associates with the lipid's polar head group. These pores allow the transient entry of the peptide to cytoplasm, where it targets intracellular components. These peptides induce the cell death by forming toroidal shaped pores within membrane causing membrane depolarization [144].

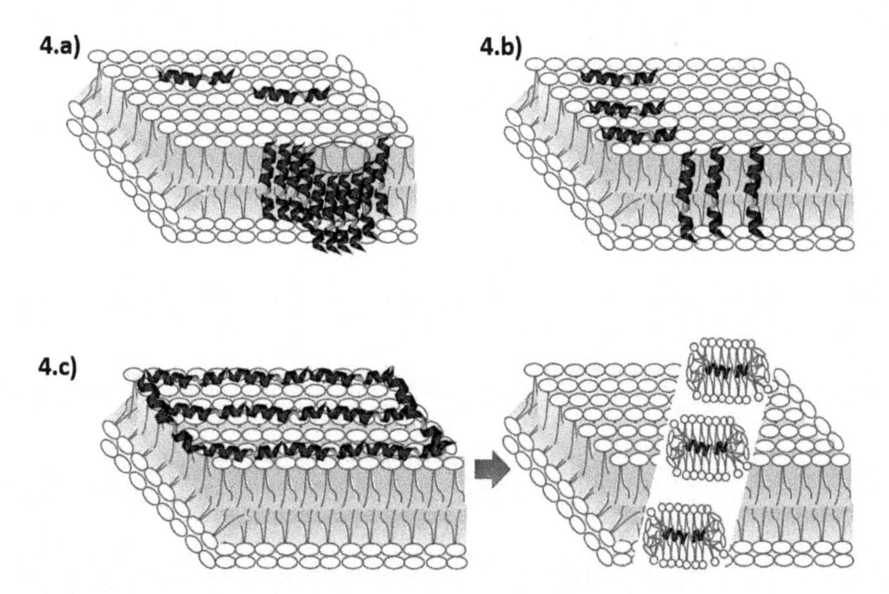

FIGURE 6.4 Models explaining the action mechanism of AMPs: (a) Barrel-Stave model, (b) Toroidal pore model, (c) Carpet model, or (c) Detergent 'like' model.

3. **Carpet Model:** AMPs peptides also act through nonspecific membrane permeabilization instead of pore formation [180]. In carpet model, the peptides initially gather on membrane surface and form a carpet-like structure locally. After reaching a critical concentration, the peptides induce membrane destabilization and permeabilization, which causes cell membrane disruption. The disruption of membrane has detergent-like effect and thus, the membrane disintegrates into micelles [128].
4. **Detergent-Like Model:** This model does not depend on the membrane-bound peptide monomer interaction and insertion (Figure 6.4c). Aurein, indolicidin, and cecropin use carpet model for their activity.

6.5.2 *INTRACELLULAR ACTIVE AMPS*

Earlier, the primary mechanism of killing by AMPs was through cell membrane permeabilization. But Later, other mechanisms for killing actions of these peptides have also been reported. Some AMPs also act intracellularly for example indolicin binds to DNA at basic sites [68, 94]. Other AMPs inhibit protein synthesis and DNA synthesis such as PR-39 acts as a proteolytic agent in bacteria that stops synthesis of protein and DNA. Apidaecin is a protein inhibitor AMP of Gram-negative bacteria and is unable to form pore in membrane and but peptide is transported inside the cell through active transport [22]. Human neutrophil defensin-1 and thrombin-induced platelet microbicidal protein-1also inhibit DNA and protein synthesis to kill broad spectrum pathogens [177]. Interestingly, some intracellular active AMPs act at defined stage of growth such as diptericine, which is active against actively growing cells of bacteria [71, 81]. Some AMPs act on multiple targets. Seminal plasmin, a protein isolated from the seminal plasma of bull, inhibits the protein synthesis completely by inhibiting RNA (ribonucleic acid) polymerase [138].

6.6 THERAPEUTIC USE OF AMPS

The excessive use of antibiotics as microbial growth inhibitors causes alarming increase of antibiotic resistance and the issue of antibiotic resistance is the most serious concern for the public health and health specialists. Therefore, an alternative is required to resolve this issue, which is able to act faster with broad-spectrum antimicrobial activity and has no side effects. AMPs peptides hold good promise as a broad-spectrum agent against commonly used antibiotics. AMPs are well established natural intrinsic defenses for most of the organisms and their distribution suggests the fundamental role as antimicrobial agents against bacteria, fungi, protozoa, and enveloped viruses, etc., which are discussed in this section.

6.6.1 INNATE IMMUNITY ACTIVATION

The synthesis and production of AMPs by all host-organisms in response to microbial infection is a conserved defense response of innate immunity. In animals, epithelial cells produce AMPs upon microbial infection and are then secreted into lymphatic system and bloodstream to target sites of infection. Paneth cells are present in small intestine and produce majority of the AMPs. The α-defensin is produced in form of pre-proteins and need MMP-7 (matrix metalloproteinase) for its activation. This mature peptide has enhanced susceptibility and lethality against *Salmonella typhimurium* and *Escherichia coli* [78]. Similarly, β-defensin clears *Homophilus influenza* and *Staphylococcus spp.* from lungs and urinary tract, respectively [104]. Plants strengthen their innate immunity against microbial invasion by several mechanisms of defense by local formation of the reactive oxygen species (ROS), proteins, and secondary metabolites having antimicrobial activity. In plants, AMPs do not circulate, but are expressed constitutively in specific organs at the site of infection, where microbes attach. Lignin and polysaccharides construct physical barriers to inhibit spreading of pathogens [157].

6.6.2 BENEFICIAL SYMBIOTIC INTERACTION

Most of the plants and animals maintain symbiosis with beneficial microbes. Number of AMPs is engaged to maintain symbiotic association among bacteria and are actively involved in the symbiosis process to yield benefits, such as nutrient acquisition, immunity development, defense against enemies, and reproduction.

1. **Controls Nitrogen-Fixation Endosymbionts:** Nodule specific cysteine rich (NCR) families of AMPs produced by *Medicago truncatula* are quite remarkable because these consist of several hundred

genes and express only in Rhizobium-infected symbiotic nodule cells. These peptides are involved in differentiation of endosymbiont *Sinorhizobium melilotin* [121]. NCRs are transported to the nodules and play effective antimicrobial activity to kill most of the bacteria present in nodules.

2. **Controls Gut Microbiota:** The symbiotic relation of microbiota with the gut provides immunity to individuals and is essential for good health and physiology. These gut microbiota eliminate the harmful pathogens and are involved in cell differentiation of intestine [166]. Regulated level of AMPs under pathogenic and healthy conditions is crucial to maintain homeostasis in the gut.

6.6.3 ANTIOXIDANT ACTIVITIES OF AMPS

Free radicals originated from lipid peroxidation cause oxidative stress if produced in excess. Antioxidants fight against oxidation therefore antioxidants are focused in many current research studies. Antioxidant peptides are present abundantly in food and other medicinal food items show the capacity to decrease oxidative damages, which are responsible for many health problems like atherosclerosis and cancer. Several studies have reported that isolated native peptides and peptides originated from protein hydrolysate act as potential antioxidants. Bioactive peptide with amino acid sequence LKQELGNLLEKQE isolated from marine oyester (*Crassosterea gigas*) has excellent effect against polyunsaturated fatty acid peroxidation. Some of the peptides and protein hydrolysates originated from fish have remarkable antioxidant to scavenge ROS and reduce ferric ions with its reducing ability [61, 183]. Gelatin peptides, FNSGPAGVL, and NGPLQAGQER, with large amount of hydrophobic amino acids have significant emulsification ability and are potent antioxidative peptides. Therefore, it has been universally accepted that oxidation-induced the oxidant agents and peptides overcome these radical-mediated cellular oxidative stresses.

6.6.4 ROLE IN PROBIOSIS ASSOCIATION

The production of AMPs by probiotic microbiota is an important attribute for strain selection of closely-related species in the complex microbial niche of gastrointestinal tract (GIT). The contribution of AMPs towards positive effect on host health is important. It allows competition and colonization of probiotics with indigenous gut microbiota, eliminates pathogens, and also plays an important role in signaling to activate host immune system [125]. Nearly, 50 *Lactobacillus* species have been identified from the gut, which produced bacteriocins having wide spectrum of antimicrobial activities against most of the pathogen [111, 165, 176]. Bacteriocins are also being used as agents in aquaculture to improve the health performance of aquatic microbiome [54].

6.6.5 ANTITUMOR ACTIVITY

Antitumor drugs exhibit different response in absorption and metabolism in the target tissue within each patient [160]. Antitumor treatment therapies by drugs are limited because of impaired drug penetration in restricted environments such as tumor vasculatures and hydrostatic pressure in the tissues. Additionally, the acquired or intrinsic drug resistance is a common cause for tumor recurrence [162]. Due to these limitations, AMPs and their manufactured synthetic analogs have identified alternate therapies to cure cancer. Peptides from *Clitoria ternatea* (like CT2, CT4, CT7, CT10, and CT12 (cyclotides)) exhibited cytotoxicity and reduced 2 to 4-fold IC_{50} value when compared with mitotic inhibitor used in cancer chemotherapy against human lung cancer cells (A549) [148]. Another peptide isolated from *Cocos nucifera* (coconut: Cn-AMP1) was examined on human epithelial colorectal adenocarcinoma cells (Caco-2), which led to 13% decline in cell viability. A peptide (lunasin) was capable of suppressing tumorigenes that is induced by chemical carcinogen. Peptide Cr-ACP and its acetylated analogs arrest cell at G_0–G_1 phase. Some other AMPs (such as Lf, BMAP-27 and BMAP-28) cause cytotoxicity by enhancing permeability of the lipid membrane, Ca^{2+} influxing along with fragmentation of the DNA [129].

6.6.6 ANTIMICROBIAL ATTRIBUTES

The potential of AMPs along with their less toxicity and without any cross-resistance supports their use as agents in clinical studies. Several antimicrobial bioactive AMPs have been identified, isolated, prepared with different methods like protein hydrolysis and synthetic process, and are used as potential agents, such as antibacterial, antiviral, antifungal, and anti-protozoans activities. AMPs are very active against multidrug-resistant bacterium, such as *Pseudomonas aeruginosa, Streptococcus pneumoniae, Streptococcus aureus, and Klebsiella pneumoniae.* A cathelicidin-derived peptide (SMAP-29 from sheep) has the excellent potential of antimicrobial activity against *S. aureus, Enterococcus faecium,* and *Pseudomonas aeruginosa* [75, 149, 158]. Human β-defensins and cathelicidins decrease the attachment of *Pseudomonas aeruginosa* and *Staphylococcus aureus* [66]. Some naturally occurring and synthetic cationic and amphipathic molecules up to 50 amino acids exhibit antifungal activity. These antimycotic peptides are cyclic lipoproteins, which inhibit β-glucan synthase enzyme, vital for integrity of the cell wall. The β-Glucan synthase has a widespread distribution in fungi [19]. Pneumocandins and caspofungin have broader spectrum of fungicidal activities [139]. In addition, some of other antifungal peptides (AFPs) like nikkomycins from *Streptomyces tendae* and *Streptomyces ansochromogenes,* polyoxins from *Streptomyces cacaoi* and aureobasidins from *Aureobasidium pullulans* are widely used as peptidyl nucleoside inhibitors of chitin synthase [63]. Some of the AMPs are now being used as alternative agents to eliminate the viral pathogens with strong potential. In 1986, α-, β-, and θ-defensin were used to inhibit viruses [30]. LfcinB exhibits inhibition against HSV-1, HSV-2, adenovirus, hepatitis B, and hepatitis C and human cytomegalovirus (HCMV). Activity of LfcinB has been ascribed to the specificity for viral binding cell membrane carbohydrates, which block the viral entry [2]. AMPs are also able to act against wide range of protozoans. Peptide dermaseptin-S4 (DRS-S4) [43] and anti-plasmodial [82] have effective hemolytic activity; and peptide acylation of AMPs increases the hydro-phobocity to enhance the penetration capability of AMPs. Therefore, antimicrobial activity of AMPs is an important concern for microbial infection.

6.7 SUMMARY

AMPS are generally produced by the host as a defense mechanism to kill the target. The AMPs interact with the plasma membrane of the target, self-oligomerize, insert into the plasma membrane forming pore, which causes leakage of molecules and metabolites and finally kills the target. The activity of AMPs depends on size, sequence, structure, presence of hydrophobic and hydrophilic amino acids and covalent and non-covalent interaction between the peptides. These AMPs form different structures depending on its amino acid sequence such as β-sheet, α-helical and extended AMPs, which affect the antimicrobial activity. AMPs kill the targets either by acting on the membrane or by interfering with the intracellular molecules. The action mechanisms of active membrane binding peptides were explained by barrel-stave, carpet, and toroidal models. These AMPs exhibit antibacterial activity, innate immunity in animals and plants, beneficial symbiotic interaction by controlling gut microbiota in animals, nitrogen fixation process in plants, proboscis association, antioxidant, antibacterial, antifungal, antiviral, antiprotozoal, and antitumor activities. AMP analogs are also being prepared to potentiate their therapeutic effects. Since AMPs occur naturally and safe to use, therefore nowadays these are being majorly focused in pharmaceuticals and food industries.

KEYWORDS

- **antimicrobial peptides (AMPs)**
- **barrel-stave model**
- **carpet model**
- **membrane disruption**
- **therapeutic use of AMPs**
- **toroidal model**

REFERENCES

1. Almasia, N. I., Molinari, M. P., Maroniche, G. A., Nahirnak, V., Baron, M. P. B., Taboga, O. A., & Rovere, C. V., (2017). Successful production of the potato antimicrobial peptide Snakin-1 in baculovirus-infected insect cells and development of specific antibodies. *BMC Biotechnology*, *17*(1), 401–412.
2. Andersen, J. H., Jenssen, H., & Gutteberg, T. J., (2003). Lactoferrin and lactoferricin inhibit Herpes simplex 1 and 2 infection and exhibit synergy when combined with acyclovir. *Antiviral Research*, *58*, 209–215.
3. Andreu, D., & Rivas, L., (1998). Animal antimicrobial peptides: An overview. *Biopolymers, 47*, 415–433,
4. Arihara, K., (2006). Strategies for designing novel functional meat products. *Meat Science, 74*, 219–229.
5. Auvin-Guette, C., Rebuffat, S., Prigent, Y., & Bodo, B., (1992). Trichogin A IV, an 11-residue lipopeptaibol from *Trichoderma longibrachiatum*. *Journal of the American Chemical Society, 114*, 2170–2174.
6. Bals, R., Weiner, D. J., Moscioni, A. D., Meegalla, R. L., & Wilson, J. M., (1999). Augmentation of innate host defense by expression of a cathelicidin antimicrobial peptide. *Infect. Immun., 67*, 6084–6089.
7. Banerjee, P., & Shanthi, C., (2012). Isolation of novel bioactive regions from bovine Achilles tendon collagen having angiotensin I-converting enzyme-inhibitory properties. *Process Biochemistry, 47*, 2335–2346.

8. Barra, D., & Simmaco, M., (1995). Amphibian skin: Promising resource for antimicrobial peptides. *Trends in Biotechnology*, *13*(6), 205–209.
9. Barra, D., & Simmaco, M., (2000). Structure-function relationships of temporins, small antimicrobial peptides from amphibian skin. *Eur. J. Biochem.*, *267*(5), 1447–1454.
10. Bauchart, C., Remond, D., Chambon, C., Mirand, P. P., Savary-Auzeloux, I., Reynes, C., & Morzel, M., (2006). Small peptides (<5 kDa) found in ready-to-eat beef meat. *Meat Science, 74*, 658–666.
11. Baumann, G., & Mueller, P., (1974). Molecular model of membrane excitability. *Journal of Supramolecular Structure*, *2*, 538–557.
12. Bergsson, G., Agerberth, B., Jornvall, H., & Gudmundsson, G. H., (2005). Isolation and identification of antimicrobial components from the epidermal mucus of Atlantic cod (*Gadusmorhua*). *The FEBS Journal*, *272*, 4960–4969.
13. Berthelot, K., Peruch, F., & Lecomte, S., (2016). Highlights on *Hevea brasiliensis* (pro) hevein proteins. *Biochimie, 127*, 258–270.
14. Birkemo, G. A., Luders, T., Andersen, O., Nes, I. F., & Nissen-Meyer, J., (2005). Hipposin, a histonederived antimicrobial peptide in Atlantic halibut (*Hippoglossus hippoglossus L.). Biochimica. Et. Biophysica. Acta*, *21*, 207–215.
15. Broekman, D. C., Frei, D. M., Gylfason, G. A., Steinarsson, A., Jornvall, H., & Agerberth, B., (2010). Cod Cathelicidin: Isolation of the mature peptide, cleavage site characterization, and developmental expression. *Developmental and Comparative Immunology*, *35*(3), 296–303.
16. Broncano, J. M., Otte, J., Petron, M. J., Parra, V., & Tim, M. L., (2012). Isolation and identification of low molecular weight antioxidant compounds from fermented "chorizo" sausages. *Meat Science, 90*, 494–501.
17. Brouwer, C. P. J. M., Bogaards, S. J. P., Wulferink, M., Velders, M. P., & Welling, M. M., (2006). Synthetic peptides derived from human antimicrobial peptide Ubiquicidin accumulates at sites of infection and eradicate (multi-drug resistant) *Stephylococcus aureus* in mice. *Peptides*, *27*(11), 2585–2591.
18. Budic, M., Rijavec, M., Petkovsek, Z., & Zgur-Bertok, D., (2011). *Escherichia coli* bacteriocins: Antimicrobial efficacy and prevalence among isolates from patients with bacteraemia. *PLoS One*, *6*(12), e28769.
19. Bulet, P., (1999). Antimicrobial peptides in insects: Structure and function. *Developmental and Comparative Immunology*, *23*, 329–344.
20. Caipang, C. M., Lazado, C. C., Brinchmann, M. F., & Kiron, V., (2010). Infection-induced changes in expression of antibacterial and cytokine genes in the gill epithelial cells of Atlantic cod, *Gadus morhua* during incubation with bacterial pathogens. *Comparative Biochemistry and Physiology-Part B: Biochemistry and Molecular Biology*, *156*, 319–325.
21. Cao, H., Ke, T., Liu, R., Yu, J., Dong, C., Cheng, M., Huang, J., & Liu, S., (2015). Identification of a novel proline-rich antimicrobial peptide from *Brassica napus*. *PloS One*, *10*(9), E-article: 0137414.
22. Castle, M., Nazarian, A., Yi, S. S., & Tempst, P., (1999). Lethal effects of apidaecin on *Escherichia coli* involve sequential molecular interactions with diverse targets. *Journal of Biological Chemistry*, *274*, 32555–32564.
23. Ceotto, H., Santos-Nascimento, J. D., Vasconcelos-Paiva-Brito, M. A., & De Freire-Bastos, M. C., (2009). Bacteriocin production by *Staphylococcus aureus* involved in bovine mastitis in Brazil. *Research in Microbiology*, *160*(8), 592–599.

24. Chang, C. I., Zhang, Y. A., Zou, J., Nie, P., & Secombes, C. J., (2006). Two cathelicidin genes are present in both rainbow trout (*Oncorhynchus mykiss*) and Atlantic salmon (*Salmo salar*). *Antimicrobial Agents and Chemotherapy, 50*, 185–195.
25. Chen, S. L., Li, W., Meng, L., Sha, Z. X., Wang, Z. J., & Ren, G. C., (2007). Molecular cloning and expression analysis of a hepcidin antimicrobial peptide gene from turbot (*Scophthalmus maximus*). *Fish and Shellfish Immunology, 22*, 172–181.
26. Chen, S. L., Xu, M. Y., Ji, X. S., Yu, G. C., & Liu, Y., (2005). Cloning, characterization, and expression analysis of hepcidin gene from red sea bream (*Chrysophrys major*). *Antimicrobial Agents and Chemotherapy*, *49*, 1608–1612.
27. Cheng, Q., Shi, X., Liu, Y., Liu, X., Dou, S., Ning, C., Liu, Z., et al., (2018). Production of nisin and lactic acid from corn stover through simultaneous saccharification and fermentation, *Biotechnology and Biotechnological Equipment*, *32*(2), 420–426.
28. Chikindas, M. L., García-Garcera, M. J., & Driessen, A. J., (1993). Pediocin PA-1, a bacteriocin from *Pediococcus acidilactici* PAC1.0, forms hydrophilic pores in the cytoplasmic membrane of target cells. *Applied and Environmental Microbiology*, *59*(11), 3577–3584.
29. Concha, M. I., Smith, V. J., Castro, K., Bastias, A., Romero, A., & Amthauer, R. J., (2002). Apolipoproteins A-I and A-II are potentially important effectors of innate immunity in the teleost fish *Cyprinus carpio*. *European Journal of Biochemistry*, *271*, 2984–2990.
30. Daher, K. A., Selsted, M. E., & Lehrer, R. I., (1986). Direct inactivation of viruses by human granulocyte defensins. *Journal of Virology*, *60*, 1068–1074.
31. De-Mello, E., Dos-Santos, I. S., Carvalho, A. O., De-Souza, L. S., De-Souza-Filho, G. A., Do-Nascimento, V. V., Machado, O. L. T., & Zottich, U., (2014). Functional expression and activity of the recombinant antifungal defensin PvD1r from *Phaseolus vulgaris L.* (common bean) seeds. *BMC Biochemistry*, *15*, 7–11.
32. Dhople, V., Krukemeyer, A., & Ramamoorthy, A., (2006). The human beta-defensin-3, an antibacterial peptide with multiple biological functions. *Biochimica et Biophysica Acta-Biomembranes*, *1758*, 1499–1512.
33. Di Luca, M., Maccari, G., Maisetta, G., & Batoni, G., (2015). BaAMPs: The database of biofilm-active antimicrobial peptides. *Biofouling*, *31*(2), 193–199.
34. Dimarcq, J. L., (1998). Cysteine-rich antimicrobial peptides in invertebrates. *Biopolymers*, *47*, 465–477.
35. Dorosky, R. J., Yu, J. M., Pierson, L. S., & Pierson, E. A., (2017). *Pseudomonas chlororaphis* produces two distinct R-tailocins that contribute to bacterial competition in biofilms and on roots. *Applied and Environmental Microbiology*, *83*(15), e-article 00706–00717.
36. Douglas, S. E., Gallant, J. W., Liebscher, R. S., Dacanay, A., & Tsoi, S. C., (2003). Identification and expression analysis of hepcidin-like antimicrobial peptides in bony fish. *Developmental and Comparative Immunology*, *27*, 589–601.
37. Ebner, P., Reichert, S., Luqman, A., Krismer, B., Popella, P., & Gotz, F., (2018). Lantibiotic production is a burden for the producing staphylococci. *Scientific Reports, 7471*(8), 132–137.
38. Ehrenstein, G., & Lecar, H., (1977). Electrically gated ionic channels in lipid bilayers. *Quarterly Reviews of Biophysics, 10*, 1–34.
39. El-Karim, I. A., Linden, G. J., Orr, D. F., & Lundy, F. T., (2008). Antimicrobial activity of neuropeptides against a range of micro-organisms from skin, oral, respiratory, and gastrointestinal tract sites. *Journal of Neuroimmunology*, *200*, 11–16.

40. Ellen, A. F., Rohulya, O. V., Fusetti, F., Wagner, M., Albers, S. V., & Driessen, A. J. M., (2011). The Sulfolobicin genes of *Sulfolobus acidocaldarius* encode novel antimicrobial proteins. *Journal of Bacteriology*, *193*(17), 4380–4387.
41. Enoki, S., & Suzuki, S., (2016). Pathogenesis-related proteins in grape. In: *Grape and Wine Biotechnology*. https://www.intechopen.com (accessed on 25 May 2020). doi: 10.5772/64873.
42. Essig, A., Hofmann, D., Munch, D., Gayathri, S., Kunzler, M., Kallio, P. T., Sahl, H. G., et al., (2014). Copsin, a novel peptide-based fungal antibiotic interfering with the peptidoglycan synthesis. *The Journal of Biological Chemistry*, *89*(50), 34953–34964.
43. Feder, R., Dagan, A., & Mor, A., (2000). Structure reactivity relationship study of antimicrobial dermaseptin S4 showing the consequences of peptide oligomerization on selective cytotoxicity, *The Journal of Biological Chemistry, 275*, 4230– 4238.
44. Fleming, A., (1922). On a remarkable bacteriolytic element found in tissues and secretions. *Proceedings of the Royal Society B: Biological Sciences, 93*, 306–317.
45. Foulston, L. C., & Bibb, M. J., (2010). Microbisporicin gene cluster reveals unusual features of lantibiotic biosynthesis in actinomycetes. *Proceedings of National Academy of Sciences, 107*(30), 13461–13466.
46. Fu, Y., Jette, F. Y., & Therkildsen, M., (2017). Bioactive peptides in beef: Endogenous generation through postmortem aging. *Meat Science, 123*, 134–142.
47. Fu, Y., Young, J. F., Løkke, M. M., Lametsch, R., Aluko, R. E., & Therkildsen, M., (2016). Re- valorization of bovine collagen as a potential precursor of angiotensin I-converting enzyme (ACE) inhibitor peptides based on *in silico* and *in vitro* protein digestions. *Junctional of Functional Foods, 24*, 196–206.
48. Gagnaire, V., Moll, M., Herrouin, M., & Leonil, J., (2001). Peptides identified during cheese ripening: Origin and proteolytic systems involved. *Journal of Agricultural and Food Chemistry, 49*, 4402–4413.
49. Ganz, T., & Lehrer, R. I., (1999). Antibiotic peptides from higher eukaryotes: Biology and applications. *Molecular Medicine Today*, *5*, 292–297.
50. Gao, G. H., Liu, W., Dai, J. X., Wang, J. F., Hu, Z., Zhang, Y., & Wang, D. C., (2001). Solution structure of PAFP-S: A new knottin-type antifungal peptide from the seeds of *Phytolacca americana*. *Biochemistry*, *40*(37), 10973–10978.
51. Gee, D. G., Gabard-Durnam, L. J., Flannery, J., Goff, B., Humphreys, K. L., Telzer, E. H., Hare, T. A., et al., (2013). Early developmental emergence of human amygdale-prefrontal connectivity after maternal deprivation. *Proceedings of the National Academy of Sciences*, *110*(39), 15638–15643.
52. Gifford, J. L., Hunter, H. N., & Vogel, H. J., (2005). Lactoferricin: A lactoferrin-derived peptide with antimicrobial, antiviral, antitumor, and immunological properties. *Cellular and Molecular Life Sciences*, *62*, 2588–2598.
53. Gogoladze, G., Grigolava, M., Vishnepolsky, B., Chubinidze, M., Duroux, P., & Lefranc, M. P., (2014). DBAASP: Database of antimicrobial activity and structure of peptides. *FEMS Microbiology Letter*, *357*, 63–68.
54. Gomez, E. A., Giraldo, P., & Orduz, S., (2017). InverPep: A database of invertebrate antimicrobial peptides. *Journal of Global Antimicrobial Resistance*, *8*, 13–17.
55. Gomez-Ruiz, J. A., Ramos, M., & Recio, I., (2004). Angiotensin-converting enzyme-inhibitory activity of peptides isolated from Manchego cheese. Stability under simulated gastrointestinal digestion. *International Dairy Journal, 14*, 1075–1080.

56. Gorbach, S. L., (2001). Antimicrobial use in animal feed: IME to stop. *The New England Journal of Medicine, 345*, 1202–1203.
57. Gotz, F., Perconti, S., Popella, P., Werner, R., & Schlag, M., (2014). Epidermin and gallidermin: *Staphylococcal lantibiotics*. *International Journal of Medical Microbiology, 304*(1), 63–71.
58. Groroos, L., Matto, J., Saarela, M., Luoma, A. R., Luoma, H., Jousimies-Somer, H., Pyhala, L., et al., (1995). Chlorhexidine susceptibilities of mutans streptococcal serotypes and ribotypes. *Antimicrobial Agents and Chemotherapy, 39*, 894–898.
59. Guilhelmelli, F., Vilela, N., Albuquerque, P., Derengowski, L., Da, S., Silva-Pereira, I., & Kyaw, C. M., (2013). Antibiotic development challenges: The various mechanisms of action of antimicrobial peptides and of bacterial resistance. *Frontiers in Microbiology, 4*, 353–357.
60. Hammami, R., Ben, H. J., Vergoten, G., & Fliss, I., (2009). PhytAMP: A database dedicated to antimicrobial plant peptides. *Nucleic Acids Research, 37*, 963–968.
61. Hammami, R., Zouhir, A., Ben, H. J., & Fliss, I., (2007). BACTIBASE: A new web-accessible database for bacteriocin characterization. *BMC Microbiology, 7*, 89.
62. Hancock, R. E. W., & Rozek, A., (2002). Role of membranes in the activities of antimicrobial cationic peptides. *FEMS Microbiology Letters, 206*, 143–149.
63. Hawser, S., (1999). Mulundocandin, an echinocandin-like lipopeptide, antifungal agent: Biological activities *in vitro*. *The Journal of Antibiotics, 52*(3), 305–310.
64. Helmerhorst, E. J., Troxler, R. F., & Oppenheim, F. G., (2001). The human salivary peptide histatin 5 exerts its antifungal activity through the formation of reactive oxygen species. *Proceedings of the National Academy of Sciences, 98*(25), 14637–14642.
65. Hirono, I., Hwang, J. Y., Ono, Y., Kurobe, T., Ohira, T., & Nozaki, R., (2005). Two different types of hepcidins from the Japanese flounder *Paralichthys olivaceus*. *The FEBS Journal, 272*, 5257–5264.
66. Hogg, A. C., Sripo, T., Beecher, B., Martin, J. M., & Giroux, M. J., (2004). Wheat puroindolines interact to form friabilin and control wheat grain hardness. *Theoretical and Applied Genetics, 108*(6), 1089–1097.
67. Holo, H., Nilssen, O., & Nes, I. F., (1991). Lactococcin-A, a new bacteriocin from *Lactococcus lactiscremoris*: Isolation and characterization of the protein and its gene. *Journal of Bacteriology, 173*(12), 3879–3887.
68. Hsu, C. H., Chen, C., Jou, M. L., Lee, A. Y. L., Lin, Y. C., & Yu, Y. P., (2005). Structural and DNA-binding studies on the bovine antimicrobial peptide, indolicidin: Evidence for multiple conformations involved in binding to membranes and DNA. *Nucleic Acids Research, 33*, 4053–4064.
69. Huang, E., & Yousef, A. E., (2015). Biosynthesis of paenibacillin, a lantibiotic with N-terminal acetylation, by *Paenibacillus polymyxa*. *Microbioloical Research, 181*, 15–21.
70. Iijima, N., Tanimoto, N., Emoto, Y., Morita, Y., Uematsu, K., & Murakami, T., (2003). Purification and characterization of three isoforms of chrysophsin, a novel antimicrobial peptide in the gills of the red sea bream, *Chrysophrys major*. *European Journal of Biochemistry, 270*, 675–686.
71. Ishikawa, M., Kubo, T., & Natori, S., (1992). Purification and characterization of a diptericin homologue from *Sarcophaga peregrina* (flesh fly). *Biochemical Journal, 287*, 573–578.

72. Jang, A., Jo, C., Kang, K. S., & Lee, M., (2008). Antimicrobial and human cancer cell cytotoxic effect of synthetic angiotensin-converting enzyme (ACE) inhibitory peptides. *Food Chemistry, 107*, 327–336.
73. Jaskiewicz, M., Orlowska, M., Olizarowicz, G., Migon, D., Grzywacz, D., & Kamysz, W., (2016). Rapid Screening of antimicrobial synthetic peptides. *International Journal of Peptide Research and Therapeutics, 22*, 155–161.
74. Jennings, A., West, J., Waine, C., Craik, D., & Anderson, M., (2001). Biosynthesis and insecticidal properties of plant cyclotides: The cyclic knotted proteins from *Oldenlandia affinis*. *Proceedings of the National Academy of Sciences*, *98*(19), 10614–10619.
75. Jenssen, H., Hamill, P., & Hancock, R. E., (2006). Peptide antimicrobial agents, *Clinical Microbiology Reviews, 19*, 491–511.
76. Joly, S., Maze, C., McCray, P. B., & Guthmiller, J. M., (2004). Human beta-defensins 2 and 3 demonstrate strain-selective activity against oral microorganisms. *Journal of Clinical Microbiology*, *42*, 1024–1029.
77. Jones, D. E., & Bevins, C. L., (1993). Defensin-6 mRNA in human paneth cells: Implications for antimicrobial peptides in host defense of the human bowel. *FEBS Letters, 315*, 187–192.
78. Joseph, S., Karnik, S., Nilawe, P., Jayaraman, V. K., & Idicula-Thomas, S., (2012). Class AMP: A prediction tool for classification of antimicrobial peptides. *IEEE/ACM Transactions on Computational Biology and Bioinformatics*, *9*, 1535–1538.
79. Kaji, T., Sugiyama, N., Ishizaki, S., Nagashima, Y., & Shiomi, K., (2006). Molecular cloning of grammistins, peptide toxins from the soapfish *Pogonoperca punctata*, by hemolytic screening of a cDNA library. *Peptides, 27*, 3069–3076.
80. Kan, A., Ozcelik, B., Kartal, M., Ozdemir, Z. A., & Ozgen, S., (2010). *In vitro* antimicrobial activities of *Cicer arietinum* L. (chickpea). *Tropical Journal of Pharmaceutical Research*, *9*(5), 475–481.
81. Keppi, E., Pugsley, A. P., Lambert, J., Wicker, C., Dimarcq, J. L., & Hoffmann, J. A., (1989). Mode of action of diptericin A, a bactericidal peptide-induced in the hemolymph of *Phormia terranovae* larvae. *Archives of Insect Biochemistry and Physiology*, *10*, 229–239.
82. Krugliak, M., Feder, R., Zolotarev, V. Y., Gaidukov, L., Dagan, A., Ginsburg, H., & Mor, A., (2000). Antimalarial activities of dermaseptin S4 derivatives. *Antimicrobial Agents and Chemotherapy*, *44*, 2442–2451.
83. Lai, Y., & Gallo, R. L., (2009). AMPed up immunity: How have antimicrobial peptides multiple roles in immune defense? *Trends in Immunology*, *30*, 131–141.
84. Lamberty, M., (2001). Solution structures of the antifungal heliomicin and a selected variant with both antibacterial and antifungal activities. *Biochemistry*, *40*, 11995–12003.
85. Lata, S., Mishra, N. K., & Raghava, G. P., (2010). AntiBP2: Improved version of antibacterial peptide prediction. *BMC Bioinformatics, 11*(1), 11–19.
86. Leah, A., Visscher, M., Van-Belkum, M. J., & Garneau-Tsodikova, S., (2008). Isolation and characterization of Carnocyclin A, a novel circular bacteriocin produced by *Carnobacterium maltaromaticum* UAL307. *Applied and Environmental Microbiology*, *74*(15), 4756–4763.
87. Li, B., Chen, F., Wang, X., Ji, B., & Wu, Y., (2007). Isolation and identification of antioxidative peptides from porcine collagen hydrolysate by consecutive chromatography and electrospray ionization-mass spectrometry. *Food Chemistry, 102*, 1135–1143.

88. Li, Y., Xiang, H., Liu, J., Zhou, M., & Tan, H., (2003). Purification and biological characterization of halocin C8, a novel peptide antibiotic from *Halobacterium strain* AS7092. *Extremophiles*, *7*(5), 401–407.
89. Liu, F., Li, J. L., Yue, G. H., Fu, J. J., & Zhou, Z. F., (2010). Molecular cloning and expression analysis of the liver-expressed antimicrobial peptide 2 (LEAP-2) gene in grass carp. *Veterinary Immunology and Immunopathy, 133*, 133–143.
90. Londelle, S. E., & Lohner, K., (2010). Optimization and high-throughput screening of antimicrobial peptides. *Current Pharmaceutical Design*, *16*, 3204–3211.
91. Lopez-Exposito, I., Gomez-Ruiz, J. A., Amigo, L., & Recio, I., (2006). Identification of antibacterial peptides from ovine alpha-s2 casein. *International Dairy Journal, 16*, 1072–1080.
92. Madani, F., Lindberg, S., Langel, U., Futaki, S., & Graslund, A., (2011). Mechanisms of cellular uptake of cell-penetrating peptides. *Journal of Biophysi.*, E-pub. ID 414729. doi: 10.1155/2011/414729.
93. Maier, V. H., Dorn, K. V., Gudmundsdottir, B. K., & Gudmundsson, G. H., (2008). Characterization of cathelicidin gene family members in divergent fish species. *Molecular Immunology*, *45*, 3723–3730.
94. Marchand, C., Krajewski, K., Lee, H. F., Antony, S., Johnson, A., & Amin, R., (2006). Covalent binding of the natural antimicrobial peptide indolicidin to DNA abasic sites. *Nucleic Acids Research*, *34*, 5157–5165.
95. Marcus, J. P., Green, J. L., Goulter, K., & Manners, J. M., (1999). Family of antimicrobial peptides is produced by processing of a 7S protein in *Macadamia integrifola* kernels. *The Plant Journal, 19*(6), 699–710.
96. Martinez, B., Rodriguez, A., & Suarez, E., (2016). Antimicrobial peptides produced by bacteria: The bacteriocins. In: Tomas, G. V., & Miguel, V., (eds.), *New Weapons to Control Bacterial Growth* (pp. 15–38). New York: Springer Nature.
97. Matsuzaki, K., Murase, O., Fujii, N., & Miyajima, K., (1995). Translocation of a channel-forming antimicrobial peptide, Magainin 2, across lipid bilayers by forming a pore. *Biochemistry*, *34*, 6521–6256.
98. Mavri, J., & Vogel, H. J., (1996). Ion pair formation of phosphorylated amino acids and lysine and arginine side chains: A theoretical study. *Proteins: Structure, Function, and Genetics, 24*, 495–501.
99. McAuliffe, O., Ryan, M. P., Ross, R. P., Hill, C., Breeuwer, P., & Abee, T., (1998). Lecticin 3147, a broad spectrum bacteriocin, which selectively dissipates the membrane potential. *Applied and Environmental Microbiology*, *64*(2), 439–445.
100. Meister, M., Lemaitre, B., & Hoffmann, J. A., (1997). Antimicrobial peptide defense in drosophila. *Bio Essays*, *19*, 1019–1026.
101. Melo, M. N., Dugourd, D., & Castanho, M. A. R. B., (2006). Omiganan pentahydrochloride in the front line of clinical applications of antimicrobial peptides. *Recent Patents on Anti-Infective Drug Discovery, 1*, 201–207.
102. Minamikawa, M., Kawai, Y., Inoue, N., & Yamazaki, K., (2005). Purification and characterization of Warnericin RB4, anti-Alicyclobacillus bacteriocin, produced by *Staphylococcus warneri* RB4. *Current Microbiology*, *51*(1), 22–26.
103. Minervini, F., Algaron, F., Rizzello, C. G., Fox, P. F., Monnet, V., & Gobbetti, M., (2003). Angiotensin I-converting-enzyme-inhibitory and antibacterial peptides from *Lactobacillus helveticus*PR-4 proteinase-hydrolyzed casein of milk from 6 species. *Applied and Environmental Microbiology, 69*, 5297–5305.

104. Minkiewicz, P., Dziuba, J., Iwaniak, A., Dziuba, M., & Darewicz, M., (2008). BIOPEP database and other programs for processing bioactive peptide sequences. *Journal of AOAC International*, *91*, 965–980.
105. Minkiewicz, P., Slangen, C. J., Dziuba, J., Visser, S., & Mioduszewska, H., (2000). Identification of peptides obtained via hydrolis of bovine casein using HPLC and mass spectrometry. *Milchwissenschaft, 55*, 14–17.
106. Moretro, T., Aasen, I. M., Storro, I., & Axelsson, L., (2000). Production of sakacin P by *Lactobacillus sakei* in a completely defined medium. *Journal of Applied Microbiology, 88*(3), 536–545.
107. Muller, I., Lurz, R., & Geider, K., (2012). Tasmancin and lysogenic bacteriophages induced from *Erwinia tasmaniensis* strains. *Microbiological Research*, *167*(7), 381–387.
108. Nilsen, T., Nes, I. F., & Holo, H., (2003). Enterolysin A, cell wall-degrading bacteriocin from *Enterococcus faecalis* LMG 2333. *Applied and Environmental Microbiology, 69*(5), 2975–2984.
109. Nissen-Meyer, J., & Nes, I. F., (1997). Ribosomally synthesized antimicrobial peptides: Their function, structure, biogenesis, and mechanism of action. *Archives of Microbiology*, *167*, 67–77.
110. Noga, E. J., Borron, P. J., Hinshaw, J., Gordon, W. C., Gordon, L. J., & Seo, J. K., (2011). Identification of histones as endogenous antibiotics in fish and quantification in rainbow trout (*Oncorhynchus mykiss*) skin and gill. *Fish Physiology and Biochemistry*, *37*(1), 135–152.
111. Novkovic, M., Simunic, J., Bojovic, V., Tossi, A., & Juretic, D., (2012). DADP: The database of anuran defense peptides. *Bioinformatics*, *28*, 1406–1407.
112. Oscariz, J. C., Lasa, I., & Pisabarro, A. G., (1999). Detection and characterization of cerein 7, a new bacteriocin produced by *Bacillus cereus* with a broad spectrum of activity. *FEMS Microbiology Letters*, *178*(2), 337–341.
113. Ostroumova, O. S., & Malev, V. V., (2015). Modifiers of membrane dipole potentials as tools for investigating ion channel formation and functioning. *International Review of Cell and Molecular Biology*, *315*, 245–297.
114. Pan, C. Y., Chen, J. Y., Cheng, Y. S., Chen, C. Y., Ni, I. H., & Sheen, J. F., (2007). Gene expression and localization of the epinecidin-1 antimicrobial peptide in the grouper (*Epinephelus coioides*), and its role in protecting fish against pathogenic infection. *DNA and Cell Biology, 26*, 403–413.
115. Parenti, F., Pagani, H., & Beretta, G., (1976). Gardimycin, a new antibiotic from *Actinoplanes*, I: Description of the producer strain and fermentation studies. *The Journal of Antibiotics*, *29*(5), 501–506.
116. Park, C. B., Lee, J. H., Park, I. Y., Kim, M. S., & Kim, S. C., (1997). A novel antimicrobial peptide from the loach, *Misgurnus anguillicaudatus*. *FEBS Letters*, *411*, 173–178.
117. Park, C. J., Park, C. B., Hong, S. S., Lee, H. S., Lee, S. Y., & Kim, S. C., (2000). Characterization and cDNA cloning of two glycine- and histidine-rich antimicrobial peptides from the roots of shepherd's purse, *Capsella bursa-pastoris*. *Plant Molecular Biology*, *44*(2), 187–197.
118. Park, C. B., (1998). Mechanism of action of the antimicrobial peptide buforin II: Buforin II kills microorganisms by penetrating the cell membrane and inhibiting cellular functions. *Biochemical and Biophysical Research Communication*, *244*, 253–257.

119. Pasupuleti, M., Schmidtchen, A., & Malmsten, M., (2012). Antimicrobial peptides: Key components of the innate immune system. *Critical Reviews in Biotechnology*, *32*, 143–171.
120. Penterman, J., Singh, P. K., & Walker, G. C., (2014). Biological cost of pyocin production during the SOS response in *Pseudomonas aeruginosa*. *Journal of Bacteriology*, *18*, 3351–3359.
121. Piotto, S. P., Sessa, L., Concilio, S., & Iannelli, P., (2012). YADAMP: Yet another database of antimicrobial peptides. *International Journal of Antimicrobial Agents*, *39*, 346–351.
122. Pridmore, D., Pittet, N. R., Suri, B., & Mollet, B., (1996). Variacin-new lanthionine-containing bacteriocin produced by *Micrococcus varians*: Ccomparison to lacticin 481 of *Lactococcus lactis*. *Applied and Environmental Microbiology*, *62*(5), 1799–1802.
123. Priya, S., Santhiya, S., & Jancy, B., (2011). Vibriocin production by marine prawn associated vibrio spp. *Biomedical and Pharmacology Journal*, *4*(1), 227–229.
124. Quiros, A., Hernandez-Ledesma, B., Ramos, M., Amigo, L., & Recio, I., (2005). Angiotensin-converting enzyme inhibitory activity of peptides derived from caprine kefir. *Journal of Dairy Science, 88*, 3480–3487.
125. Qureshi, A., Thakur, N., & Kumar, M., (2013). HIPdb: A database of experimentally validated HIV inhibiting peptides. *PLoS One, 8*, p. 8, e-article ID 54908.
126. Qureshi, A., Thakur, N., Tandon, H., & Kumar, M., (2014). AVPdb: A database of experimentally validated antiviral peptides targeting medically important viruses. *Nucleic Acids Research*, *42*, 1147–1153.
127. Ramamoorthy, A., Thennarasu, S., Tan, A., Gottipati, K., Sreekumar, S., & Heyl, D. L., (2006). Deletion of all cysteines in Tachyplesin I abolishes hemolytic activity and retains antimicrobial activity and lipopolysaccharide selective binding. *Biochemistry, 45*, 6529–6540.
128. Ren, Z., & Shai, Y., (1998). Mode of action of linear amphipathic α-helical antimicrobial peptides. *Biopolymers*, *47*, 451–463.
129. Risso, A., Braidot, E., Sordano, M. C., Vianello, A., Macri, F., & Skerlavaj, B., (2012). BMAP-28, an antibiotic peptide of innate immunity, induces cell death through opening of the mitochondrial permeability transition pore. *Molecular and Cellular Biology, 22*, 1926–1935.
130. Rival, S. G., Fornaroli, S., Boeriu, C. G., & Wichers, H. J., (2001). Caseins and casein hydrolysates, I: Lipoxygenase inhibitory properties. *Journal of Agricultural and Food Chemistry, 49*, 287–294.
131. Rogozhin, E., Ryazantsev, D., Smirnov, A., & Zavriev, S., (2018). Primary structure analysis of antifungal peptides from cultivated and wild cereals. *Plants, 7*(3), 74–80.
132. Roudot-Algaron, F., LeBars, D., Kerhoas, L., Einhorn, J., & Gripon, J. C., (1994). Phosphopeptides from Comte cheese: Nature and origin. *Journal of Food Science, 59*, 544–547.
133. Saiga, A., Iwai, K., Hayakawa, T., Takahata, Y., Kitamura, S., Nishimura, T., & Morimatsu, F., (2008). Angiotensin I converting enzyme: Inhibitory peptides obtained from chicken collagen hydrolysate. *J. Agric. Food Chem., 56*, 9586–9591.
134. Saito, R., Nakamura, T., Kitazawa, H., Kawai, Y., & Itoh, T., (2000). Isolation and structural analysis of antihypertensive peptides that exist naturally in Gouda cheese. *Journal of Dairy Science, 83*, 1434–1440.

135. Salas, C. E., Badillo-Corona, J. A., Ramírez-Sotelo, G., & Oliver-Salvador, C., (2015). Biologically active and antimicrobial peptides from plants. *Bio. Med. Research International*, *11*, 102–129.
136. Salerno, G., Parrinello, N., Roch, P., & Cammarata, M., (2007). cDNA sequence and tissue expression of an antimicrobial peptide, dicentracin: A new component of the moronecidin family isolated from head kidney leukocytes of sea bass, *Dicentrarchus labrax*. *Comparative Biochemistry and Physiology- Part B: Biochemistry and Molecular Biology*, *146*, 521–529.
137. Scannell, A. G., Hill, C., Ross, R., Marx, S., Hartmeier, W., & Arendt, E. K., (*2000*). Development of bioactive food packaging materials using immobilized Bacteriocins Lacticin 3147 and Nisaplin. *International Journal of Food Microbiology, 60*, 241–249.
138. Scheit, K. H., Reddy, E. S., & Bhargava, P. M., (1979). Seminaplasmin is a potent inhibitor of *E. coli* RNA polymerase *in vivo*. *Nature*, *279*, 728–731.
139. Schmatz, D. M., (1990). Treatment of *Pneumocystis carinii* pneumonia with 1,3-beta-glucan synthesis inhibitors. *Proceedings of the National Academy of Sciences*, *87*(15), 5950–5954.
140. Schroeder, B. O., Wu, Z., Nuding, S., Groscurth, S., Marcinowski, M., & Beisner, J., (2011). Reduction of disulfide bonds unmasks potent antimicrobial activity of human β-defensin-1. *Nature, 469*, 419–423.
141. Scocchi, M., Pallavicini, A., Salgaro, R., Bociek, K., & Gennaro, R., (2009). The salmonid cathelicidins: A gene family with highly varied C-terminal antimicrobial domains. *Comparative Biochemistry and Physiology-Part B: Biochemistry and Molecular Biology, 152*, 376–381.
142. Seebah, S., Suresh, A., Zhuo, S., Choong, Y. H., Chua, H., Chuon, D., Beuerman, R., & Verma, C., (2007). Defensins knowledgebase: A manually curated database and information source focused on the defensins family of antimicrobial peptides. *Nucleic Acids Research*, *35*, 265–268.
143. Segura, A., Moreno, M., Molina, A., & Garcia-Olmedo, F., (1998). Novel defensin subfamily from spinach (*Spinacia oleracea*) Author links open overlay panel. *FEBS Letters, 435*(2–3), 159–162.
144. Sengupta, D., Leontiadou, H., Mark, A. E., & Marrink, S. J., (2008). Toroidal pores formed by antimicrobial peptides show significant disorder. *Biochimica. Et. Biophysica. Acta-Biomembranes*, *1778*, 2308–2317.
145. Sharma, S., Verma, H., & Sharma, N. K., (2014). Cationic bioactive peptide from the seeds of *Benincasahispida. International Journal of Peptides, 3*(12), 8, e-article ID 156060.
146. Shenkarev, Z. O., Gizatullina, A. K., Finkina, E. I., Alekseeva, E. A., Balandin, S. V., Mineev, K. S., Arseniev, A. S., & Ovchinnikova, T. V., (2014). Heterologous expression and solution structure of defensin from *Lentil Lens culinaris*. *Biochemical and Biophysical Research Communications*, *451*(2), 252–257.
147. Shike, H., Shimizu, C., Lauth, X., & Burns, J. C., (2004). Organization and expression analysis of the zebrafish hepcidin gene, an antimicrobial peptide gene conserved among vertebrates. *Developmental and Comparative Immunology*, *28*, 747–754.
148. Silva, O. N., Porto, W. F., Migliolo, L., Mandal, S. M., Gomes, D. G., & Holanda, H. H., (2012). Cn-AMP1: A promiscuous peptide with potential for microbial infection treatment. *Biopolymers, 98*, 322–331.

149. Skerlavaj, B., Benincasa, M., Risso, A., Zanetti, M., & Gennaro, R., (1999). SMAP-29: A potential antibacterial and antifungul peptide from sheep leukocytes. *FEBS Letters, 463*(1–2), 58–62.
150. Smacchi, E., & Gobbetti, M., (1998). Peptides from several Italian cheeses inhibitory to proteolytic enzymes of lactic acid bacteria, *Pseudomonas fluorescens* ATCC 948 and to the angiotensin I-converting enzyme. *Enzyme and Microbial Technology, 22*, 687–694.
151. Soscia, S. J., Kirby, J. E., Washicosky, K. J., Tucker, S. M., Ingelsson, M., Hyman, B., Burton, M. A., et al., (2010). Alzheimer's disease-associated amyloid beta-protein is an antimicrobial peptide. *PLoS One, 5*(9), e-article ID 9505.
152. Stein, T., Dusterhus, S., Stroh, A., & Entian, K. D., (2004). Subtilosin production by two *Bacillus subtilis* subspecies and variance of the sbo-alb cluster. *Applied and Environmental Microbiology, 70*(4), 2349–2353.
153. Stepek, G., Buttle, D. J., Duce, I. R., & Behnke, J. M., (2005). Assessment of the anthelmintic effect of natural plant cysteine proteinases against the gastrointestinal nematode, *Heligmosomoides polygyrus*. *Parasitology, 130*(2), 203–211.
154. Swithenbank, L., & Morgan, C., (2017). The role of antimicrobial peptides in lung cancer therapy. *International Journal of Antimicrobial Agents, 3*(134), 1–6.
155. Szweda, P., Gorczyca, G., Filipkowski, P., Zalewska, M., & Milewski, S., (2014). Efficient production of *Staphylococcus simulans* lysostaphin in a benchtop bioreactor by recombinant *Escherichia coli*. *Preparative Biochemistry and Biotechnology, 44*(4), 370–381.
156. Theolier, J., Fliss, I., Jean, J., & Hammami, R., (2014). MilkAMP: A comprehensive database of antimicrobial peptides of dairy origin. *Dairy Science and Technology, 94*, 181–193.
157. Thomas, S., Karnik, S., Barai, R. S., Jayaraman, V. K., & Idicula-Thomas, S., (2010). CAMP: A useful resource for research on antimicrobial peptides. *Nucleic Acids Research, 38*, 774–780.
158. Todorov, S. D., Wachsman, M. B., Knoetze, H., Meincken, M., & Dicks, L. M. T., (2005). An antibacterial and antiviral peptide produced by *Enterococcus mundtii* ST4V isolated from soya beans. *International Journal of Antimicrobial Agents, 25*, 10, ID 508e13.
159. Ullal, A. J., Litaker, R. W., & Noga, E. J., (2008). Antimicrobial peptides derived from hemoglobin are expressed in epithelium of channel catfish (*Ictalurus punctatus, Rafinesque*). *Developmental and Comparative Immunology, 32*, 1301–1312.
160. Usmani, S. S., Kumar, R., Kumar, V., Singh, S., & Raghava, G. P. S., (2018). AntiTbPdb: A knowledgebase of anti-tubercular peptides. *Database, 2018*, 1–8.
161. Uzzell, T., Stolzenberg, E. D., Shinnar, A. E., & Zasloff, M., (2003). Hagfish intestinal antimicrobial peptides are ancient cathelicidins. *Peptides, 24*, 1655–1667.
162. Van, H. A. J., De Jong, A., Song, C., Viel, J. H., Kok, J., & Kuipers, O. P., (2018). BAGEL4: A user-friendly web server to thoroughly mine RiPPs and bacteriocins. *Nucleic Acids Research, 46*, 278–281.
163. van-Hemert, S., Meijerink, M., Molenaar, D., Bron, P. A., de-Vos, P., Kleerebezem, M., Wells, J. M., & Marco, M. L., (2010). Identification of *Lactobacillus plantarum* genes modulating the cytokine response of human peripheral blood mononuclear cells. *BMC Microbiology, 10*, 293–299.

164. Vizioli, J., (2001). Gambicin: A novel immune responsive antimicrobial peptide from the malaria vector Anopheles gambiae. *Proceedings of the National Academy of Sciences, 98*, 12630–12635.
165. Wang, C. K. L., Kaas, Q., Chiche, L., & Craik, D. J., (2007). CyBase: A database of cyclic protein sequences and structures, with applications in protein discovery and engineering. *Nucleic Acids Research, 36*, 206–210.
166. Wang, G., Li, X., & Wang, Z., (2009). APD2: The updated antimicrobial peptide database and its application in peptide design. *Nucleic Acids Research, 37*, 933–937.
167. Wang, L., Liu, Q., Chen, J. C., Cui, Y. X., Zhou, B., Chen, Y. X., Zhao, Y. F., & Li, Y. M., (2012). Antimicrobial activity of human islet amyloid polypeptides: An insight into amyloid peptides connection with antimicrobial peptides. *Biological Chemistry, 393*, 641–646.
168. Wang, Y. C., Wu, D. C., Liao, J. J., Wu, C. H., Li, W. Y., & Weng, B. C., (2009). *In vitro* activity of *Impatiens balsamina* L. against multiple antibiotic-resistant *Helicobacter pylori. The American Journal Chinese Medicine, 37*(4), 713–722.
169. Wang, Y. D., Kung, C. W., & Chen, J. Y., (2010). Antiviral activity by fish antimicrobial peptides of epinecidin-1 and hepcidin 1–5 against nervous necrosis virus in medaka. *Peptides, 31*, 1026–1033.
170. Whitmore, L., & Wallace, B. A., (2004). The peptaibol database: A database for sequences and structures of naturally occurring peptaibols. *Nucleic Acids Research, 32*, 593–594.
171. Wilhelm, C., Anke, H., Flores, Y., & Sterner, O., (2004). New peptaibols from *Mycogone cervina. Journal of Natural Products, 67*(3), 466–468.
172. Wimley, W. C., (2010). Describing the mechanism of antimicrobial peptide action with the interfacial activity model. *ACS Chemical Biology, 5*, 905–917.
173. Wong, J. H., & Ng, T. B., (2005). Lunatusin, a trypsin-stable antimicrobial peptide from lima beans (*Phaseolus lunatus* L.). *Peptides, 26*(11), 2086–2092.
174. Wong, J. H., & Ng, T. B., (2005). Vulgarinin, a broad-spectrum antifungal peptide from haricot beans (*Phaseolus vulgaris*). *The International Journal of Biochemistry and Cell Biology, 37*(8), 1626–1632.
175. Wong, J. H., Ng, T. B., & Randy, C., (2010). Proteins with antifungal properties and other medicinal applications from plants and mushrooms. *Applied Microbiology and Biotechnology, 87*(4), 1221–1235.
176. Wu, H., Lu, H., Huang, J., Li, G., & Huang, Q., (2012). Enzy base: Novel database for enzybiotic studies. *BMC Microbiology, 2012*, 12–54.
177. Xiong, Y. Q., Yeaman, M. R., & Bayer, A. S., (1999). *In vitro* antibacterial activities of platelet microbicidal protein and neutrophil defensin against *Staphylococcus aureus* are influenced by antibiotics differing in mechanism of action. *Antimicrobial Agents and Chemotherapy, 43*, 1111–1117.
178. Yamamoto, N., Akino, A., & Takano, T., (1994). Antihypertensive effect of the peptides derived from casein by an extracellular proteinase from *Lactobacillus helveticus* CP790. *Journal of Dairy Science, 77*, 917–922.
179. Yang, L., Harroun, T. A., Weiss, T. M., Ding, L., & Huang, H. W., (2001). Barrel-stave model or toroidal model? A case study on melittin pores. *Biophysical Journal, 81*, 1475–1485.
180. Yeaman, M. R., & Yount, N. Y., (2003). Mechanisms of antimicrobial peptide action and resistance. *Pharmacological Review, 55*, 27–55.

181. Zasloff, M., (2002). Antimicrobial peptides of multi cellular organisms. *Nature*, *415*, 389–395.
182. Zhang, Y. A., Zou, J., Chang, C. I., & Secombes, C. J., (2004). Discovery and characterization of two types of liver-expressed antimicrobial peptide 2 (LEAP-2) genes in rainbow trout. *Veterinary Immunology and Immunopathy*, *101*, 259–269.
183. Zhao, X., Wu, H., Lu, H., Li, G., & Huang, Q., (2013). LAMP: A database linking antimicrobial peptides. *PLoS One,* 8, 7, e-article ID 66557.

CHAPTER 7

APPLICATIONS OF BDELLOVIBRIO BACTERIA AS BIOCONTROL AGENTS: FOOD SAFETY AND MITIGATING CLINICAL PATHOGENS

VALERIE D. ZAFFRAN, GABRIELLE KIRSHTEYN, and PRASHANT SINGH

ABSTRACT

Bdellovibrio is a Gram-negative bacteria that possess the ability to prey on other larger Gram-negative microorganisms. They are generally spread in numerous natural marine and soil conditions. Upon recognition, *Bdellovibrio* penetrates Gram-negative bacterial cells and multiply inside by forming bdelloplasts, which are finally released as new *Bdellovibrio* cells through lysis of the prey organism to continue their predatory life-cycle. The attachment and penetration of *Bdellovibrio* to Gram-negative bacteria is non-specific, which confers them the capacity to prey on multiple Gram-negative genera. Strains of *Bdellovibrio* have been shown to lyse multiple foodborne pathogens (e.g., *E. coli*, *Salmonella*, and *Shigella*, *Yersinia*) in pure culture, biofilms, and food models. Additionally, they have been reported to prey on antibiotic-resistant strains of pathogens. Further, they have been tested to treat various bacterial infections (e.g., ocular, and lung infections) in animal models. The safety of *Bdellovibrio* has been extensively evaluated using human cell lines, human epithelial cells, immunological assays, and at the gut microbiome level. *Bdellovibrio* is a unique bacterial genus that can be utilized for mitigating foodborne pathogens and enhancing food safety.

7.1 INTRODUCTION

According to the Food and Agriculture Organization (FAO), around 800 million people are affected by hunger with global food security continuing to be threatened by climate change and building pressure on natural resources [56]. Further, with an estimated increase in the global population to 9.7 billion persons by 2050, there is uncertainty as to whether the current food and agricultural systems will be able to meet the anticipated demand. Based on the projected population increase, Overall food production will need to increase by 50% between 2012 and 2050 [56]. Food safety and food loss due to microbial spoilage are major factors that continue to pose challenges to meet this food production demand.

An increase in fresh produce intake has paralleled with an increase in fresh produce-associated outbreaks [12]. In the United States, from 1973 to 2012, fresh leafy greens were associated with a total of 606 outbreaks, causing 20,000 illnesses and many deaths, making fresh produce one of the most commodities for transferring foodborne illness [26]. Each year, the number of produce-associated outbreaks range from 23 to 60 [5]. Multiple approved chemical sanitizers (e.g., chlorine, peracetic acid (PAA)) are used at pre- and post-harvest steps to enhance food safety. However, their efficacy has been shown to be limited in commercial processing facilities. The currently utilized chemical sanitizers can act only on foodborne pathogens present on the outer surface of fresh produce, resulting in 0.5–2 log CFU (colony forming unit) reductions after sanitizer treatment. Similarly, physical methods (e.g., cold plasma, UV irradiation) for enhancing food safety have shown positive results in laboratory conditions but their practical application is still limited [34].

Plant pathogens pose a major challenge towards increasing global fresh produce production. Infection by phytopathogens and pests results in loss of yield, reduced quality, and faster decay of produce and are responsible for a 20–40% loss of global food production [19]. In the absence of proper refrigeration, Gram-negative phytopathogens (e.g., *Pseudomonas*, *Erwinia*) and other saprophytic bacteria can cause rot and facilitate biofilm formation, which can promote colonization and survival of other pathogens.

Antibiotic resistance is another major challenge. Antibiotics are commonly used as prophylactic agents in animal husbandry. Over time, the widespread use of antibiotics for production of livestock has exhorted a selective pressure, resulting in an increased prevalence of antibiotic-resistant bacterial (ARB) strains that can have serious public health implications. Foodborne ARBs strains can spread through consumption of food harboring ARB strains and/or contact with domesticated food animals and their manure. In the United States, over 2 million people are infected by ARB each year, causing at least 23,000 deaths [8]. With limited pharmaceutical options available for drug development against ARB strains, predatory bacteria may be utilized as an alternative option.

Bdellovibrio belongs to the Delta-proteobacteria that was first discovered by Hans Stolp in the 1960s [54]. Later, it was described as a tiny bacterium with the ability to attack and lyse *Pseudomonas*. *Bdellovibrio* is a small, Gram-negative, motile bacteria that undergo a predatory lifestyle. They are found in aquatic and marine ecosystems and can also be isolated from soil and sewage, where they commonly prey on Gamma-proteobacteria. They attack larger Gram-negative organisms by penetrating inside the cells; digesting their periplasm and killing, the host cells from inside [42]. *Bdellovibrio* cells grow and multiply rapidly within 2–3 hours after invasion by dividing inside the prey cell. The process ends with lysis of target cells and the release of manifold swimming *Bdellovibrio* offspring. The motile progeny seeks other large gram-negative cells to attack, continuing the predatory life cycle [41, 42, 54]. In the absence of any large Gram-negative cells, *Bdellovibrio* cells can undergo a quiescent state [29].

This chapter focuses on the genetic makeup, mode of action, and potential applications of *Bdellovibrio* as a biocontrol agent.

7.2 BDELLOVIBRIO GENOME

The *Bdellovibrio bacteriovorus* HD100 strain genome was sequenced in 2004 [47]. Despite its small cell size (0.35 × 1.1 μm), the genome of *B. bacteriovorus* HD100 is considerably large (3.8 Mbps) and lacks any plasmid or phage-derived elements. The *B. bacteriovorus* HD100 genome encodes for 20 DNAses and nine RNAses. In addition to these nucleases, the HD100 genome encodes for 15 lipases, 10 glycanases, 150 proteases, and 89 other hydrolytic genes. The relatively large number of hydrolytic genes, compared to other closely related bacteria, is essential for *Bdellovibrio's* predatory behavior. The hydrolytic gene products are employed for degrading host/prey cell structures; and their breakdown products are used as a raw material for synthesis of new *Bdellovibrio* cell components.

Karunker et al. [31] performed a RNA-sequence analysis and compared the gene expression profile during attack-phase versus growth-phase of

Bdellovibrio cells. Interestingly, during the attack-phase only 33% of genes were transcribed compared to 85% transcribed during the predator-growth phase. The upregulation of genes during the growth phase was largely comprised of ribosomal protein-encoding genes. Many of these genes are involved in transcriptional regulation, nucleic acid degradation/modification/ synthesis, peptidoglycan modification/degradation, and various proteases that assist in degradation of prey cell components [32].

7.3 *BDELLOVIBRIO*: MECHANISM OF PREY ENCOUNTER

The *Bdellovibrio* are very motile and their flagellar motility drives their chemotactic response and bringing them close to the prey cells. The flagella, gliding organelles, and pili belonging to the Type-IV secretion system are other important *Bdellovibrio* components that are involved in the predation process. The predator flagellums, which provide high-speed motility to the cells, are primarily responsible for bringing the predator close to the prey cells (predator-prey collision). A specific mutation in the flagellar genes has been shown to reduce (but not completely eliminate) predation potential of the predator. Genetically modified predator cells completely lacking flagella can still prey on other Gram-negative cells using gliding motility [33].

Bdellovibrio cells have shown to possess a chemotactic response towards a high concentration of bacteria, which increases predation efficiency [41]. Medina et al. [40] identified methyl-accepting chemotactic receptor proteins that are involved in enhancing the predation processes. Recently, Avidan et al. [2] postulated that the deletion of several pilin-like genes can abolish the predation process. They further proposed that these proteins may be components of signaling pathways that are required for the predator entry into the prey cells.

7.4 BIOCONTROL POTENTIAL OF PREDATORY BACTERIA

Although the mode of action of predatory bacteria appears to be similar to that of bacteriophages, yet the predatory bacteria mechanism is unique. Unlike a bacteriophage, which has a very specific host-range, predatory bacteria bind to other Gram-negative prey in a non-specific fashion. The wide host-range of predatory bacteria makes them suitable for multiple applications, such as: enhancing food safety, biocontrol against phytopathogens, water clean-up, clinical therapy, and for combating ARB strains. Predatory bacteria have the ability to grow at a broad temperature range (19–37°C) [21], similar to the growth temperature range of common foodborne pathogens. Additionally, predatory bacteria are very motile and can rapidly swim towards target organisms, enabling them to target Gram-negative foodborne pathogens located within a biofilm [29]. In the absence of any prey/host cells, predatory bacteria can survive for longer periods by converting into axenic host-independent (HI) phase [54].

Further, while some Gram-negative bacteria are able to develop genetic resistance against bacteriophage attack, no such adaptation has been observed against predatory bacteria with the exception of some plastic resistance, where cell-surface protein receptors were lost on prey [16]. Furthermore, predatory bacteria also possess ability to attack actively growing, stationary, as well as metabolically inactive cells [39]. The metabolically inactive cells are often generated after sanitizer and antibiotic treatments. Ability of predatory bacteria to attack metabolically inactive cells further enhances their efficacy to eliminate Gram-negative pathogens.

7.5 POTENTIAL OF *BDELLOVIBRIO* IN AQUACULTURE

By 2030, it is projected that almost 2/3 of the fish supply will be provided by aquaculture production [11, 18]. Despite significant potential for contribution to the global protein supply, one of the major threats to aquaculture production is the spread of diseases through densely-populated fish farms [58]. Application of antibiotics may control spread of pathogenic bacteria through aquatic communities; however, antibiotic-use may not be affordable to all farmers and may pose a risk to the environment [24]. *Bdellovibrio*'s ability to lyse host bacterial cells offers a potential alternative for elimination of pathogenic bacteria in aquaculture.

Aeromonas hydrophilia is one example of a prominent aquaculture pathogen that may be alleviated by *Bdellovibrio* application. Cao et al. [7] isolated a *Bdellovibrio* strain F16 which showed ability to invade *A. hydrophilia* strain S1 cells. Compared to the control group, the 39.31% reduction in *A. hydrophilia* was observed after inoculation with *Bdellovibrio* strain F16 after eight days of incubation. In addition to *A. hydrophilia*, *Bdellovibrio* strain F16 demonstrated lysis ability in thirteen other tested pathogenic *Aeromonas* strains, indicating the potential role of the isolate as a probiotic agent for sturgeon farming.

Prevalence of contaminated fish and shellfish resulting in foodborne illness have been extensively documented [4]. *Vibrio* species-specifically *V. alginolyticus, V. fluvialis, V. cholerae, V. hollisae, V. minicus, V. vulnificus,* and *V. parahaemolyticus*—are most important foodborne pathogen species affecting fish and shellfish. Due to these limitations of conventional control methods, *Bdellovibrio* and like organisms (BALO) have also been used for control of *Vibrio* species. Cai et al. [4] reported lysis-ability of four BALO strains against 41 strains of *Vibrio*. Lysis rates for individual BALO strain applications were 27.3–63.6% (*V. alginolyticus*), 33.3–66.7% (*V. cholera*), 11.1–88.9% (*V. fluvialis*), 25.0–100% (*V. minicus*), and 33.3–66.7% (*V. parahaemolyticus*). When used in combination, susceptibility of *Vibrio* to BALOs ranged from 81.8–100%. *V. parahaemolyticus* and *V. vulnificus* are notably implicated in foodborne illness due to the consumption of raw oysters [49].

While, it was previously accepted that *Vibrio* presence in oysters was primarily dependent on physical parameters of the water source, Richards et al. [48] showed that presence of predatory bacteria *Bdellovibrio* in seawater is a major factor leading to inhibition of *V. vulnificus and V. parahaemolyticus* in shellfish. Among BALO, *Bacteriovorax* are exclusively saltwater predators and preferentially target *V. parahaemolyticus* [44]. Seasonal patterns for *Bacteriovorax* presence in the U.S. Atlantic, Gulf, and Pacific seawaters differed greatly depending on the location [49].

B. bacteriovorus potential for *V. cholerae* elimination was evident after *in vitro* inoculation with *B. bacteriovorus* H16 strain and resulted in around 3 log CFU *V. cholerae* reduction (99.9%) after eight days incubation [6]. Further, the predatory potential of *B. bacteriovorus* H16 was observed in a range of other tested pathogenic *Vibrio* strains. *In vivo* predation of *B. bacteriovorus* H16 was also recorded for mitigating infection caused by *V. Cholerae* on *Litopenaeus vannamei* (whiteleg shrimp). Relative survival of *P. vannamei* in the presence of QH isolate was 63.3% after addition of 10,000 PFU/mL of H16 for seven days.

It is estimated that roughly 2.8% of domestic U.S. production and 10% of imported raw seafood is contaminated with *Salmonella*. Among the imported seafood, *S. Typhimurium* was ranked 12th serotype detected [25]. Lu and Cai [36] characterized four *Bdellovibrio*-and-like organisms (BDXGL, BD5GL, BD2GS, and BD2GL) isolated from freshwater pond in China. The four isolates had differential lytic ability towards 32 tested Gram-negative bacterial strains, which were spread across nine genera (*Aeromonas*, *Erwinia*, *Escherichia*, *Pseudomonas*, *Providencia*, *Salmonella*, *Serratia*, *Shewanella*, *Vibrio*).

The BD2GS had the highest reported lytic rate against five tested *Salmonella* strains and its efficacy was further assessed for enhancing safety of tilapia fish fillets contaminated with *S. Typhimurium*, where a time-dependent ability for BD2GS was observed. Within 3–12 h, BD2GS significantly reduced growth of *S. Typhimurium*. However, *S. Typhimurium* count resembled that of the control (no BD2GS) at 24 and 36 h. The authors reported that the tilapia matrix may not be suitable for BD2GS survival, which led to the reduced efficacy over time.

Chu and Zhu [10] isolated 14 *Bdellovibrio*-like organisms (BALO) strains from fish ponds. Of the 14 strains tested, *Bdellovibrio* C-1 formed large plaques after incubation with six fish pathogens (*A. hydrophilia, E. coli, Edwardsiella tarta, V. parahaemolyticus, V. harveyi* and *V. alginolyticus*). Plaque formation was not observed following incubation with tested Gram-positive bacteria, *Staphylococcus aureus,* and *Bacillus subtilis*. *Bdellovibrio* C-1 showed a protective effect in fish immersed with *A. hydrophilia* J-1 cells and significantly decreased the mortality rate by 75%. These results suggest that *Bdellovibrio* treatment may inhibit bacterial growth in water and can control transmission of pathogens.

7.6 *BDELLOVIBRIO* APPLICATION FOR MITIGATING *SALMONELLA IN POULTRY*

Atterbury et al. [1] reported the first study to assess *Bdellovibrio* strain HD100 effects on live animals, using a poultry model. Poultry are frequently infected with *Salmonella enterica* Enteritidis, a primary source of foodborne illness in developed countries [37]. Prior to administering of *Bdellovibrio* to the poultry, the authors first performed *in vitro* tests in simulated gastric conditions to assess viability of *Bdellovibrio*. Although *Bdellovibrio* strains are typically cultured at 29–32°C under aerobic conditions, yet ability of *B. bacteriovorus* HD100 to survive was reported up to 2 days at 42°C under aerobic, microaeorbic, and anaerobic conditions, suggesting ability of *Bdellovibrio* to survive under poultry gastric conditions. Oral administration of 100 μl of 1.9×10^7 PFU/ml of *Bdellovibrio* HD100 to live chickens showed no adverse effect on animal health or change in behavior during the 28-day assessment.

Further, *Bdellovibrio* did not spread from animal to animal or drinking water in closely housed environment. Assessment of cecal bacterial populations suggested that *Bdellovibrio* colonization in the intestine is transient and does not readily alter the gut microbiota. Further, compared to the control groups, *Bdellovibrio*-treated birds had significantly lower cecal counts of *S. enterica* Enteritidis (phage type 4), reduced cecal inflammation, and reduced defects in cecum suggesting effective predatory action of *Bdellovibrio* on *Salmonella* strains in the gut. This study concludes that strains of *Bdellovibrio* can be used as a biocontrol agent for treating infection caused by Gram-negative bacteria in live food animals.

7.7 *BDELLOVIBRIO* APPLICATION FOR MITIGATING DISCOLORATION IN MUSHROOMS

In addition to aquatic species and live animals, *Bdellovibrio* application has been assessed in mushrooms to control the growth of *Pseudomonas tolaasii*, a Gram-negative soil-borne pathogen responsible for causing brown discoloration [50]. The cause of discoloration is melanin production by the mushrooms in response to the tolaasin toxin produced by *P. tolaasii*. Tolaasin toxin disrupts host cell plasma membranes, which allow access of the bacteria to cell nutrients [3]. *P. tolassii* contamination of mushrooms ultimately results in lower crop yield and economic loss due to qualitative spoilage [55]. Mushroom growth relies on several non-spoilage microbes present in soil. For this reason, sterilization is not a feasible option to control *P. tolassii* [59].

Therefore, *B. bacteriovorus* strain HD100 was investigated as a potential candidate for reducing *P. tolassii* growth on button mushroom (*Agaricus bisporus*). After 48-hour incubation of *P. tolaasii* alone at 29°C, the mushrooms exhibited the traditional symptoms (dark surface lesions). Application of *B. bacteriovorus* HD100 30 minutes prior to or 30 minutes post-inoculation with *P. tolaasii* both resulted in mushrooms that developed significantly lighter lesions with a concurrent reported reduction in the number of *P. tolaasii* cells. Predation of *P. tolaasii* by *B. bacteriovorus* HD100 was confirmed using scanning electron microscope imaging (SEMI). Reduction was slightly higher when *Bdellovibrio* was added before *P. tolaasii* inoculation, indicating greater efficacy of *Bdellovibrio* HD100 as a preventative agent to control *P. tolaasii* on mushroom.

7.8 *BDELLOVIBRIO* APPLICATION FOR TREATING SOYBEAN BACTERIAL BLIGHT

Because of globally important crops for protein and oil, protection of soybean crops from bacterial blight is of utmost significance [57]. It has been estimated that 4–40% of soybean yield in the U.S. is lost due to the aforementioned bacterial blight [46, 57]. Scherff [51] isolated multiple *B. bacteriovorus* strains from soybean roots and evaluated their efficacy for inhibiting *P. glycinea* in soybeans. Significant reductions in disease severity were observed after soybean leaves were inoculated with Bd-17 strain at *B. bacteriovorus*: *P. glycinia* ratios of 9:1 and 99:1. The development of necrotic lesions and systemic toxemia (characterized by chlorosis) were inhibited by Bd-17 application on the soybean leaves. *B. bacteriovorus* may be used to alter the microbial diversity on plant surfaces and can be used for fighting phytopathogens and increasing food production. Applications of *Bdellovibrio* to control foodborne and phytopathogens are listed in Table 7.1.

TABLE 7.1 Applications of *Bdellovibrio* to Control Foodborne and Phytopathogens

Food Source	Pathogen	BALO Strain	References
Carp	*Aeromonas hydrophilia*	BdC-1	[10]
Chicken	*Salmonella enterica*	HD100	[1]
Mushroom	*Pseudomonas tolaasii*	HD100	[50]
Shrimp	*Vibrio cholerae*	H16	[6]
Soybean	*Pseudomonas glycinea*	Bd-17	[51]
Sturgeon	*Aeromonas hydrophilia*	F16	[7]
Tilapia	*Salmonella* Typhimurium	BD2GS	[36]

7.9 APPLICATION OF *BDELLOVIBRIO* FOR MITIGATING BIOFILMS

Poor cleaning of food contact surfaces and processing equipment promotes soil and moisture buildup, resulting in the development of biofilms. Pathogenic microorganisms can survive in these biofilms formed on food contact surfaces for an extended period and can lead to cross-contamination of food [9]. As the biofilm matures, its bio transfer potential also increases as does its resistance to sanitizers [22]. Chlorine-based sanitizers, PAA, ozonated water, quaternary ammonium compounds, and iodophors are commonly used for controlling biofilm. However, the main limitation of these sanitizers against biofilm is their inability to penetrate within the biofilm. In the absence of physical aberration, sanitizers are effective only on the surface, and pathogens may survive inside the biofilm or may enter a metabolically inactive state.

The predation potential of *B. bacteriovorus* 109J against *Pseudomonas fluorescens* and *E. coli* strains in biofilms was evaluated by Kadouri and

O'Toole [29]. The stable biofilm was formed in 96-well microtiter plates with ~10^8 CFU cell concentration per well. After the application of *B. bacteriovorus* 109J to the biofilm, the reduction of *E. coli* cells in biofilm biomass was observed 3 h after exposure. After a 24 h incubation period, 87% reduction (4-log) was achieved. An initial concentration of 10^2 PFU *B. bacteriovorus* 109J per well and a contact time of 30 min was adequate to initiate infection of *E. coli* cells in the biofilm. Scanning electron micrographs revealed application of *B. bacteriovorus* resulted in extensive cellular damage of *E. coli* cells in biofilm, leaving behind cellular residue and matrix [29].

Fratamico and Cooke [20] also isolated *Bdellovibrio* strain 45k (soil isolate) and 88e (sewage isolate), which showed high predation potential against *Salmonella* and *E. coli* O157. These *Bdellovibrio* isolates (45k and 88e) resulted in 2.5 to 7.9 CFU/mL *E. coli* O157 and *Salmonella* reductions after 7 h incubation. The *Bdellovibrio* 45k strain also showed a high rate of predation on dried *E. coli* cells on a stainless-steel surface resulting in a 3.6 log CFU/cm^2 reduction after 24 h incubation. Results from the study supportability of predators to target prey bacteria inside the biofilms as well as on food contact surfaces.

7.10 THE POTENTIAL ROLE OF *BDELLOVIBRIO* IN DISEASE MANAGEMENT

This section illustrates the potential role of *Bdellovibrio* species to improve food security. In addition to controlling foodborne pathogens, predatory *Bdellovibrio* bacteria may also have a role in the control of bacteria associated with human disease. Disease treatment with conventional antimicrobial agents is becoming increasingly less effective due to the development of antibiotic-resistant among pathogenic bacterial strains. Single-cell planktonic bacteria and surface-attached bacteria present in biofilms are capable of developing resistance in response to selective pressures (e.g., antibiotics, and sanitizers) [13]. Surface-attached bacteria in biofilms are particularly challenging to eliminate as they can be resistant up to 1000 times the concentrated sanitizer treatment [27]. *Acinetobacter baumannii, E. coli, Klebsiella pneumonia,* and *Pseudomonas aeruginosa* are the most frequently involved antibiotic-resistant Gram-negative bacteria involved in human infection [53].

Dashiff et al. [13] investigated the host-specificity and predation potential of two predatory bacteria, *B. bacteriovorus* 109J and *Micavibrio aeruginosavorus* ARL-13, using 83 medically-relevant bacterial species. The results demonstrated that *B. bacteriovorus* 109J and *Micavibrio aeruginosavorus* ARL-13 when co-cultured with test pathogen were able to effectively prey on 68 out of 83 tested bacteria (82%). These medically-relevant bacteria belonged to 17 different genera. Both predatory bacteria showed similar predation potential against pathogens in pure culture and in multi-species mixed cultures. When pathogenic bacterial lawns were used to simulate biofilm, lytic halos were observed after inoculation with *B. bacteriovoruson* 56 (*n* = 56/67) tested bacteria (84%), exhibiting its wide predation potential, and host-range. Assessment of *B. bacteriovorus's* predatory action in a simulated multilayer biofilm showed a 2–3 log reduction and 6–7 log CFU reduction after 24 h and 48 h incubations, respectively. Efficacy of *B. bacteriovorus* to prey on pathogenic bacteria was again similar when single-species and multi-species biofilms were compared.

Application of *B. bacteriovorus* to parasitize pathogens contributing to dental disease has also been explored [14]. Periodontitis involves numerous Gram-negative bacteria that form a biofilm (dental plaque) in the oral cavity [43]. Oral pathogens are capable of causing systemic illness beyond the oral cavity and therefore have broad clinical significance. Due to the reported challenges in removal of oral plaques or biofilms by traditional methods, Dashiff, and Khadouri [14] evaluated the predatory action of *B. bacteriovorus* 109J against oral pathogens. The authors observed that all tested serotypes of *A. actinomycetemcomitans* were prone to *B. bacteriovorus* attack. Predation was also observed in *Ei. corrodens* and *F. nucleatum* PK1594, but not in *F. nucleatum* ATCC 10953, *Pr. intermedia, T. forsythia,* or *Po. gingivalis.*

Significant reduction (>81%) of *A. actinomycetemcomitans (*Gram-negative) biofilms prepared in a 96-well plate was observed after inoculation with *B. bacteriovorus* for 48 h. In addition, *A. actinomycetemcomitans* CU1000 biofilms prepared on hydroxyapatite squares were also substantially reduced by *B. bacteriovorus* after 72 h. Presence of saliva during incubation of *B. bacteriovorus* with *A. actinomycetemcomitans* biofilms did not interfere with predation. Similarly, *Ei. corrodens* biofilms developed in-plate were reduced (45–81%) upon 48 h inoculation with *B. bacteriovorus.*

In a separate study by Van Essche et al. [17], six BALO strains were assessed for their predation efficiency against common periodonto-pathogens. Out of six tested strains (*Bacteriovorax stolpii* Uki2, *Bdellovibrio* BEP2, *B. bacteriovorus* 109J, *B. bacteriovorus* HD100, *Bacteriovorax* FCE, and *Peredibacter starrii* A3.12), *B. bacteriovorus* HD100 was the most versatile and was effective in reducing four out of the six tested oral pathogens (*A.*

actinomycetemcomitans, Po. gingivalis, P. intermedia, F. nucleatum, C. sputigena, and *E. corrodens*). Major differences in predation efficiency were observed across the tested strains due to specific predator-prey interactions.

The most susceptible prey bacterium was *F. nucleatum,* which was targeted by four of the six tested BALO strains [17]. In addition to *F. nucleatum*, another anaerobic bacterium *P. intermedia* was also subjected to attack, demonstrating ability of BALO to predate anaerobic bacteria. The authors also assessed whether the presence of non-BALO-targeted bacteria influences predation efficiency of BALO. Although predation speed was reduced by the presence of non-prey bacteria (possibly due to steric hindrance), yet the final reduction of prey viability was identical to the control (no decoy cells).

In addition to oral pathogens, *Bdellovibrio* can be also applied for attacking pathogens associated with cystic fibrosis-associated infections. Cystic fibrosis patients are susceptible infections caused by *P. aeruginosa* and *S. aureus,* which form resistant biofilms leading to inflammatory clinical symptoms [38]. *B. bacteriovorus* strain HD100 was evaluated for predatory activity against bacterial strains (*P. aeruginosa* and *S. aureus*) commonly implicated in cystic fibrosis [28]. In TSB broth, *B. bacteriovorus* HD100 reduced cell viability in *P. aeruginosa* and unexpectedly, *S. aureus* was also susceptible to predation. Biofilms of the same two pathogens were also assessed following 24 h incubation with *B. bacteriovorus*. The biofilms were significantly reduced by 76% and 74%, respectively, for *P. aeruginosa* and *S. aureus*.

To further evaluate the predatory behavior of HD100 towards Gram-positive *S. aureus*, field emission scanning electron microscopy was used to visualize the interaction using a *S. aureus* biofilm. After 24 h, the biofilm was sufficiently reduced and high magnification revealed HD100 cells attacking *S. aureus* cells in an epibiotic manner, distinctly different from the peri-plasmic predation mechanism used against Gram-negative bacteria. *S. aureus* destruction appeared to occur through membrane destruction and subsequent release of cellular contents. The *B. bacteriovorus* HD100 remained attached throughout prey destruction. When assessed under dynamic settings using a Bioflux microfluidic system, which is used to mimic physiological shear-stress, a similar reduction was observed after 20 h in *P. aeruginosa* and *S. aureus* (~38% reduction). Compared to static conditions, the flow conditions reduced the efficacy of *B. bacteriovorus* to attack prey by 38% and 28% for *P. aeruginosa* and *S. aureus*, respectively.

Multiple studies have evaluated safety of predatory bacteria using various cell lines and animal models. Gupta et al. [23] accessed effects of three predatory bacterial strains (e.g., *B. bacteriovorus* HD100 and *B. bacteriovorus* 109 J, and *Micavibrioaeruginosavorus* ARL-13) using five human cell lines (e.g., human keratinocytes, two human liver epithelial cells and two human kidney epithelial cells). The study concluded that predatory bacteria are not cytotoxic to these cell lines. Predatory bacteria when inoculated at higher concentration did not elicit any pro-inflammatory cytokines markers, thus showing their non-cytotoxic and non-immunogenic. In another study, safety of predatory bacteria (e.g., *M. aeruginosavorus* ARL-13 and *B. bacteriovorus* 109 J) was evaluated by intrarectal inoculations of predators in rats [52].

Intrarectal inoculations of predators did not result in any visual histopathological abnormalities. However, a slight increase in IL-13 cytokines level was observed in the serum samples, which later came to baseline on day seven. Gut microbiome analysis showed overgrowth of *Prevotella* in the fecal samples. The study reported that introduction of predatory bacteria to gut has no adverse effect on rat health and predatory bacteria can be considered as a viable option for treating Gram-negative bacterial infections [52].

Although the reports described within this section suggest efficacy of live predatory bacteria for treatment of medically relevant pathogens, yet further large-scale assessment is required to determine their *in vivo* efficacy.

7.11 APPLICATION OF PREDATORY BACTERIA AGAINST ANTIBIOTIC RESISTANT BACTERIA (ARB)

Multiple drug resistant (MDR) bacteria are big threat to human health. These MDR Gram-negative bacterial strains produce potent enzymes, which are capable of inactivating ß-lactam as well as extended-spectrum ß-lactam antibiotics. Polymyxins, including colistin, are considered as last resort antibiotics and are used to treat severe cases of infections caused by highly antibiotic resistant strains. The severity of the antibiotic resistance problem has prompted the development of alternative strategies for combating the problem.

Kadouri et al. [30] tested predation potential of three predatory bacteria (*B. bacteriovorus* HD100, *B. bacteriovorus* 109J, and *M. aeruginosavorus* ARL-13) against 14 MDR strains of *K. pneumonia, E. coli*, *A. baumannii*, and *Pseudomonas* spp. The strains used in this study were positive for AmpC-type ß-lactamase, extended-spectrum ß-lactamase (ESBL), KPC-type carbapenemase and metallo-ß-lactamase enzymes. The results showed the potential of predators to prey on MDR clinical pathogens.

Colistin is a very strong antibiotic, which was initially introduced in the market in 1959. It is infrequently prescribed due to its renal and neuro toxicities [45]. However, colistin is commonly used for animal husbandry purposes, especially in pig farms. Liu et al. [35] for the first time reported isolation of a plasmid-mediated colistin-resistant *E. coli* strain isolated from a pig model. Predation potential of *B. bacteriovorus* HD100, 109J and *M. aeruginosavorus* ARL-13 strains against *mcr-1*-positive, colistin-resistant isolates (*K. pneumoniae* ATCC 13883, *E. coli* ATCC 25922, *A. baumannii* ATCC 17978, and *P. aeruginosa* ATCC 47085) were evaluated by Dharani et al. [15]. The *Bdellovibrio* strains when co-cultured with colistin-resistant isolates resulted in 2.5–5 log CFU reductions after 24 h incubation, demonstrating ability of predatory bacteria to reduce viability of antibiotic-resistant pathogens. In cases where antibiotic treatment is ineffective, predatory bacteria may be useful as a therapeutic agent for treating infections.

7.12 SUMMARY

Predatory bacteria are a group of small-sized bacteria spread across multiple genera, which possess the ability to prey on a wide range of other larger Gram-negative bacteria. They are commonly found in soil and marine environments. Due to their predatory behavior, they have been explored as biocontrol agents against foodborne pathogens, other disease-causing bacterial strains, MDR pathogens and to treat plant infection caused by various Gram-negative bacteria. Predatory bacteria-based biocontrol methods appear to be safe and effective strategies for the mitigation of Gram-negative pathogens.

KEYWORDS

- **antimicrobial resistance**
- ***Bdellovibrio***
- **extended-spectrum ß-lactamase**
- **living antibiotics**
- **predation**
- **scanning electron microscope imaging**

REFERENCES

1. Atterbury, R. J., Hobley, L., & Till, R., (2011). Effects of orally administered *Bdellovibrio* bacteriovorus on the well-being and *Salmonella* colonization of young chicks. *Appl. Environ. Microbiol.*, *77*(16), 5794–5803.
2. Avidan, O., Petrenko, M., & Becker, R., (2017). Identification and characterization of differentially-regulated type IVb Pilin genes necessary for predation in obligate bacterial predators. *Sci. Rep.*, *7*(1), 1013–1017.
3. Brodey, C. L., Rainey, P. B., Tester, M., & Keith, J., (1991). Bacterial blotch disease of the cultivated mushroom is caused by an ion channel forming lipo depsipeptide toxin. *Mol. Plant Microbe. Interact.*, *4(4)*, 407–411.
4. Cai, J., Zhao, J., Wang, Z., Zou, D., & Sun, L., (2008). Lysis of vibrios by *Bdellovibrio*-and-like organisms (balos) isolated from marine environment. *J. Food Safety*, *28(2)*, 220–235.
5. Callejón, R. M., Rodríguez-Naranjo, M. I., Ubeda, C., Hornedo-Ortega, R., Garcia-Parrilla, M. C., & Troncoso, A. M., (2015). Reported foodborne outbreaks due to fresh produce in the United States and European Union: Trends and causes. *Food Borne Pathog. Dis.*, *12(1)*, 32–38.
6. Cao, H., An, J., Zheng, W., & He, S., (2015). Vibrio cholerae pathogen from the freshwater-cultured white leg shrimp *Penaeus vannamei* and control with *Bdellovibrio* bacteriovorus. *J. Invertebr. Pathol.*, *130*, 13–20.
7. Cao, H., He, S., Wang, H., Hou, S., Lu, L., & Yang, X., (2012). *Bdellovibrios*-potential biocontrol bacteria against pathogenic *Aeromonas hydrophila*. *Vet. Microbiol.*, *154*(3–4), 413–418.
8. CDC. (2013). *The Biggest Antibiotic-Resistant Threats in the US.* Centers for Disease Control and Prevention (CDC). https://www.cdc.gov/drugresistance/biggest_threats.html (accessed on 25 May 2020).
9. Chmielewski, R., & Frank, F., (2003). Biofilm formation and control in food processing facilities. *Compr. Rev. Food Sci. Food Safety*, *2(1)*, 22–32.
10. Chu, W. H., & Zhu, W., (2010). Isolation of bdellovibrioas biological therapeutic agents used for the treatment of aeromonas hydrophila infection in fish. *Zoonoses Public Health*, *57(4)*, 258–264.
11. Contributing to Food Security and Nutrition for All (2016). The State of World Fisheries and Aquaculture, FAO, Rome. http://www.fao.org/3/a-i5555e.pdf (accessed on 25 May 2020).
12. Cui, Y., Liu, D., & Chen, J., (2018). Fate of various Salmonella enterica and enterohemorrhagic *Escherichia coli* cells attached to alfalfa, fenugreek, lettuce, and tomato seeds during germination. *Food Control*, *88*, 229–235.
13. Dashiff, A., Junka, R., Libera, M., & Kadouri, D. E., (2011). Predation of human pathogens by the predatory bacteria *Micavibrio aeruginosavorus* and *Bdellovibrio* bacteriovorus. *J. Appl. Microbiol.*, *110*(2), 431–444.
14. Dashiff, A., & Kadouri, D., (2011). Predation of oral pathogens by *Bdellovibrio* bacteriovorus 109J. *Mol. Oral Microbiol.*, *26*(1), 19–34.
15. Dharani, S., Kim, D. H., Shanks, R. M. Q., Doi, Y., & Kadouri, D. E., (2018). Susceptibility of colistin-resistant pathogens to predatory bacteria. *Res. Microbiol.*, *169*(1), 52–55.

16. Dwidar, M., Monnappa, A. K., & Mitchell, R. J., (2012). The dual probiotic and antibiotic nature of *Bdellovibrio* bacteriovorus. *BMB Rep.*, *45*(2), 71–78.
17. Essche, M. V., Quirynen, M., & Sliepen, I., (2011). Killing of anaerobic pathogens by predatory bacteria. *Mol, Oral Microbiol.*, *26*(1), 52–61.
18. FAO, (2013). *Fish to 2030: Prospects for Fisheries and Aquaculture* (p. 102). http://www.fao.org/docrep/019/i3640e/i3640e.pdf (accessed on 25 May 2020).
19. FAO, (2017). *Plant Health and Food Security* (p. 2). Rome: FAO. http://www.fao.org/3/a-i7829e.pdf (accessed on 25 May 2020).
20. Fratamico, P. M., & Cooke, P. H., (1996). Isolation of *Bdellovibrios* that prey on *Escherichia Coli* O157:h7 and *Salmonella* species and application for removal of prey from stainless steel surfaces. *J. Food Safety*, *16*(2), 161–173.
21. Fratamico, P. M., & Whiting, R. C., (1995). Ability of *Bdellovibrio* bacteriovorus 109J to lyse gram-negative food-borne pathogenic and spoilage bacteria. *J. Food Prot.*, *58*(2), 160–164.
22. Garrett, T. R., Bhakoo, M., & Zhang, Z., (2008). Bacterial adhesion and biofilms on surfaces. *Prog. Nat. Sci.*, *18*(9), 1049–1056.
23. Gupta, S., Tang, C., Tran, M., & Kadouri, D. E., (2016). Effect of predatory bacteria on human cell lines. *PLoS One*, *11*(*8*), 5, e-article ID 0161242.
24. Harikrishnan, R., Balasundaram, C., & Heo, M., (2010). Effect of probiotics enriched diet on Paralichthys olivaceus infected with lymphocystis disease virus (LCDV). *Fish Shellfish Immunol.*, *29*(*5*), 868–874.
25. Heinitz, M. L., Ruble, R. D., Wagner, D. E., & Tatini, S. R., (2000). Incidence of Salmonella in fish and seafood. *J. Food Prot.*, *63*(5), 579–592.
26. Herman, K. M., Hall, A. J., & Gould, L. H., (2015). Outbreaks attributed to fresh leafy vegetables, United States, 1973–2012. *Epidemiol Amp. Infect.*, *143*(14), 3011–3021.
27. Høiby, N., Bjarnsholt, T., Givskov, M., Molin, S., & Ciofu, O., (2010). Antibiotic resistance of bacterial biofilms. *Int. J. Antimicrob Agents*, 35(*4*), 322–332.
28. Iebba, V., Santangelo, F., & Totino, V., (2013). Higher prevalence and abundance of *Bdellovibrio* bacteriovorus in the human gut of healthy subjects. *Plos One*, 8(*4*), e61608.
29. Kadouri, D., & O'Toole, G. A., (2005). Susceptibility of Biofilms to *Bdellovibrio* bacteriovorus attack. *Appl. Environ. Microbiol.*, *71*(*7*), 4044–4051.
30. Kadouri, D. E., To, K., Shanks, R. M. Q., & Doi, Y., (2013). Predatory bacteria: A potential ally against multidrug-resistant gram-negative pathogens. *PLoS One*, *8*(*5*), e63397.
31. Karunker, I., Rotem, O., Dori-Bachash, M., Jurkevitch, E., & Sorek, R., (2013). Global transcriptional switch between the attack and growth forms of *Bdellovibrio* bacteriovorus. *Plos One*, *8*(*4*), e61850.
32. Lambert, C., Chang, C. Y., Capeness, M. J., & Sockett, R. E., (2010). The first bite: Profiling the Predatosome in the bacterial pathogen *Bdellovibrio*. *PLoS One*, *5*(1), e-article ID 8599, pages 6.
33. Lambert, C., Fenton, A. K., Hobley, L., & Sockett, R. E., (2011). Predatory *Bdellovibrio* bacteria use gliding motility to scout for prey on surfaces. *J. Bacteriol.*, *193*(12), 3139–3141.
34. Li, X., & Farid, M., (2016). A review on recent development in non-conventional food sterilization technologies. *J. Food Eng.*, *182*, 33–45.

35. Liu, Y., Wang, Y., & Walsh, T. R., (2016). Emergence of plasmid-mediated colistin resistance mechanism MCR-1 in animals and human beings in China: A microbiological and molecular biological study. *Lancet Infect. Dis.*, *16*(2), 161–168.
36. Lu, F., & Cai, J., (2010). The protective effect of *Bdellovibrio*-and-like organisms (BALO) on tilapia fish fillets against *Salmonella enterica* ssp. *Enterica serovar typhimurium*. *Letters Appl. Microbiol.*, *51*(6), 625–631.
37. Lu, S., Killoran, P. B., & Riley, L. W., (2003). Association of *Salmonella enterica serovar enteritidis* yard with resistance to chicken egg albumen. *Infect. Immun.*, *71*(12), 6734–6741.
38. Lyczak, J. B., Cannon, C. L., & Pier, G. B., (2002). Lung infections associated with cystic fibrosis. *Clin. Microbiol. Rev.*, *15*(2), 194–222.
39. Markelova, N. Y., & Kerzhentsev, A. S., (1998). Isolation of a new strain of the genus *Bdellovibrio* from plant rhizosphere and its lytic spectrum. *Microbiology*, *67*, 696–699.
40. Medina, A. A., Shanks, R. M., & Kadouri, D. E., (2008). Development of a novel system for isolating genes involved in predator-prey interactions using host independent derivatives of *Bdellovibrio* bacteriovorus 109J. *BMC Microbiol.*, *8*(1), 33–37.
41. Negus, D., Moore, C., Baker, M., Raghunathan, D., Tyson, J., & Sockett, R. E., (2017). Predator versus pathogen: How does predatory *Bdellovibrio* bacteriovorus interface with the challenges of killing gram-negative pathogens in a host setting? *Annu. Rev. Microbiol.*, *71*(1), 441–457.
42. Olanya, O. M., & Lakshman, D. K., (2015). Potential of predatory bacteria as biocontrol agents for foodborne and plant pathogens. *J. Plant Pathol.*, *97*(3), 405–417.
43. Pihlstrom, B. L., Michalowicz, B. S., & Johnson, N. W., (2005). Periodontal diseases. *Lancet London England*, *366*(9499), 1809–1820.
44. Pineiro, S. A., Stine, O. C., Chauhan, A., Steyert, S. R., Smith, R., & Williams, H. N., (2007). Global survey of diversity among environmental saltwater *Bacteriovoracaceae*. *Environ. Microbiol.*, *9*(10), 2441–2450.
45. Poirel, L., & Nordmann, P., (2016). Emerging plasmid-encoded colistin resistance: The animal world as the culprit? *J. Antimicrob. Chemother.*, *71*(8), 2326–2327.
46. Qi, M., Wang, D., Bradley, C. A., & Zhao, Y., (2011). Genome sequence analyses of *Pseudomonas savastanoi* pv. glycinea and subtractive hybridization-based comparative genomics with nine pseudomonads. *PLoS One*, *6*(1), 5, e-article ID 16451.
47. Rendulic, S., Jagtap, P., & Rosinus, A., (2004). Predator unmasked: Life cycle of *Bdellovibrio* bacteriovorus from a genomic perspective. *Science*, *303*(5658), 689–692.
48. Richards, G. P., Fay, J. P., Dickens, K. A., Parent, M. A., Soroka, D. S., & Boyd, E. F., (2012). Predatory bacteria as natural modulators of vibrio parahaemolyticus and vibrio vulnificus in seawater and oysters. *Appl. Environ. Microbiol.*, 78(*20*), 7455–7466.
49. Richards, G. P., Watson, M. A., & Boyd, E. F., (2013). Seasonal levels of the vibrio predator bacteriovorax in Atlantic, Pacific, and Gulf coast seawater. *Int. J. Microbiol.*, *4*, e-article ID 375371.
50. Saxon, E. B., Jackson, R. W., Bhumbra, S., Smith, T., & Sockett, R. E., (2014). *Bdellovibrio* bacteriovorus HD100 guards against *Pseudomonas tolaasii* brown-blotch lesions on the surface of post-harvest *Agaricus bisporus* supermarket mushrooms. *BMC Microbiol.*, *14*, 163.
51. Scherff, R. H., (1973). Control of bacterial blight of soybean by Bdellovibrio bacteriovorus. *Phytopathology*, *63*(328), 400–402.

52. Shatzkes, K., Tang, C., & Singleton, E., (2017). Effect of predatory bacteria on the gut bacterial microbiota in rats. *Sci. Rep.*, *7*, 6, e-article ID 43483.
53. Singh, P., Pfeifer, Y., & Mustapha, A., (2016). Multiplex real-time PCR assay for the detection of extended-spectrum β-lactamase and carbapenemase genes using melting curve analysis. *J. Microbiol. Methods*, *124*, 72–78.
54. Sockett, R. E., (2009). Predatory lifestyle of Bdellovibrio bacteriovorus. *Annu. Rev. Microbiol.*, *63*(1), 523–539.
55. Soler-Rivas, C., Jolivet, S., Arpin, N., Olivier, J. M., & Wichers, H. J., (1999). Biochemical and physiological aspects of brown blotch disease of *Agaricus bisporus*. *FEMS Microbiol. Rev.*, *23*(5), 591–614.
56. *The Future of Food and Agriculture*. Rome, Food and Agriculture Organization of the United Nations FAO, (2019) The Future of Food and Agriculture. http://www.fao.org/publications/fofa/en/ (accessed on 5 June 2020).
57. USDA, (2018). *Crop Production Historical Track Records* (p. 240). https://www.nass.usda.gov/Publications/Todays_Reports/reports/croptr18.pdf (accessed on 25 May 2020).
58. WRAP, (2018). *Food Futures* (p. 81). http://www.wrap.org.uk/sites/files/wrap/Food_Futures_%20report_0.pdf (accessed on 25 May 2020).
59. Zarenejad, F., Yakhchali, B., & Rasooli, I., (2012). Evaluation of indigenous potent mushroom growth promoting bacteria (MGPB) on *Agaricus bisporus* production. *World J. Microbiol. Biotechnol.*, *28*(1), 99–104.

CHAPTER 8

SAFETY ASPECTS OF NOVEL BACTERIOCINS

TEJINDER P. SINGH and SHALINI ARORA

ABSTRACT

Increasing awareness and understanding towards the importance of healthy as well as safe food has put pressure on food business operators to satisfy customers demand for minimally processed, highly nutritious, natural, and safe food. Despite of existing good manufacturing and hygiene practices, up-gradation of the food preservation methodologies, and advanced understanding of microorganisms; food safety and security are still a major concern both for the developed and developing countries. The reason for this could be the development of new pathogens and/or contaminants, development of resistance and cross-resistance against preservatives in use. Moreover, the health implications associated with the chemical preservatives has driven consumers demand for natural food with minimal preservatives. Considering the issues discussed, the trend is shifting from chemical preservatives and antibiotics to biopreservatives. Among different biopreservatives such as antimicrobial metabolites, fermentates, bioprotective cultures, and bacteriophages, applications of bacteriocins and/or bacteriocinogenic cultures in the food sector have attracted the researchers and food producers to ensure food safety and security. Mostly, the bacteriocins have narrow spectrum of activity and can be used in combinations or as a component in Hurdle technology. Bacteriocins can be used either in purified, semi-purified form or even the bacteriocinogenic strains can be used as starters but their safety needs to be ensured prior to application in food sector. Therefore, this chapter will focus on issues related to bacteriocin resistance and cross-resistance, safety aspects of bacteriocinogenic cultures, and regulatory considerations for bacteriocins and/or bacteriocinogenic strains.

8.1 INTRODUCTION

Increasing awareness and understanding towards the importance of healthy as well as safe food has put pressure on food business operators to satisfy customers demand for specific, fresh, natural, nutritious, minimally processed, and ready to eat food without compromising with the safety concerns associated with the products. Also, the food business operators are facing challenges not only to meet the customers demand and satisfy the regulatory policies but also to make the process economical by including various means to enhance food safety and security. USDA Economic Research Service has reported that the loss of food mainly occurs at three marketing stages viz., retail, during servicing of food and customer which constitutes a quarter of edible food available for consumption in USA significant loss to food industry was attributed to microbial spoilage that results in products with inferior quality [33]. Globally around a quarter of the food is lost due to microbial spoilage [10]. According to FAO, the food spoilage results in total economic losses of approximately 1300 million tons every year [16].

For several decades, the food safety (the removal of pathogens) and food security (controlling growth of spoilage causing organisms) were considered as separate issues which now have been studied as an inseparable, intervened, and indistinguishable concepts. Despite chill chains, chemical preservatives, and a better understanding of microorganisms, food-borne diseases (FBD) represent an important health problem for developed and developing countries. Health issues and economical losses due to food-borne illnesses are comparable with the economic losses due to food deterioration [30]. Food contaminated with harmful microorganisms or chemicals may cause several diseases-ranging from diarrhea to life-threatening diseases like cancer. As per FAO, every tenth person falls ill and approx.4.2 lakh people die every year due to consumption of contaminated food. Also, foodborne diseases prevent socio-economic development by constricting health care systems and having negative impact on national economies, trade, as well as tourism. Hence, Food safety and food security are of paramount importance as it not only improves the shelf-life of product but also is key to attain good health.

Despite of existing good manufacturing and hygiene practices, up-gradation of the food preservation methodologies and advanced understanding of microorganisms; food safety and security is still a major concern both for the developed and developing countries. Due to globalization, the food supply chains have crossed multiple national borders and have led to the increasing demand for preservatives. Since long, antibiotics or other chemical/synthetic preservatives (e.g., nitrite, and sulfur dioxide) were in use to delay microbial growth, extend shelf-life, and possible corruption which may have an adverse impact on the health of the consumer. Moreover, the emergence of new pathogens and/or contaminants, development of resistance and cross-resistance against preservatives in use is also a challenge which demands the advancement in preservation techniques. Considering the issues discussed, the trend is shifting from chemical preservatives and antibiotics to biopreservatives.

Among different biopreservatives such as antimicrobial metabolites, fermentates, bioprotective cultures, and bacteriophages, applications of bacteriocins and/or bacteriocinogenic cultures in food sector have attracted the researchers and food producers to ensure food safety and security. Mostly, the bacteriocins have narrow spectrum of activity and can be used in combinations or as a component in Hurdle technology. Bacteriocins can be used either in purified, semi-purified form or even the bacteriocinogenic strains can be used as starters but their safety needs to be ensured prior to application in food sector.

This chapter focuses on issues related to bacteriocin resistance and cross resistance, safety aspects of bacteriocinogenic cultures, and regulatory considerations for bacteriocins and/or bacteriocinogenic strains.

8.2 BACTERIOCINS

The risks associated with the contaminated food carrying spoilage and pathogenic microorganisms, and with the synthetic preservatives used to control their growth in food products have diverted the interest of food scientists towards the discovery of new biopreservatives [3]. Also, among biopreservatives, bacteriocins have caught the attention and become an important alternate option because of limited availability of new drugs that could target drug-resistant pathogens [23]. Bacteriocins are peptides/proteins, ribosomally transcribed, that may kill (bactericidal effect) or halt the growth (bacteriostatic effect) of the target organisms [24, 69]. They may inhibit taxonomically related bacteria (narrow spectrum activity) or may inhibit a wide variety of bacteria (broad spectrum activity) [24]. Bacteriocins are generally cationic and hydrophobic in nature. Bacteriocins are preferred as they exhibit antimicrobial activity against drug-resistant bacteria at nanomolar concentrations; are ineffective to eukaryotic cells; negligibly affect the sensory quality of the product to which it has been incorporated [3–5, 23, 24].

The 177 bacteriocin sequences (156 from Gram-positive and 18 from Gram-negative bacteria) are available in an open-access database on bacteriocins, BACTIBASE [49]. As per the database, bacteriocins from Gram-positive bacteria may constitute 20–60 residues of amino acid, whereas, may constitute as much as 688 residues of amino acid in bacteriocins from Gram-negative bacteria. Different schemes for bacteriocins classification, based on producing strains, mode of action, molecular weight (MW), and chemical configuration, have been proposed as mentioned in Table 8.1.

TABLE 8.1 Classification of Bacteriocins

Classes	Features	Example
Class I (Lantibiotics)	Peptides undergo posttranslational modifications; Contain lanthionine and-methyl lanthionine charge; Molecular weight (MW) 2–5 kDa	Nisin A & Z Enterocin A & B
Class II	Heat stable; no posttranslational modifications in amino acids; MW < 10 kDa.	Lactococcin A & B Pediocin PA-1
Class III	Unmodified amino acids; Heat-sensitive Large peptides; MW > 30 kDa.	Lacticin A & B
Class IV	Protein in combination with carbohydrate or lipid component	
Class V	Circular bacteriocins	Gassericin, Reutericin

8.2.1 MODE OF ACTIVITY OF BACTERIOCINS

Different mechanisms explaining bacteriocins activity can be divided based on their bactericidal (cell death), or bacteriostatic (prevent cell growth) effects [27]. Most of the LAB bacteriocins target the cell envelope-associated mechanisms to exert their antagonistic effect, particularly those affecting Gram-positive bacteria [23]. Several bacteriocins (lantibiotics and others) target Lipid II (a molecule involved in synthesis of peptidoglycan) and inhibit peptidoglycan synthesis [12]. Other bacteriocins use Lipid II or mannose phosphotransferase system (Man-PTS) as a docking molecule to create pores in the cell membrane with subsequent dissipation of the membrane potential and finally death of the cell [23, 24, 70]. Many bacteriocins can destroy the target organism by inhibiting gene expression [84, 103] and protein production [78].

Applications of bacteriocins can be accomplished by [15, 94]: (a) using bacteriocinogenic strain as starter culture; (b) incorporation of bacteriocin (pure or crude) as a food additive; and/or (c) supplementation of an ingredient containing preformed bacteriocins.

8.2.2 BACTERIOCIN IMMUNITY

Bacteriocins structure consists of three different domains involved in (a) recognition of specific receptor, (b) translocation, and (c) responsible for their toxic activity [26]. The bacteriocins operon may be present on the bacterial chromosome, extra-chromosomal material (plasmids), and transposons (both plasmids and chromosome carried) [32]. These operons include several genes implicated in the synthesis of functional peptide, post-translation modifications, secretion, regulation, and for self-immunity (SI) [90]. The bacteriocin producer usually synthesizes SI proteins; and genes encoding SI proteins are generally located within the bacteriocin gene clusters, which provide protection against cognate bacteriocin [8, 51, 82, 101]. SI proteins have been either attached to the surface of cell membrane [36] or embedded in the membrane [8, 18]. Some are mainly exported and remain trapped within the cell envelope [53].

These SI proteins may have either of the following mechanisms to provide self-protection to producer cells against its cognate bacteriocin:

- Degradation or modification of a receptor molecules [65];
- Expulsion of bacteriocins [8, 81, 101], (d) alterations in cell envelope [37]; and
- Scavenge bacteriocins or compete for the bacteriocin receptor [31, 36].

8.3 DEVELOPMENT OF BACTERIOCIN RESISTANCE

Every newly identified bacteriocin, which has been proven for its efficacy against pathogens and safety, must be evaluated for the risks of resistance development as well as for the frequency of developing resistance against selected bacteriocin upon its prolonged exposure [23]. Nisin with GRAS label has long been used for biopreservation, but no report has claimed nisin resistance amongst spoilage causing micro-organisms in the food industry

[9]. Therefore, the available knowledge related to bacteriocin resistance development is primarily based on *in-vitro* studies [23].

The available reports are inappropriate to give clear understanding of the bacteriocin resistance because some researchers experimented with purified preparations while others used the fermentates of the bactericinogenic culture. Hence, there is no clarity regarding low-, moderate- or high-level resistance to bacteriocins due to variability in experimental approaches and terminology [58]. Different researchers have proposed different ways to estimate development of bacteriocin resistance.

Gravesen et al. [46] suggested that 10-fold increase in MIC of nisin resistance in *Listeria monocytogenes* can be considered as a high-level resistance. Whereas, Blake et al. [9] suggested that mutants displaying susceptibility to 8-fold or less of nisin MIC exhibit low resistance, whilst highly resistant mutants will exhibit a 32-fold high/elevation/risein nisin MIC. However, the researchers considered subclass IIa bacteriocins mutants as highly resistant when:

- They survived in presence of $\geq$1600 arbitrary units (AU) bacteriocin ml^{-1} [89];
- The MICs/IC_{50}s were 1000 times higher compared to those of wild strains [47, 58]; or
- The IC_{50} was $\geq$1 mg ml^{-1} [97].

The mutants inhibited by <1600 AU bacteriocins ml^{-1} [89] or the IC_{50} was <1 $mgml^{-1}$ [97], were considered under low-level resistance. Katla et al. [58] reported several methods to determine bacterial susceptibility to bacteriocins and suggested the use of microtiter plate assay than the plating methods.

Bacteriocins generally act fast, thereby, the frequency of resistance development remains minimum in their presence among population of susceptible cells [24]. However, mutations in the absence of bacteriocins may result in development of resistance to bacteriocins, which further get selected in the presence of bacteriocin.

The bacteriocin resistance mechanisms can further be studied as innate resistance (that naturally seen in certain genera/species) and acquired resistance (that developed in wild type sensitive strain) [21]. The association between these mechanisms and genes has been revealed by gene deletions or knockouts, over-expression analysis or complementation studies. If mutations occurred in the gene renders bacteriocin sensitivity, then a said gene is liable for innate resistance; whereas, if the mutation in the genes results in bacteriocins resistance then it is said to be related with acquired resistance [98].

8.3.1 INNATE RESISTANCE

Very few studies have compared variety of strains for innate susceptibility to bacteriocins [57, 58]. However, the studies have suggested that innate bacteriocin resistance may develop due to: (a) immunity mimicry, (b) bacteriocin degradation, (c) growth conditions or (d) changes in the target cell envelope.

8.3.1.1 IMMUNITY MIMICRY

This mechanism is specific for bacteriocins. It has been observed that non-bacteriocinogenic strains carry genes, designated as 'orphan immunity genes,' encoding functional analogs to the bacteriocin immunity organizations. The heterologous expression particularly of these genes converses defense against associated bacteriocin [35]. This phenomenon has been observed mainly in the case of lantibiotics and class II bacteriocins [34, 40].

8.3.1.2 BACTERIOCIN DEGRADATION

This mechanism suggests that the certain enzymes produced by mutants or resistant strains may degrade bacteriocins. For example, Nisinase-produced by some nisin-resistant strains of *Paenibacillus polymyxa* and *Bacillus* spp. during sporulation-degrades nisin as it disrupts the C-terminal lanthionine ring [54, 55]. Also, NSR protease, a membrane-associated protein conferring nisin resistance-has been observed in the non-nisin producing strains of *L. lactis*, which degrades the bacteriocin by removing the nisin C-terminal tail [96]. Similarly, membrane-embedded protease (YqeZ), encoded by the σ^{W}-regulated ygeZyqfABoperon, confer resistance against the sub-lancin 168 in *Bacillus* spp. [13].

8.3.1.3 RESISTANCE ASSOCIATED WITH GROWTH CONDITIONS

Several reports have described the link between growth conditions and innate bacteriocin resistance. Jydegaard et al. [56]reported higher tolerance to nisin and pediocin in stationary-phase cells compared to log phase cells of *Listeria monocytogenes.* They also reported high resistance to pediocin PA-1 in thermally-stressed cells (5°C for 60–80 min) and osmotically stressed cells (6.5% NaCl). Jydegaard et al. [56] explained that high salt concentrations affect the electrostatic interactions between bacteriocins and cell surface due to high ion concentration and also, the increased osmolarity of medium results in alterations in cell envelope.

Similarly, there are reports for association of nisin resistance of *L. monocytogenes* with acid stress [6]. The authors proposed that under acidic conditions the glutamate decarboxylase (GAD) system gets activated which perform decarboxylation of glutamate to γ-aminobutyrate and CO_2 and, contributes to the cellular ATP pool. Nisin creates pores in cell envelope that eventually releases ATP followed by cell death, whereas GadD system may reestablish the intracellular ATP pools, thereby contributing to nisin resistance.

8.3.1.4 VARIATIONS IN THE BACTERIAL CELL ENVELOPE

Bacteriocins being cationic in nature interact with the cell envelope having negative charge. It has been observed that bacteriocins resistant mutants employ various means to reduce the negative charge on the cell surface to minimize their interaction with bacteriocins and other cationic antimicrobial peptides (CAMPs). Teichoic acids (TAs) or lipoteichoic acids (LTAs) are major constituents of Gram-positive cell wall, which are highly charged by deprotonated phosphate and extends through peptidoglycan to the surface of the cell [41]. Reports suggest that dlt-operon codes for a protein, which cause esterification of TAs and LTAs backbone with D-alanine. It lowers the net negative charge on cell envelope due to introduction of basic amino group [41, 86]. The contribution of the dlt-operon in conferring innate resistance to bacteriocins and CAMPs has been experimentally proven for *Bacillus cereus* [1], *Clostridium difficile* [76] and *S. aureus* [86].

In fact, the increased expression of the *dlt-DABC* genes was observed in *C. difficile* WT cells when cultivated in the presence of bacteriocins [76]. Moreover, the *dlt-A* gene mutations increased the sensitivity resulted from defective esterification of TAs and LTAs backbone that increased negative charge on cell envelope thereby, increased interaction with bacteriocin. Another functional protein, MprF, is responsible for the lysylphosphatidylglycerols biosynthesis [85].

High concentration of lysylphosphatidylglycerols in cell membrane also contributes to bacteriocins resistance due to overall lowering of cell envelope negative charge. It has also been observed that *mprF* gene mutants exhibit sensitivity to bacteriocins and other CAMPs due to defective incorporation of lysine to membrane phospholipids [85, 93, 99]. Also, *dltA* and *mprF* in *S. aureus* and *L. monocytogenes* are regulated by response regulators GraRS and VirRS, respectively [38, 73]. Further, inactivation of these regulators resulted in increased susceptibility to bacteriocins and other CAMPs in bacteria [38, 60, 99]. Similarly, other factors that contribute towards structure and charge of the cell envelope are summarized in Table 8.2.

TABLE 8.2 Factors Affecting Cell Envelope Structure

Factor	Mechanism of Action	Organism Under Study	References
AnrAB	Regulates ATP transporter and bacteriocin, movement outside the cell; specific for bacitracin and β-lactam antibiotics	*L. monocytogenes*	[20]
braRS (nsaRS or bceRS)/vraDE	Regulates ATP transporter and bacteriocin movement outside the cell: nisin and nukacin ISK-1 specific	*S. aureus*	[52, 61]
graRS/vraFG	Regulates ATP transporter and bacteriocin efflux; confers broad-spectrum resistance against CAMPs	*S. aureus*	[60, 61]
lcrRS/lctFEG	Regulates ATP transporter and bacteriocin efflux	*Streptococcus mutans*	[60]
lmo1967	Mechanism unknown; Specific for bacteriocin gallidermin& nisin, and also to antibiotics such as bacitracin, cefotaxime & cefuroxime	*L. monocytogenes*	[22]
NsrRS	controls the expression of membrane-bound protein NsrX that modifies unidentified factor that blocks the interaction of nisin and lipid II	*Streptococcus mutans*	[60]
sigB gene	Control expression of genes associated with the bacterial cell envelope alterations, mainly those codes for penicillin-binding proteins, efflux pumps, autolysins, and many other proteins; Regulate transporters involved in bacteriocin movement outside the cell.	*L. monocytogenes*	[7]

TABLE 8.2 *(Continued)*

Factor	Mechanism of Action	Organism Under Study	References
Sigma factor: σ^M	Controls expression of gene (ltaSa) encoding for a LTA synthase under stress-conditions	*Bacillus subtilis*	[62]
Sigma factor: σ^X	promotes the phosphatidylethanolamine synthesis; reduces the net negative charge of the cell surface; regulates dlt-operon expression	*Bacillus subtilis*	[14, 62]

8.3.2 ACQUIRED RESISTANCE

Variability in the development of bacteriocin resistance due to spontaneous mutations depends on several factors [9, 45, 75, 79], such as micro-organism and strain tested, the bacteriocin verified, and the employed assay along with the ratio of bacteria/bacteriocin and growth conditions.

8.3.2.1 RESISTANCE TO LANTIBIOTICS

Different mechanisms related to acquire resistance against lantibiotics have been reported by different investigators in different microorganisms. In several reports, the resistance phenotype was linked with alterations in cell wall, synthesis unusual cell wall, and inhibition of autolysin [72], the higher amounts of altered LTAs [66, 74] and increased thickness at the septum [66] or content of rhamnose [104].

Variability in composition of cytoplasmic membrane may a possible factor for bacterial resistance to lantibiotics. Verheul et al. [102] studied a nisin-resistance *L. monocytogenes* Scott A variant, which produces more phosphatidylglycerol and less diphosphatidylglycerol incytoplasmic membrane compared to parental strain. It was concluded that nisin penetration is more in lipid monolayers of diphosphatidylglycerol than those of phosphatidylglycerol [102].

Crandall and Montville [25] reported that lower C15:C17 fatty acids ratio, as well as less phosphatidylglycerol and more phosphatidylethanolamine in cytoplasmic membrane, resulted in nisin resistance in *Listera* strains. They suggested that these variations in composition may result in increased rigidity and less fluidity in the cytoplasmic membrane, thereby, preventing nisin entry into the target. Furthermore, the less interaction between nisin and membrane is due to low phosphatidylglycerol content resulting in-reduced negative charge on lipid bilayer [25].

8.3.2.2 RESISTANCE TO CLASS II BACTERIOCINS

For class II bacteriocins, both spontaneous as well as induced mutations have been observed [58, 89]. Different mechanisms have been suggested for class II bacteriocins resistance and the resistance level varies with the mechanism involved. Several reports have suggested that subclass IIa bacteriocins and lactococcin A resistance is due to lost or reduced expression of their receptors [47, 64, 83, 87, 88]. Elevated resistance to subclass IIabacteriocins among Gram-positive bacteria has been linked to lost or reduced Man-PTS expression [47, 64, 83, 98]. The investigators have studied the mutants with reduced or no expression of *mptA*, *mptC* or *mptD* gene from Man-PTS operon in different target organisms [47, 64, 83, 97]. Furthermore, the introduction of the *La. Lactis* ptnABCD or the *L. monocytogenesm ptC* gene increased sensitivity of *La. Lactis* resistant strains for lactococcin [64] and sub-class IIabacteriocins [88], respectively.

Some investigators have linked subclass IIabacteriocins resistance with regulatory gene rpoN, encoding for different sigma factor σ^{54} (SigL), *in L. monocytogenes* and *Enterococcus faecalis* [28, 91]. SigL is associated with the mptA/BCD operon activation that encodes for EIIA, BC, and D in *L. monocytogenes* and *Enterococcus faecalis* [50], in combination with ManR (the transcriptional activator for σ^{54}) [29]. Mutations in rpoN and/or ManR resulted in downregulation of genes that are encoding for subclass IIabacteriocin receptor, which ultimately led to resistance development in the bacteria [28, 29, 31, 47, 50, 83]. Also, there are reports describing shift in metabolism from homo-fermentation in sensitive strains to mixed/hetero-fermentation in mutants [80, 83, 100].

Resistance to class II bacteriocin has further been linked to the alterations in the cell envelope [2], for instance:

- Bacteriocin efflux by twin-arginine translocases system;
- Cell membranes with higher proportion of unsaturated fatty acids along with short-to long-acyl chain species, thereby raised the membrane fluidity;

- Change in peptidoglycan structure due to higher proportion of tripeptide side-chains than pentapeptide side-chains in peptidoglycan;
- Higher the D-alanine amount in TAs resulted in the lowering of cell wall negative charge; and
- Saturated fatty acid in higher proportion reduces the membrane fluidity.

8.3.2.3 RESISTANCE TO CLASS III BACTERIOCINS

Resistance development for Lysostaphin was studied under *in vitro* as well as *in vivo* conditions [17, 68]. This resistance has shown to be linked with femAB mutations. Such mutations lead to modification of peptidoglycan cross-bridges mainly due to replacement of glycine with serine residues [43, 95]. Several authors revealed that such changes in peptidoglycan generally increase β-lactam susceptibility [17, 48, 63, 92]. Therefore, it has been recommended to use β-lactam antibiotics and lysostaphinin combination for therapeutic trials, to both restrain resistance and to stimulate synergy [67].

8.3.2.4 RESISTANCE TO CYCLIC BACTERIOCINS

Higher tolerance of *L. monocytogenes* mutant to enterocin AS-48 was linked to altered fatty acid composition in cell envelope, such as higher C15:C17 ratio and higher amount of branched-chain fatty acids [77]. Grande-Burgos et al. [44] identified the genes encoding a transcriptional regulator of PadR-type and an unidentified membrane protein that are responsible for enterocin AS-48 resistance in *Bacillus cereus* by unknown mechanism. Also, Gabrielsen et al. [42] linked the garvicin ML resistance in *La. Lactis* IL1403 mutants with the malEFG operon, encoding maltose ABC transporter, implying that maltose transporter acts possibly as a target molecule for bacteriocin.

8.4 STRATEGIES TO OVERCOME BACTERIOCIN RESISTANCE

Different mechanisms may be conferring resistance to bacteriocin in different bacterial strains of a species or maybe sharing similar resistance mechanism for different bacteriocins. The application of bacteriocin in food industry is not affected by this as various strategies can be used to control bacteriocin resistance. Hurdle technology is one such strategy used in food industries that reduces the risk of bacteriocin resistance development [3, 24]. It has been shown that the bacterial mutants resistant to bacteriocins remain equally or more sensitive to sodium chloride, low pH, sodium nitrite or potassium sorbate compared to wild type cells [59]. Therefore, bacteriocin resistance rarely influences the use of bacteriocins as a factor in hurdle technologies to improve food safety.

Another approach is to use bacteriocins with different mode of action in combination. The advantage of using bacteriocins in combination is that lower doses can be used, and their efficacy will be higher than that of either bacteriocin alone [3, 11, 45, 59]. It is advisable to use bacteriocins belonging to different classes in combination as it minimizes the chances of cross resistance development.

Macwana and Muriana [71] classified bacteriocins into functional groups, based on mechanism of action, which suggests the most appropriate combination of bacteriocins that can be used in bio-control strategies. Application of bacteriocins and conventional drugs in combination is another approach to control the development of resistant mutants [19, 23]. Due to their different mode of actions, the development of cross-resistance is minimally expected [19].

8.5 REGULATORY CONSIDERATIONS

Bacteriocins can be used in any of the following way [39]:

- Addition of purified or semi-purified forms;
- Use of bacteriocinogenic cultures as starters;
- Incorporation of products pre-fermented with bacteriocinogenic cultures as an ingredient, or (d) genetic expression of bacteriocin in food-producing organisms.

Different authorities-such as Food and Drug Administration (FDA), U.S. Department of Agriculture (USDA), Food Safety and Inspection Services (FSIS), Environment Protection Agency (EPA)-regulate the use of bacteriocins depending on its intended use and type of food in which it is used.

Bacteriocins serving the purpose of preservation in processed foods falls under the authority of FDA and regulated through Federal Food, Drug, and Cosmetic Act (FFDCA) as food ingredients. Under FFDCA, the ingredients under GRAS status are exempted from premarket approval but those that do not fall under GRAS need to be considered as food additives and require premarket approval from FDA. The GRAS status given to any substance is either on the basis of (a) their safe history in food application or (b) science-based proof related to safety concerns [39].

In the case of purified bacteriocin, or cells producing bacteriocins application in food that are intended to be used as a preservative would fall within the FDA's authority under FFDCA. Whereas, they must be considered for additional assessment under the Federal Meat Inspection Act (FMIA) and the Poultry Products Inspection Act (PPIA) of USDA authority, if must be used in meat and poultry, respectively [39]. And, if used for controlling pests in whole food and vegetables, would be considered as pesticide and assessed under Federal Insecticide, Fungicide, and Rodenticide Act (FIFRA) under jurisdiction of the EPA. Lastly, FDA Center for Veterinary Medicine is currently engaged in developing policies on the regulation of genetic modifications to animals if the bacteriocin needs to be expressed in animals for preventing disease as they may constitute "drugs" under the relevant statutory definitions [39].

FDA Center for Food Safety and Applied Nutrition (CFSAN) has produced several guidance and policy documents that include safety assessment procedures which the producers can use to assess the safety of their products. CFSAN has concluded that biologically derived products for which pathways of digestion and metabolism are understood can often be determined to be safe for consumption. In the case of bacteriocins, information that would be required to carry out a safety assessment would likely include, but not necessarily be limited to the following considerations:

1 Characterization of Substance:

- Chemical name of purified components;
- Common or trade names used;
- Properties of the substance (including biological activity and specificity, and if available, the mechanism of action);
- Protein preparations used in food are typically complex mixtures, therefore, the percentage of the preparation that is the desired component and information on any contaminating constituents will be required;
- Source of the preparation (including documentation of the taxonomic identification of the specific strain of the organism to be used and evidence that it is not toxigenic or pathogenic).

2. Details of Preparation:

    - Manufacturing process should be well described and should include considerations of culture conditions and purification methods, as well as details of procedures used to guarantee the cultural purity of the producing organism.
    - For manufacturing methods which utilize recombinant DNA techniques, additional information may be required, including descriptions of genetic manipulation sand the final form the results of such manipulations take in the production organism. Finally, descriptions of methods used to ensure the absence of toxins that could be produced will be necessary.

3. Dose and Use:

    - Intended uses and use levels;
    - Foods in which the substance is to be used;
    - If the source organism and/or the substance itself is present in foods (e.g., if bacteriocinogenic culture is used as starter) the natural level of the source organism and/or its product in food should be characterized and provided so that relative exposures can be calculated if needed.

4. Effect: The desired technical effect should be stated and documentation of the usefulness of the product should be available. For example, if a bacteriocin is used to control growth of a spoilage or pathogenic organism, its efficacy against that particular organism under typical use conditions should be demonstrated.
5. Ingredient Standardization: Data on multiple batches should be available to ensure that the ingredient can be reproducibly manufactured to meet the proposed specifications and that the lots of ingredient that are tested are representative of the material that will enter the food supply.
6. Control of Function: Evidence of controls adequate to ensure that viable cells or functional genetic material of the production strain will not be present in the final ingredient should be available if such considerations are relevant to safety.
7. Metabolism: Documentation of the metabolic fate of the protein(s), including in vitro digestion studies to determine if the protein(s) is easily digested in the mammalian gastrointestinal tract (GIT).

8. Other Pertinent Studies: If necessary, results of other studies required to demonstrate safety.

8.6 SUMMARY

Bacteriocins have important applications in food industry because of their numerous advantages over synthetic or chemical preservatives. However, the development of bacteriocin resistance may raise an issue, though the resistance does not have significant impact. It is believed that the resistance development makes cells weak and, in that case, wild type sensitive cells are expected to grow at high rate and ultimately outnumber/eliminate mutants/resistant cells in the medium culture or in the food matrix. Therefore, the mutants/bacteriocin-resistant cells will rarely hamper the efficacy of bacteriocins. It is hard to predict the frequency of resistance development against bacteriocin in food matrix and all the existing data is based on laboratory experiments. Therefore, there is a need to study bacteriocin resistance development in food model systems to estimate the frequency and influence on microbial control. Lastly, it is important to continue research on the bacteriocin resistance, because the clear understanding of mechanisms involved can facilitate in improving effectiveness of bacteriocins-based strategies to improve food safety and security.

KEYWORDS

- **bacteriocin resistance**
- **bacteriocins**
- **cationic antimicrobial peptides**
- **food-borne diseases**
- **glutamate decarboxylase**
- **regulatory agencies**

REFERENCES

1. Abi, K. Z., Rejasse, A., Destoumieux-Garzon, D., Escoubas, J. M., Sanchis, V., Lereclus, D., Givaudan, A., et al., (2009). The DLT operon of *Bacillus cereus* is required for

resistance to cationic antimicrobial peptides and for virulence in insects. *Journal of Bacteriology*, *191*, 7063–7073.
2. Bastos, M. C., Coelho, M. L., & Santos, O. C., (2015). Resistance to bacteriocins produced by Gram-positive bacteria. *Microbiology, 161*, 683–700.
3. Bastos, M. C. F., & Ceotto, H., (2011). Bacterial antimicrobial peptides and food preservation. In: *Natural Antimicrobials in Food Safety and Quality* (pp. 62–76). Wallingford: CAB International.
4. Bastos, M. C. F., Ceotto, H., Coelho, M. L. V., & Nascimento, J. S., (2009). Staphylococcal antimicrobial peptides: Relevant properties and potential biotechnological applications. *Current Pharmaceutical Biotechnology*, *10*, 38–61.
5. Bastos, M. C. F., Coutinho, B. G., & Coelho, M. L. V., (2010). Lysostaphin: A staphylococcal bacteriolysin with potential clinical applications. *Pharmaceuticals (Basel), 3*, 1139–1161.
6. Begley, M., Cotter, P. D., Hill, C., & Ross, R. P., (2010). Glutamate decarboxylase-mediated nisin resistance in *Listeria monocytogenes*. *Applied and Environmental Microbiology*, *76*, 6541–6546.
7. Begley, M., Hill, C., & Ross, R. P., (2006). Tolerance of *Listeria monocytogenes* to cell envelope-acting antimicrobial agents is dependent on SigB. *Applied and Environmental Microbiology*, *72*, 2231–2234.
8. Bierbaum, G., & Sahl, H. G., (2009). Lantibiotics: Mode of action, biosynthesis, and bioengineering. *Current Pharmaceutical Biotechnology*, *10*, 2–18.
9. Blake, K. L., Randall, C. P., & O'Neill, A. J., (2011). In vitro studies indicate a high resistance potential for the lantibioticnisin in *Staphylococcus aureus* and define a genetic basis for nisin resistance. *Antimicrobial Agents and Chemotherapy, 55*, 2362–2368.
10. Bondi, M., Messi, P., Halami, P. M., Papadopoulou, C., & De Niederhausern, S., (2014). Emerging microbial concerns in food safety and new control measures. *Bio. Med. Research International*. Article ID 251512.
11. Bouttefroy, A., & Milliere, J., (2000). Nisin-curvaticin 13 combinations for avoiding the regrowth of bacteriocin resistant cells of *Listeria monocytogenes* ATCC 15313. *International Journal of Food Microbiology*, *62*, 65–75.
12. Breukink, E., & De Kruijff, B., (2006). Lipid-II as a target for antibiotics. *Nature Reviews Drug Discovery, 5*, 321–323.
13. Butcher, B. G., & Helmann, J. D., (2006). Identification of *Bacillus subtilis* sW-dependent genes that provide intrinsic resistance to antimicrobial compounds produced by Bacilli. *Molecular Microbiology*, *60*, 765–782.
14. Cao, M., & Helmann, J. D., (2004). The *Bacillus subtilis* extracytoplasmic- function sX factor regulates modification of the cell envelope and resistance to cationic antimicrobial peptides. *Journal of Bacteriology*, *186*, 1136–1146.
15. Chen, H., & Hoover, D., (2003). Bacteriocins and their food applications. *Comprehensive Reviews in Food Science and Food Safety*, *2*, 82–100.
16. Cichello, S. A., (2015). Oxygen absorbers in food preservation: A review. *Journal of Food Science and Technology*, *52*, 1889–1895.
17. Climo, M. W., Ehlert, K., & Archer, G. L., (2001). Mechanism and suppression of lysostaphin resistance in oxacillin-resistant *Staphylococcus aureus*. *Antimicrobial Agents and Chemotherapy*, *45*, 1431–1437.
18. Coelho, M. L. V., Coutinho, B. G., Santos, O. C. S., Nes, I. F., & Bastos, M. C. F., (2014). Immunity to the *Staphylococcus aureus* leaderless four-peptide bacteriocinaureocin A70

is conferred by AurI, an integral membrane protein. *Research in Microbiology, 165*, 50–59.

19. Coelho, M. L. V., Nascimento, J. S., Fagundes, P. C., Madureira, D. J., Oliveira, S. S., Brito, M. A. V. P., & Bastos, M. C. F., (2007). Activity of staphylococcal bacteriocins against *Staphylococcus aureus* and *Streptococcus agalactiae* involved in bovine mastitis. *Research in Microbiology, 158*, 625–630.
20. Collins, B., Curtis, N., Cotter, P. D., Hill, C., & Ross, R. P., (2010). The ABC transporter AnrAB contributes to the innate resistance of Listeria monocytogenes to nisin, bacitracin, and various beta-lactam antibiotics. *Antimicrobial Agents and Chemotherapy, 54*, 4416–4423.
21. Collins, B., Guinane, C. M., Cotter, P. D., Hill, C., & Ross, R. P., (2012). Assessing the contributions of the LiaS histidine kinase to the innate resistance of *Listeria monocytogenes* to nisin, cephalosporins, and disinfectants. *Applied and Environmental Microbiology, 78*, 2923–2929.
22. Collins, B., Joyce, S., Hill, C., Cotter, P. D., & Ross, R. P., (2010). TelA contributes to the innate resistance of *Listeria monocytogenes* to nisin and other cell wall-acting antibiotics. *Antimicrobial Agents and Chemotherapy, 54*, 4658–4663.
23. Cotter, P. D., Ross, R. P., & Hill, C., (2013). Bacteriocins: A viable alternative to antibiotics? *Nature Reviews Microbiology, 11*, 95–105.
24. Cotter, P. D., Hill, C., & Ross, R. P., (2005). Bacteriocins: Developing innate immunity for food. *Nature Reviews Microbiology, 3*, 777–788.
25. Crandall, A. D., & Montville, T. J., (1998). Nisin resistance in *Listeria monocytogenes* ATCC 700302 is a complex phenotype. *Applied and Environmental Microbiology, 64*, 231–237.
26. Cursino, L., Smatda, J., Charton-Souza, E., & Nascimento, A. M. A., (2002). Recent updates aspects of colicins of *Entero-bacteriaceae. Brazilian Journal of Microbiology, 33*, 185–195.
27. Da Silva, S. S., Vitolo, M., González, J. M. D., & De Souza, O. R. P., (2014). Overview of *Lactobacillus plantarum* as a promising bacteriocin producer among lactic acid bacteria. *Food Research International, 64*, 527–536.
28. Dalet, K., Briand, C., Cenatiempo, Y., & Hechard, Y., (2000). The rpoN gene of *Enterococcus faecalis* directs sensitivity to subclass IIa bacteriocins. *Current Microbiology, 41*, 441–443.
29. Dalet, K., Cenatiempo, Y., Cossart, P., & Hechard, Y., (2001). European listeria genome consortium: A s54-dependent PTS permease of the mannose family is responsible for sensitivity of *Listeria monocytogenes* to mesentericin Y105. *Microbiology, 147*, 3263–3269.
30. Di Renzo, L., Colica, C., Carraro, A., Cenci, G. B., Marsella, L. T., Botta, R., Colombo, M. L., et al., (2015). Food safety and nutritional quality for the prevention of non-communicable diseases: The nutrient, hazard analysis, and critical control point process (NACCP). *Journal of Translational Medicine, 13*, 128.
31. Diep, D. B., Skaugen, M., Salehian, Z., Holo, H., & Nes, I. F., (2007). Common mechanisms of target cell recognition and immunity for class II bacteriocins. *Proceedings of the National Academy of Sciences of the United States of America, 104*, 2384–2389.
32. Dimov, S., Ivanova, P., & Harizanova, N., (2005). Genetics of bacteriocins biosynthesis by lactic acid bacteria. *Biotechnology and Biotechnological Equipment, 19*(2), 4–10.

33. Dousset, X., Jaffre, E., & Zagorec, M., (2016). Spoilage: Bacterial spoilage. In: Caballero, B., Finglas, P. M., & Toldra, F., (eds.), *Encyclopedia of Food and Health* (Vol. 5, pp. 106–112). Academic Press, Oxford.
34. Draper, L. A., Grainger, K., Deegan, L. H., Cotter, P. D., Hill, C., & Ross, R. P., (2009). Cross-immunity and immune mimicry as mechanisms of resistance to the lantibiotic lacticin 3147. *Molecular Microbiology*, *71*, 1043–1054.
35. Draper, L. A., Tagg, J. R., Hill, C., Cotter, P. D., & Ross, R. P., (2012). The spiFEG locus in *Streptococcus infantarius* subsp. infantarius BAA- 102 confers protection against nisin. *Antimicrobial Agents and Chemotherapy*, *56*, 573–578.
36. Dubois, J. Y. F., Kouwen, T. R. H. M., Schurich, A. K., Reis, C. R., Ensing, H. T., Trip, E. N., Zweers, J. C., & Van, D. J. M., (2009). Immunity to the bacteriocinsublancin 168 is determined by the SunI (YolF) protein of *Bacillus subtilis*. *Antimicrobial Agents and Chemotherapy*, *53*, 651–661.
37. Ehlert, K., Tschierske, M., Mori, C., Schroder, W., & Berger, B. B., (2000). Site-specific serine incorporation by Lif and Epr into positions 3 and 5 of the Staphylococcal peptidoglycan interpeptide bridge. *Journal of Bacteriology*, *182*, 2635–2638.
38. Falord, M., Mader, U., Hiron, A., Debarbouille, M., & Msadek, T., (2011). Investigation of the Staphylococcus aureus GraSR regulon reveals novel links to virulence, stress response, and cell wall signal transduction pathways. *PLoS One*, *6*, p. 6 e-article ID21323.
39. Fields, F. O., (1996). Use of bacteriocins in food: Regulatory considerations. *Journal of Food Protection*, *59*(13), 72–77.
40. Fimland, G., Eijsink, V. G., & Nissen-Meyer, J., (2002). Comparative studies of immunity proteins of pediocin-like bacteriocins. *Microbiology*, *148*, 3661–3670.
41. Fischer, W., (1988). Physiology of lipoteichoic acids in bacteria. *Advances in Microbial Physiology*, *29*, 233–302.
42. Gabrielsen, C., Brede, D. A., Hernandez, P. E., Nes, I. F., & Diep, D. B., (2012). The maltose ABC transporter in *Lactococcus lactis* facilitates high-level sensitivity to the circular bacteriocingarvicin ML. *Antimicrobial Agents and Chemotherapy*, *56*, 2908–2915.
43. Gargis, S. R., Heath, H. E., LeBlanc, P. A., Dekker, L., Simmonds, R. S., & Sloan, G. L., (2010). Inhibition of the activity of both domains of lysostaphin through peptidoglycan modification by the lysostaphin immunity protein. *Applied and Environmental Microbiology*, *76*, 6944–6946.
44. Grande, B. M. J., Kovacs, A. T., Mironczuk, A. M., Abriouel, H., Galvez, A., & Kuipers, O. P., (2009). Response of *Bacillus cereus* ATCC 14579 to challenges with sub lethal concentrations of enterocin AS-48. *BMC Microbiology*, *9*, 227–234.
45. Gravesen, A., Jydegaard-Axelsen, A. M., Mendes, D. S. J., Hansen, T. B., & Knøchel, S., (2002). Frequency of bacteriocin resistance development and associated fitness costs in *Listeria monocytogenes*. *Appl. Environ. Microbiol.*, *68*, 756–764.
46. Gravesen, A., Kallipolitis, B., Holmstrøm, K., Høiby, P. E., Ramnath, M., & Knøchel, S., (2004). The pbp2229-mediated nisin resistance mechanism in *Listeria monocytogenes* confers cross-protection to class IIa bacteriocins and affects virulence gene expression. *Applied and Environmental Microbiology*, *70*, 1669–1679.
47. Gravesen, A., Ramnath, M., Rechinger, K. B., Andersen, N., Jansch, L., Hechard, Y., Hastings, J. W., & Knøchel, S., (2002). High-level resistance to class IIa bacteriocins is

associated with one general mechanism in *Listeria monocytogenes*. *Microbiology*, *148*, 2361–2369.

48. Guignard, B., Entenza, J. M., & Moreillon, P., (2005). Beta-lactams against methicillin-resistant *Staphylococcus aureus*. *Current Opinion in Pharmacology*, *5*, 479–489.
49. Hammami, R., Zouhir, A., LeLay, C., Ben, H. J., & Fliss, I., (2010). BACTIBASE second release: A database and tool platform for bacteriocin characterization. *BMC Microbiology*, *10*, 22–28.
50. Hechard, Y., Pelletier, C., Cenatiempo, Y., & Frere, J., (2001). Analysis of s54-dependent genes in *Enterococcus faecalis*: A mannose PTS permease (EIIMan) is involved in sensitivity to a bacteriocin, mesentericin Y105. *Microbiology*, *147*, 1575–1580.
51. Heng, N. C. K., Wescombe, P. A., Burton, J. P., Jack, R. W., & Tagg, J. R., (2007). The diversity of bacteriocins in gram positive bacteria. In: Riley, M. A., & Chavan, M. A., (eds.), *Bacteriocins: Ecology and Evolution* (pp. 45–92). Springer, New York.
52. Hiron, A., Falord, M., Valle, J., Debarbouille, M., & Msadek, T., (2011). Bacitracin and nisin resistance in *Staphylococcus aureus*: A novel pathway involving the BraS/BraR two-component system (SA2417/SA2418) and both the BraD/BraE and VraD/VraE ABC transporters. *Molecular Microbiology*, *81*, 602–622.
53. Hoffmann, A., Schneider, T., Pag, U., & Sahl, H. G., (2004). Localization and functional analysis of PepI, the immunity peptide of Pep5-producing *Staphylococcus epidermidis* strain 5. *Applied and Environmental Microbiology, 70*, 3263–3271.
54. Jarvis, B., (1967). Resistance to nisin and production of nisin inactivating enzymes by several Bacillus species. *Journal of General Microbiology*, *747*, 33–48.
55. Jarvis, B., & Farr, J., (1971) Partial purification, specificity and mechanism of action of the nisin-inactivating enzyme from Bacillus cereus. *Biochemicaet. Biophysica. Acta*, *1227*, 232–240.
56. Jydegaard, A. M., Gravesen, A., & Knøchel, S., (2000). Growth condition-related response of *Listeria monocytogenes* 412 to bacteriocin inactivation. *Letters in Applied Microbiology*, *31*, 68–72.
57. Katla, T., Møretrø, T., Sveen, I., Aasen, I. M., Axelsson, L., Rørvik, L. M., & Naterstad, K., (2002). Inhibition of *Listeria monocytogenes* in chicken cold cuts by addition of sakacin P and sakacin P-producing lactobacillus sakei. *Journal of Applied Microbiology*, *93*, 191–196.
58. Katla, T., Naterstad, K., Vancanneyt, M., Swings, J., & Axelsson, L., (2003). Differences in susceptibility of *Listeria monocytogenes* strains to sakacin P, sakacin A, pediocin PA-1, and nisin. *Applied and Environmental Microbiology*, *69*, 4431–4437.
59. Kaur, G., Singh, T. P., & Malik, R. K., (2013). Antibacterial efficacy of nisin, pediocin 34 and enterocin FH99 against *Listeria monocytogenes* and cross resistance of its bacteriocin resistant variants to common food preservatives. *Brazilian Journal of Microbiology*, *44*, 63–71.
60. Kawada-Matsuo, M., Oogai, Y., Zendo, T., Nagao, J., Shibata, Y., Yamashita, Y., Ogura, Y., et al., (2013b). Involvement of the novel two-component NsrRS and LcrRS systems in distinct resistance pathways against nisin A and nukacin ISK-1 in *Streptococcus mutans*. *Applied and Environmental Microbiology*, *79*, 4751–4755.
61. Kawada-Matsuo, M., Yoshida, Y., Zendo, T., Nagao, J., Oogai, Y., Nakamura, Y., Sonomoto, K., et al., (2013). Three distinct two-component systems are involved in resistance to the class I bacteriocins, nukacin ISK-1 and nisin A, in *Staphylococcus aureus*. *PLoS One*, *8*, 5, e-article ID 69455.

62. Kingston, A. W., Liao, X., & Helmann, J. D., (2013). Contributions of the sW, sM and sX regulons to the lantibioticresistance of *Bacillus subtilis*. *Molecular Microbiology*, *90*, 502–518.
63. Kiri, N., Archer, G., & Climo, M. W., (2002). Combinations of lysostaphin with b-lactams are synergistic against oxacillin-resistant *Staphylococcus epidermidis*. *Antimicrobial Agents and Chemotherapy*, *46*, 2017–2020.
64. Kjos, M., Nes, I. F., & Diep, D. B., (2011). Mechanisms of resistance tobacteriocins targeting the mannose phosphotransferase system. *Applied and Environmental Microbiology*, *77*, 3335–3342.
65. Kjos, M., Snipen, L., Salehian, Z., Nes, I. F., & Diep, D. B., (2010). The Abi proteins and their involvement in bacteriocin self-immunity. *Journal of Bacteriology*, *192*, 2068–2076.
66. Kramer, N. E., Hasper, H. E., Van, D. B. P. T. C., Morath, S., De Kruijff, B., Hartung, T., Smid, E. J., et al., (2008). Increased D-alanylation of lipoteichoic acid and a thickened septum are main determinants in the nisin resistance mechanism of *Lactococcus lactis*. *Microbiology*, *154*, 1755–1762.
67. Kusuma, C. M., Jadanova, A., Chanturiya, T., & Kokai-Kun, J. F., (2007). Lysostaphin-resistant variants of *Staphylococcus aureus* demonstrate reduced fitness *in vitro* and *in vivo*. *Antimicrobial Agents and Chemotherapy*, *51*, 475–482.
68. Kusuma, C. M., & Kokai-Kun, J. F., (2005). Comparison of four methods for determining lysostaphin susceptibility of various strains of Staphylococcus aureus. *Antimicrobial Agents and Chemotherapy*, *49*, 3256–3263.
69. Leroy, F., & De Vuyst, L., (2004). Lactic acid bacteria as functional starter cultures for the food fermentation industry. *Trends in Food Science and Technology*, *15*, 67–78.
70. Machaidze, G., & Seelig, J., (2003). Specific binding of cinnamycin (Ro 09–0198) to phosphatidylethanolamine: Comparison between micellar and membrane environments. *Biochemistry*, *42*, 12570–12576.
71. Macwana, S., & Muriana, P. M., (2012). Spontaneous bacteriocin resistance in *Listeria monocytogenes* as a susceptibility screen for identifying different mechanisms of resistance and modes of action by bacteriocins of lactic acid bacteria. *Journal of Microbiological Methods*, *88*, 7–13.
72. Maisnier-Patin, S., & Richard, J., (1996). Cell wall changes in nisinresistant variants *of Listeria innocua* grown in the presence of high nisin concentrations. *FEMS Microbiology Letters*, *140*, 29–35.
73. Mandin, P., Fsihi, H., Dussurget, O., Vergassola, M., Milohanic, E., Toledo-Arana, A., Lasa, I., Johansson, J., & Cossart, P., (2005). VirR: A response regulator critical for *Listeria monocytogenes* virulence. *Molecular Microbiology*, *57*, 1367–1380.
74. Mantovani, H. C., & Russell, J. B., (2001). Nisin resistance of *Streptococcus bovis*. *Applied and Environmental Microbiology*, 67, 808–813.
75. Mazzotta, A. S., Crandall, A. D., & Montville, T. J., (1997). Nisin resistance in *Clostridium botulinum* spores and vegetative cells. *Applied and Environmental Microbiology*, *63*, 2654–2659.
76. McBride, S. M., & Sonenshein, A. L., (2011a). The DLT operon confers resistance to cationic antimicrobial peptides in *Clostridium difficile*. *Microbiology*, *157*, 1457–1465.
77. Mendoza, F., Maqueda, M., Galvez, A., Martinez-Bueno, M., & Valdivia, E., (1999). Anti-listerial activity of peptide AS-48 and study of changes induced in the cell envelope properties of an AS-48-adapted strain of *Listeria monocytogenes*. *Applied and Environmental Microbiology*, *65*, 618–625.

78. Metlitskaya, A., Kazakov, T., Kommer, A., Pavlova, O., Praetorius, M., & Ibba, M., (2006). Aspartyl-tRNA synthetase is the target of peptide nucleotide antibiotic Microcin C. *Journal of Biological Chemistry*, *281*, 18033–18042.
79. Ming, X., & Daeschel, M. A., (1993). Nisin resistance of food borne bacteria and the specific resistance responses of *Listeria monocytogenes* scott A. *Journal of Food Protection*, *56*, 944–948.
80. Naghmouchi, K., Kheadr, E., Lacroix, C., & Fliss, I., (2007). Class I/ Class IIa bacteriocin cross-resistance phenomenon in *Listeria monocytogenes*. *Food Microbiology*, *24*, 718–727.
81. Nascimento, J. S., Coelho, M. L. V., Ceotto, H., Potter, A., Fleming, L. R., Salehian, Z., Nes, I. F., & Bastos, M. C. F., (2012). Genes involved in immunity to and secretion of aureocin A53, an atypical class II bacteriocin produced by *Staphylococcus aureus* A53. *Journal of Bacteriology*, *194*, 875–883.
82. Nissen-Meyer, J., Rogne, P., Oppegard, C., Haugen, H. S., & Kristiansen, P. E., (2009). Structure-function relationships of the non-lanthionine-containing peptide (class II) bacteriocins produced by gram-positive bacteria. *Current Pharmaceutical Biotechnology*, *10*, 19–37.
83. Opsata, M., Nes, I. F., & Holo, H., (2010). Class IIabacteriocin resistance in *Enterococcus faecalis* V583: The mannose PTS operon mediates global transcriptional responses. *BMC Microbiology*, *10*, 224–227.
84. Parks, W. M., Bottrill, A. R., Pierrat, O. A., Durrant, M. C., & Maxwell, A., (2007). The action of the bacterial toxin, microcin B17, on DNA gyrase. *Biochem.*, *89*, 500–507.
85. Peschel, A., Jack, R. W., Otto, M., Collins, L. V., Staubitz, P., Nicholson, G., Kalbacher, H., et al., (2001). A. *Staphylococcus aureus* resistance to human defensins and evasion of neutrophil killing via the novel virulence factor MprF is based on modification of membrane lipids with L-lysine. *Journal of Experimental Medicine*, *193*, 1067–1076.
86. Peschel, A., Otto, M., Jack, R. W., Kalbacher, H., Jung, G., & Gotz, F., (1999). Inactivation of the dlt operon in *Staphylococcus aureus* confers sensitivity to defensins, protegrins, and other antimicrobial peptides. *Journal of Biological Chemistry*, *274*, 8405–8410.
87. Ramnath, M., Arous, S., Gravesen, A., Hastings, J. W., & Hechard, Y., (2004). Expression of mptC of *Listeria monocytogenes* induces sensitivity to class IIa bacteriocins in *Lactococcuslactis*. *Microbiology*, *150*, 2663–2668.
88. Ramnath, M., Beukes, M., Tamura, K., & Hastings, J. W., (2000). Absence of a putative mannose-specific phosphotransferase system enzyme IIAB component in a leucocin A-resistant strain of *Listeria monocytogenes*, as shown by two-dimensional sodium dodecyl sulfate polyacrylamide gel electrophoresis. *Applied and Environmental Microbiology*, *66*, 3098–3101.
89. Rasch, M., & Knøchel, S., (1998). Variations in tolerance of *Listeria monocytogenes* to nisin: Pediocin PA-1 and bavaricin A. *Letters in Applied Microbiology*, *27*, 275–278.
90. Riley, M. A., & Wertz, J. E., (2002). Bacteriocins: Evolution, ecology, and application. *Annual Review of Microbiology, 56*, 117–137.
91. Robichon, D., Gouin, E., Debarbouille, M., Cossart, P., Cenatiempo, Y., & Hechard, Y., (1997). The rpoN (s54) gene from *Listeria monocytogenes* is involved in resistance to mesentericin Y105, anantibacterial peptide from *Leuconostocmes enteroides*. *Journal of Bacteriology*, *179*, 7591–7594.

92. Rohrer, S., & Berger-Bachi, B., (2003). FemABX peptidyl transferases: A link between branched-chain cell wall peptide formation and blactam resistance in gram-positive cocci. *Antimicrobial Agents and Chemotherapy*, *47*, 837–846.
93. Samant, S., Hsu, F. F., Neyfakh, A. A., & Lee, H., (2009). The Bacillus anthracis protein MprF is required for synthesis of lysylphosphatidylglycerols and for resistance to cationic antimicrobial peptides. *Journal of Bacteriology*, *191*, 1311–1319.
94. Silva, C. C. G., Silva, S. P. M., & Ribeiro, S., (2018). Application of bacteriocins and protective cultures in dairy food preservation. *Frontiers in Microbiology*, *9*, 294–297.
95. Stranden, A. M., Ehlert, K., Labischinski, H., & Berger-Bachi, B., (1997). Cell-wall monoglycine cross-bridges and methicillin hypersusceptibility in a femAB null mutant of methicillin-resistant *Staphylococcus aureus*. *Journal of Bacteriology*, *179*, 9–16.
96. Sun, Z., Zhong, J., Liang, X., Liu, J., Chen, X., & Huan, L., (2009). Novel mechanism for nisin resistance via proteolytic degradation of nisin by the nisin resistance protein NSR. *Antimicrobial Agents and Chemotherapy*, *53*, 1964–1973.
97. Tessema, G. T., Møretrø, T., Kohler, A., Axelsson, L., & Naterstad, K., (2009). Complex phenotypic and genotypic responses of *Listeria monocytogenes* strains exposed to the class IIa bacteriocinsakacin P. *Applied and Environmental Microbiology*, *75*, 6973–6980.
98. Tessema, G. T., Møretrø, T., Snipen, L., Axelsson, L., & Naterstad, K., (2011). Global transcriptional analysis of spontaneous sakacinresistant mutant strains of *Listeria monocytogenes* during growth on different sugars. *PLoS One*, *6*(4), e-article ID 1619.
99. Thedieck, K., Hain, T., Mohamed, W., Tindall, B. J., Nimtz, M., Chakraborty, T., Wehland, J., & Jansch, L., (2006). The MprF protein is required for lysinylation of phospholipids in listerial membranes and confers resistance to cationic antimicrobial peptides (CAMPs) on *Listeria monocytogenes*. *Molecular Microbiology*, *62*, 1325–1339.
100. Vadyvaloo, V., Arous, S., Gravesen, A., Hechard, Y., Chauhan-Haubrock, R., Hastings, J. W., & Rautenbach, M., (2004). Cell-surface alterations in class IIa bacteriocin-resistant *Listeria monocytogenes* strains. *Microbiology*, *150*, 3025–3033.
101. vanBelkum, M. J., Martin-Visscher, L. A., & Vederas, J. C., (2011). Structure and genetics of circular bacteriocins. *Trends in Microbiology*, *19*, 411–418.
102. Verheul, A., Russell, N. J., Vanthof, R., Rombouts, F. M., & Abee, T., (1997). Modifications of membrane phospholipid composition in nisin-resistant *Listeria monocytogenes* Scott A. *Applied and Environmental Microbiology*, *63*, 3451–3457.
103. Vincent, P. A., & Moreno, R. D., (2009). The structure and biological aspects of peptide antibiotic microcin J25. *Current Medicinal Chemistry*, *16*, 538–549.
104. Xuanyuan, Z., Wu, Z., Li, R., Jiang, D., Su, J., Xu, H., Bai, Y., Zhang, X., Saris, P. E. J., & Qiao, M., (2010). Loss of IrpT function in *Lactococcus lactis* subsp. lactis N8 results in increased nisin resistance. *Current Microbiology*, *61*, 329–334.

Part III

Potential of Novel Technologies for Food Preservation

CHAPTER 9

POTENTIAL OF NONTHERMAL PLASMA TECHNOLOGY IN FOOD PRESERVATION

SUJIT DAS and SUBROTA HATI

ABSTRACT

Preserving food against food-borne pathogens/spoilage microorganisms is of primary concern for food industries and related agencies across the globe. Non-thermal plasma (NTP) is an emerging technology for food processing and preservation and has the potential to battle decontamination of water, food packaging, and disinfecting food surfaces and related equipment's by disrupting the microbial contaminations and spores present on the surface of liquid/solid foods. It is often considered as the fourth state after liquid, solid, and gas. For preserving the organoleptic qualities and improvising the food production sustainability, NTP is considered as an eco-friendly green novel technology with variations, such as pulsed electric fields (PEF), dielectric barrier discharges (DBDs), corona plasma discharges, atmospheric pressure plasma (APP) jet, high hydrostatic pressure (HHP); and high voltage arc discharge followed by cold plasma. This novel technology is related to: lower thermal loads, no formation of toxic by-products, and no follow-up treatment requirements, which are the major elements for extension of the storage period of freshly processed foods. This chapter highlights on the application of various NTP technologies for microbial inactivation, food preservation, and related future perspectives.

9.1 INTRODUCTION

Since the prehistoric period, food processing techniques were prevalently used by the hunter-gather human population. During those earlier times, the requirement for food preservation was not very intense because the consumption of fresh food was common. Gradually in due course of time, the food preservation became a necessity. Few of the house-hold poplar processing techniques were sun-drying under the sun, fermentation, grinding of cereals, and oven baking. However, the earliest food preservation technology was thermal food processing, during modern times. These food processing and preserving methods provided few specific desirable changes, such as coagulation of proteins, swelling-up of starch, textural softening of food, and releasing of aromatic elements. However, few undesirable changes like loss of vitamins, reduction of concentration of minerals, freshness, taste, and flavor were witnessed relating to which the consumers are aware of few major elements followed by hunting of fresh foods with suitable organoleptic properties. Therefore, the food scientists across the globe began to prepare for new methodologies to assist in well-maintained protocol of food preservation without any loss of major nutrients from the food [23].

The recent advanced technologies for food biopreservation are categorized as in the shadows of non-thermal processing of foods, which are termed as minimum process technologies and these processing methods possess the efficacy of preserving of food products without any substantial heat and simultaneously retrieving their nutritional impacts/features and organoleptic attributes in the meantime thereby also contributing by increasing storage-life of respective food items by inhibition of food-borne microorganisms with pathogenic characteristics. Hence, the non-thermal technology provides food products that taste fresh and are more nutritious in nature, without any employment of warm or synthetic chemical compounds. Improvised novel non-thermal methodologies have laid an impact and gained the attention of many food manufacture ring companies and research institutes, in search of new food processing methods [13].

Plasma is basically a quasi-neutral electrified gas with a chemically reactive medium that consists of various species as depicted in Figure 9.1 and various states of molecules that also possess more than 99% matter in the universe (Figure 9.1). The name was first coined by a chemist named Irving Langmuir in 1923, who witnessed distinctive oscillations in ionizing gas that are based on mass and density; and collectively these oscillations consisting of various charged elements were termed as "plasma oscillations." It is often termed as the fourth state, followed by acquainting states of solid, liquid, and gas altogether [24]. By connecting energy and gaseous modes together via various technologies (mechanical, thermal, chemical or nuclear methods or application of high currents) or by injection of electromagnetic waves are followed by combining amongst themselves to break down the complex gas molecules into simpler gas molecules and related species at a lower atmospheric pressure [17].

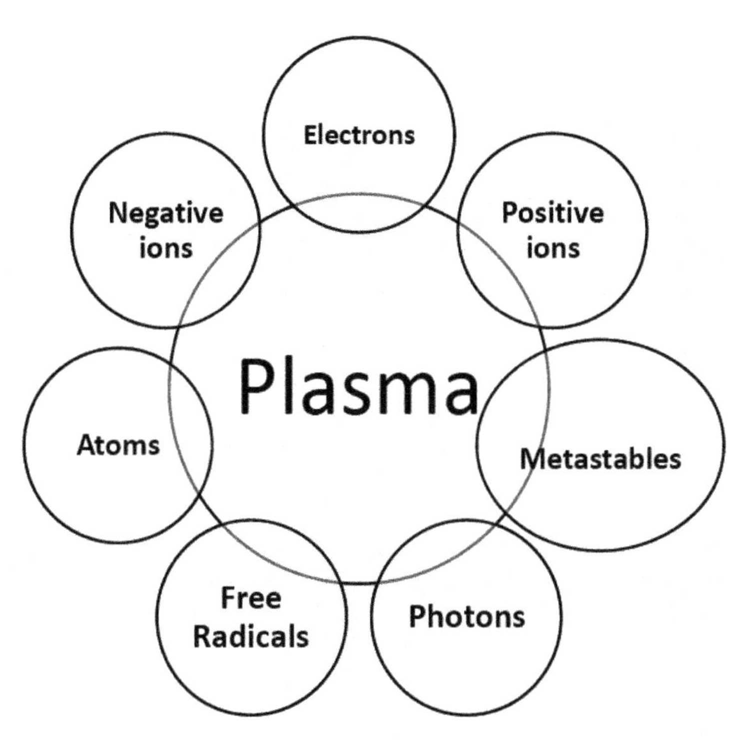

FIGURE 9.1 Various species that constitute the formation of plasma.

For developing modified starch followed by substituting its physicochemical features as well, the non-thermal plasma (NTP) process is considered as a conventional process that is modern in nature. Generally, this technique is employed for the disruption of microorganisms present on the surface of various food substrates, such as fruits, meat products, cheese, dairy products, etc. However, it is also applied for altering the rate of seed germination, endogenous enzymes, modifications of starch, and few limiting factors due to its potentiality at the food sector and packaging as novel technology. As an alternative to general techniques, it is an economical compatible method that is employed in food preservation and several related applications. In near future ahead, one of the major objective linked along with NTP technology that promises high impact on inactivation of microbes while maintenance of organoleptic parameters that ensure proper texture and quality at the same time [63].

The principle of NTP technology consists of discharge of electricity into air or liquid to release uniform mix of energetic atoms, radicals that are highly reactive, ozone, etc., which disrupt microbial cells. Electrical energy liberates NTP species despite heat [10]; and therefore, this technology is efficient in energy and results in minimal damages due to heat induction applied to foods. Impactful parameters consist of pasteurization of foods mainly liquid in nature; disinfection of equipment's employed in processing, floors of the specific plant, wastewater treatment, materials used for packing and controlling of pollutions, etc. Processing of plasma ionization is a primary parameter along with reaction rate, the mean of the free path engaged, and the distribution of energies of electrons [24]. The highlighted features for processing of plasma as an important tool for inactivating microorganisms have been globally considered as: (a) reduced thermal denaturation of nutritional content organoleptic characteristics; (b) energy reduction required for adequate processing/packaging; and (c) efficacy for treating of foods contained inside a flexible pack of film.

The chapter explores modes of action of non-thermal processes and broadly attempts to highlight on the primary methods of NTP technologies for long-term food and nutrition storage and preservation.

9.2 TYPES OF NON-THERMAL PLASMA (NTP)

9.2.1 ATMOSPHERIC PRESSURE PLASMA (APP)

Low-pressure glow discharge plasmas were considered one of the booming research interests in microelectronic industries, but their limitation lies in the vacuum equipment. As alternative atmospheric pressure plasma (APP) was considered as a major tool in NTP technology to improve the safety and preservation of foods. Power sources for generating APP include pulsed current, radiofrequency, alternating, and direct current (DC) [28]. Tools developed under the principle of APP for generating NTP are [45, 55]:

- One atmospheric uniform glow discharge;
- Micro hollow cathode discharges;
- Discharge for gliding arc;
- Dielectric plasma needle barrier discharge (DBDs);
- APP, which is usually employed for lightening, surface modification, etching, and deposition and several other industrial applications.

9.2.2 CORONA DISCHARGES (CDS)

Corona generates near edges or points of atmospheric pressure in which the electric fields are larger. These types are weakly luminous in nature that may undergo ignition with relatively maximum voltage, which generally consumes area nearby each electrode specifically [15]. The major features of corona discharges (CDs) make them unique and stable to be applied as chemical reactors that are non-equilibrium. The stable nature with easier modes of operation includes a wider concentration of gases with pressures that includes atmosphere pressure, specific ionizing areas, where hot electrons that are hot in nature gets uniformly mixed with the cooler gas thereby proliferating reactions with the absence of any backward reactions thereby extending their portions further with lower field drift that functions as electrolytes (gaseous) further incorporating electrochemical reactions on the surface a rear of food products [46].

9.2.3 DIELECTRIC BARRIER DISCHARGES (DBDS)

These kinds of NTP s are released in coupling with an insulating element fit and placed amidst the electrodes that are responsible for a self-dependent function. Dielectric barrier discharges (DBDs) are generally categorized under non-thermal atmospheric pressure gas discharges. Primarily, these discharges were employed for ozone liberation and today these are considered as important device not only in present on-thermal plasma technology but also as agate way for fundamental studies. Another interest within this area is partial discharges that are fitted in maximum voltage set up that may be explored further in the near future. The complete disruption methods and the structure formation including the action of surface processes are presently undergoing further investigation. These discharges with cold plasma are a better strategy for the food decontamination as a substitution method for those products that are unable to sanitize via conventional procedures. The inactivation of pathogenic microbes via cold plasma treatment can be correlated along with various synergistically modes, such as generating ultraviolet (UV) rays, ozone, elements that are charged, and radicals of oxygen with addition to various other reactive species [47].

9.2.4 RADIO FREQUENCIES (RFS)

Radio frequencies (RFs) are generated, when the specific gaseous element is placed in an electromagnetic field in a continuous oscillation manner. The inductive discharge field is created with the help of a coil (inducing in nature) completely masking the specific reactor or via creating a capacitive discharge field by employing different electrodes that are placed uniformly on the outer surface area of the reactor [29].

9.3 PROCESSING METHODS IN NON-THERMAL PLASMA (NTP) TECHNOLOGY

9.3.1 PULSED ELECTRIC FIELD (PEF)

Recently, the pulsed electric field (PEF) technology has an immense impact in the study for purifying fluidic foods and to define transferring objective sat larger scale in the food industry. PEF involves the principle of NTP that employs an electric field of high-voltage, which is applied at a very low temperature on a short-term basis for the food preservation and preserving the parameters, such as taste, flavor, and various nutritional components and packaging.

This process mainly targets to disrupt microbes thereby resulting in little or no alteration in the features of shelf life increment [69]. Here, a higher capacity of electrical forces is achieved by energy storage into the bank of capacitors from the power source of DC that further releases pulses of higher voltages. The fluidic foods are set amidst the electrodes that possess higher electrical capacitance in the range of 22–82 kV/cm that are utilized in form of pulses (short) resulting in the drop of microorganism (vegetative) in food gradually [34]. The widely prevalent applications under PEF are: rectangle pulsating model and logarithmic model compared to the pulse beat models [5].

The handling time applied in this process is analyzed via unifying the pulse number with the actual pulsating period of actual pulses. In this process, the following two mechanisms have been documented to wipe out microorganisms with specific electric fields:

1. Electrical Break on Cells: Here, the PEF is generated from the strength of an electric field that largely exceeds the critical value that might further induce a potential (transmembrane) higher to 1 V (volt) for vegetative cells, thus leading to urgent discharging and decomposing of the membrane [26].
2. Electroporation: This very technique claims to aim at the rise of the absorptivity of the membrane since there is compression and poration [48]. Due to the result of the following aspects of the electroporation process, microbial inactivation occurs:

   - The operating parameters might result in the intense nature of the electrical field, frequencies engaged, and temperature time or during the treatment time.
   - Based on features of food preservation, the electrical field (pulsed in nature) is very often considered for disruption of vegetative cells

at a higher rate with the greater electric field of intense nature that is often employed in combination with maximum pulse time [48].

9.3.2 ULTRASONICATION

Ultrasonication is a substituting technology compared to thermal processing techniques. It is based on the principle of food pasteurization followed by food preservation along with the disruption of microbial enzymes at lower temperatures [23]. The ultrasound food processing industry employs the following two types of variations:

- Ultrasound with lower intense nature (≤ 1 W cm^{-2}); and
- Ultrasound with higher intense nature (Range 10–1000 W cm^{-2}).

Ultra-sonication method states "vibrating each second along with sound waves of ≥20,000 to conduct development of energies" [62]. Tools prepared for generating ultrasonic waves are specifically at a specific frequency ranging from 21 kHz to 11 MHz [31]. The most major parameter of ultra-sonication that gives a cutting edge towards the processing of plasma is the distortion in temperature for processing. It must be low in concentration, when providing unwanted and pathogens for enzyme activation, which are engaged for contamination purposes. Although, the flavor, aroma, texture, and nutritional content of the products are obtained after undergoing better preservation and pure-products that are quite close to the state properties [14, 39, 62].

9.3.3 IONIZING RADIATION

The irradiation of food products is a process technology mainly used in food preservation industry for increasing the storage-life and for improving, the microbial safety engaged in the food and the mode of action of this process is carried out without any induction of any specific chemical pesticide by showcasing the specific food to be exposed to ionized radiation. Ionizing radiation has the potential in freezing the electrons, thereby isolating from their respective bonds (atomic) without encountering the food particles. This methodology is based on the disruption of pathogenic/spoilage microbes thereby contributing in the minimal risk of causing any sort of foodborne diseases. The mechanism involved in this process also possesses potential for causing delay, elimination of sprouting and ripening of food products. The sensory as well as the nutritional parameters remains constant since heat treatment is not involved in this mode [26]. Radioactive substances release various rays on the environment during the atomic fragmentation in a continuous manner and this ionizing radiation is the root cause for the organization of ions that are electrically charged in their own strike material, respectively. For this to happen, Gamma, and X-rays along with beams of electron (accelerated) are majorly employed for food preservation [20, 27].

9.3.4 HIGH HYDROSTATIC PRESSURE (HPP)

The high hydrostatic pressure (HPP) is a methodology under non-thermal technology for preserving mainly foods falling under hard as well as fluidic category, where packed or unpacked sample is projected to a pressure ranging from 150 to 2000 MPa [30]. The major mechanism for processing of this pressure rests on the theory of application of compressive fluid force to the surrounding fluid of the product surface [6]. Presently, this method has scored for itself as source of broad application in the food preservation industry [25]. Examples of applications of this method are: inactivating the spoilage microorganisms, denaturation of proteins and enzyme activation [32], the formation of gel [37], taste-aroma, and color, and protection of sensory quality parameters [35] for increasing the extraction yield.

This process includes immersion of the food packages in the liquid followed by releasing out the pressure uniformly throughout the food that the pressure is transferred uniformly on entire and all area of the food. Although HPP technique is thought to be of non-thermal manner but when pressure is imparted to food products, the temperature rises due to adiabatic heating, which depends on density of water/food elements. In this process, there is a rise in temperature@ 3°C per 100 MPa and this can go beyond only if the specific food sample consists of fat. The temperature turns back due to adiabatic cooling after the depressurization is achieved [26]. The following three basic theories involved in high pressure non-thermal technology are:

- Principle of Le Chatelier: Any sort of specific reaction, which deals in volume reduction, is expanded. However, reactions which maximize the particular volume are undergoing repression.

- Isostatic Theory: A constant pressure gets projected on food products from each and every angle and the mechanism causes food compression; and the specific food product bounce back to the original morphology after the pressure dispersion.
- Theory of Microscopic Ordering: If an optimum temperature is fixed and the pressure is maximized, then the degree of molecules is ordered that is engaged in specific food product thereby rises to a great extent [9].

9.3.5 PLASMA STERILIZATION

Various physicists consider gas plasma as a matter of the fourth state and it is generally elaborated as a gas made up of ions and free electrons [33]. There exist few particles that are non-charged (atoms, molecules, free radicals) in plasma, which is generated when a specific gas is left out on a direct mode of current amidst two electrodes, and these particles are usually applied for sterilization purpose. Electrons and ions are released during generation of plasma, which targets the cell walls of the microbes resulting in their metabolism inactivation thereby accomplishing sterilization. Previous results indicate that it results with impactful aspects in microbial basis in various food products. [36, 54].

Generally, plasma is divided into thermal and non-thermal components. The thermal plasma demands for higher pressure and optimum temperature along with turbid quantity of electrons. The NTP is released at atmospheric conditions that range from 32°C–62°C acquainted by very minimal energy. Cold atmospheric plasma is considered as a novel parameter in food processing and preservation technology for disruption of airborne microbes; eliminating chemical compounds, those are organic in nature or bacterial disruption along with decontamination of food surface, instrument packaging, and working surfaces [63].

9.4 APPLICATIONS OF NON-THERMAL PLASMA (NTP) TECHNOLOGY IN FOOD PRESERVATION AND PACKAGING INDUSTRY

Plasma consists of numerous applications in the food preservation industry (Figure 9.2) with various efficacies and challenges that are briefly elaborated in this section.

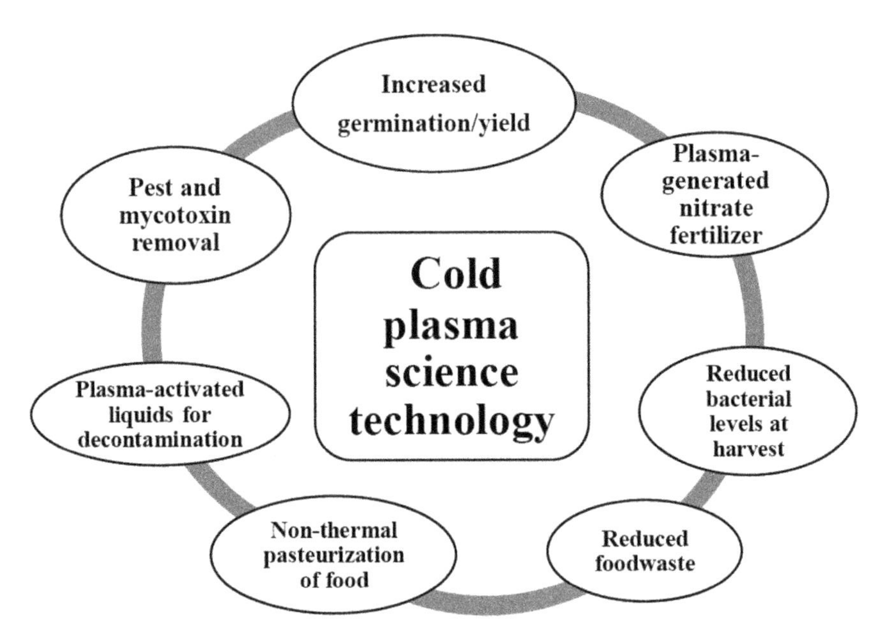

FIGURE 9.2 Potential applications of cold plasma for increasing food safety and quality at the post-harvesting stages.

9.4.1 NON-THERMAL PLASMA (NTP) APPLICATION FOR MICROBIAL DISRUPTION

Application of plasma for the proper sterilizing method was first patented in 1968 [38] and plasma procured from oxygen was first put into use in 1989 and just after that advanced research studies have been achieved on the mode of mechanisms of inactivating microbes via agents of plasma. Interaction with agents of plasma along with matter (biological) results in a fatal action in which hooking up with plasma may effectively disrupt a wider variation of microbial flora consisting of spores and viruses [60]. Application of plasma on various microorganisms can be completely specific in nature leading to disrupt the pathogenic/spoilage microorganisms without harming the host. It can also target and activate different pathways in different organisms to carry out their objectives [19].

The effect of treating microbial cells with plasma is primarily because of the ions associated with the ions of plasma and few cellular interactions. The species that are considered as reactive in plasma have been broadly linked up with straight oxidation effects taking place on the exterior surface area of the cells of microbes. The various effects of plasma rely upon water's presence, the maximum effect was noticed in a moist organism when subjected to difference with lower in the dried organism [19]. The major principle of plasma for inactivating purpose completely implies that species of plasma of reactive in nature cause disruption of the deoxyribonucleic acid (DNA) in the chromosomes.

With the use of radiobiology, Wiseman, and Halliwell [67] interpreted that mode of action of plasma on a specific cell is via the development of reactive oxygen species (ROS) that includes directly into the vicinity of a DNA molecule inside the cell nucleus. Generally, applying of NTP involves formation of malondialdehyde (MDA) in cells of microbes that participate in the DNA adducts development resulting in the cell damage [19]. Species that are reactive in nature react along with water resulting with liberation of OH* ions [70], which are highly reactive and toxic in nature to the cells. It must be mentioned that OH* (hydroxyl) radical is considered important since such radicals are generated in the layer of hydration rotating around the surface of DNA molecule that are held responsible for damage of 90% DNA. Hydroxyl radicals may then mix up with the nearest organic molecules resulting in chain oxidation thus leading to complete degradation of not only DNA molecules but also cellular membranes and related components [19].

During applying of cold plasma, microbes are projected for maximum bombarding via free radicals causing induction of lesions present in the surface so that the cell (living) is unable to maintain a steady rate and this entire process is regarded as "etching" (Figure 9.3).

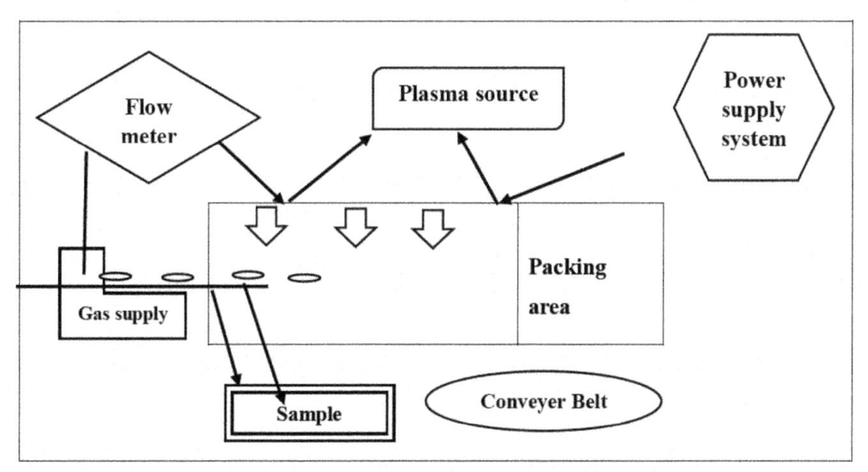

FIGURE 9.3 Schematic representation of continuous plasma system for decontamination of food products.

Plasma etching lies on the interacting with ions that are relatively energetic in nature and activated species with related molecular conformations belonging to the substrate. Hence, the overall aggregation of particular charges delivers electrostatic pressure towards the external cell surface of the membranes leading to a technique known as electro-permeabilization that possesses the efficacy of rupturing cells justifying the mechanism that occurs in PEFs [41, 64]. While implicating with plasma treatment-where plasma undergoes initialization, catalyzation, or helps to resist a response that is complexly biological-a membrane structure (that is compromised and alteration in proteins (membrane bound) and enzymes as well resulting with complex cellular responses that might impact many specific cells) gives direction to channel others also [19, 70].

Plasma possesses both efficacy of inactivation of both cells (vegetative) and dormant endospores of bacteria. The major technical objective consists of representing various food products at a specific manner to the plasma field for liberation of disrupted microbes. The derivatives of lower lipid and fat content like herbs (dried) and exotic spices, condiments, and related other horticultural-based products in which exposure of UV and radicals will produce minimal action on oxidation or other chemical changes that stands chances of offering the best opportunities to implicate lower temperature plasma on foods [66].

9.4.2 MODIFICATION OF PROPERTIES OF FOOD PRODUCTS

Novel research in food technology is the implication of NTP to modify particular properties of food varieties in which advanced and suitable bio-functional parameters are incorporated and improved further. Thirumdas et al. [63] reviewed the treatment of plasma in starch to explore its functional parameters. The changes in the characteristics are because of depolymerizing and cross-linking of mesh formed due to branching of side chains of amylose and amylopectin. NTP application was reported for causing reduction in weight (molecular), viscous nature, and temperature for gelatinizing. Etching of plasma maximizes energy on the surface and enriches the hydrophilic properties of granulated starch [63]. The treatment with flour may possibly possess maximum positive impact on the overall morphological features of resulting dough that are formed.

Misra et al. [42] stated the rheological characteristics of the flour wheat treatment indicated modification in strengthening of dough at the optimal time of mixing and NTP was also reported to incorporate alterations in the protein's secondary structure present in the flour. They interpreted that non-thermal atmospheric plasma could be possibly further explored as suitable means for modulating biofunctional attributes of wheat flour. Yepez and Keener [68] incorporated a discharge of cold plasma contained within an atmosphere of hydrogen gas for partial hydrogenation of soy oil with complete absence of fatty acid (trans) formation and they interpreted that on-thermal plasma is an advanced process technology that has potential to substitute the primary traditional hydrogenation (catalytic) process technology.

Misra et al. [43] reported efficacy of plasma for improvising the hydrophobic nature on the outer surface of baked biscuits thereby inducing rapid vegetable oil spreads and this method predominates the retention time of oil's functional properties with minimal oil or saturated fat requirement due to the rise in the spread leading a way for healthy and nutritional protectants. It is hypothesized that the functionalization of features of the food applying NTP may possibly possess various technical/regulatory boundary barriers than that related to the disinfection of food products.

9.4.3 MICROPLASMA ARRAYS

Microplasmas are NTP s that are designed in such a manner so that they can fit into chambers of smaller type of dimension generally varying from 10–1000 μm that deals with discharges at stable conditions at pressure (atmospheric) and could function in air (open) and hence can be considered as potential element for biological processes. The design of microplasmas permits the scaling up efficiently since they could be mended in both designs of geometrical and that of packaging density as well linked with its potentiality to function at atmospheric pressure with low voltages [7]. Nowadays, devices related to microdischarges function at atmospheric pressure are attaining popularity rapidly because of decreasing significant cost of production and processes as analyzed against their counterparts of lower pressure [7]; and these specific designs have the potential to assist in scaling up larger volume of food treatments via treating a homogeneous larger area and could be employed efficiently for processing conditions at a continuous rate. Still the morphology of micro-plasma arrays comes up with specific fluctuations when placed against the discharges liberated by large and highly advanced morphological characteristics because of the ratio of higher surface area and the entire volume of micro-plasma array systems. Boettner et al. [11] enquired an array of microplasmas possessing a high amount (50 × 50) of singular microplasma discharge and represented on the suitability of specific designs for providing a wide variety of implications in the food preservation industry.

9.4.4 IMPACT OF NON-THERMAL PLASMA (NTP) ON ENZYMES

The use of NTP has also impacted the browning of enzymes [19]. Mostly due to enzymatic browning, fruits, and vegetables suffer from spoilage due to the mode of action of few endogenous enzymes specifically, polyphenoloxidase, and peroxidase that is taken as loss (secondary) while handling and storage after harvesting since these enzymes oxidize phenolic components at the expense of hydrogen peroxide leading to off-flavors [58].

Browning of enzymes can be prevented by employing heat, blanch, and commercial sterilizing processes [22]. Thorough reaction via oxidation catalyzed by free radicals along with atomic oxygen, enzymes get deactivated [56]. The various other non-conventional methods considered for inactivating enzyme (mainly endogenous) reapplication of PEFs, irradiation rays, and high-pressure processing [61]. The most emerging trend presently for decontaminating of microbes of food and biocompounds is to use on-thermal atmospheric pressure technology. Pankaj et al. [53] mentioned the implication of NTP technology (in-package) as an advanced mode for inactivating peroxidase enzyme present in tomatoes.

Tappi et al. [59] reported use of plasma gas on fresh apples employing a DBDs at three different time periods of 10, 20 and 30 min and a significant reduction was noticed in treated samples up to about 65% for 30 min time interval after storage for 4 h compared to the control (ones in terms of areas undergoing browning); and it also reduced the tissue's metabolism and other features were slightly affected due to the treatment thereby interpreting that in-packed cold plasma technique is capable enough for stabilization of freshly cut fruits.

NTP is also considered as efficient for controlling of insects and a study was conducted for analyzing the impact of atmospheric-pressure plasma jet (APPJ) on an insect, *Plodia* spp. and few operating parameters were as follows: (a) diameter along the nozzle pipeline of APPJ (10, 12, 14 cm); (b) rate number APPJ pulses induced (2, 4, 11, 14 or 21 pulses). This research study reported that larval and pupal mortality were increased significantly whereas adult emergence suffered a reduction with increasing rate of APPJ pulses and reduction of distance away from the nozzle. Moreover, alterations to enzymes (catalase (CAT), glutathione S-transferase (GST), peroxidase) with antioxidative activities in the tissues of final larva stage were analyzed after a period of 24 h post-application by pulse number of 16 with 12 cm of distance; and no significance in peroxidase activity but significance in the maximization of concentration of lipid peroxide. The percentage of glutathione was dropped along with the occurrence of protein in larvae that was treated and compared against the control sample [1].

9.4.5 EFFECT OF NON-THERMAL PLASMA (NTP) IN PACKAGING

Materials used for packaging of food should be held responsible for protecting food products from outer surroundings while handling, transporting, and distributing from source to destination. Cold plasma is employed for decontaminating materials used for external packing where rates of shadow effect is almost null since plasma move sand covers up all surface area [4]. Plasma processing is broadly meant for altering or implementing modification of surfaces (cleaning, coating, printing, painting, and adhesive bonding) in case of packing materials [50]. Plasma sterilization with low-temperature gas promises rapid and safer sterilization of packaging elements without causing any harm or altering characteristics of the specific material.

Cold plasma can also be employed for sterilization of packaging materials (such as polythene, ethylene, and/or polycarbonate), which are considered heat sensitive due to lower temperature and the surface area of polymers meant for packaging films that are edible in nature and must show more hydrophobicity with minimal energies on the surfaces [65]. Crucial parameters, surface modification of food surface, and sterilization of packaging of food material have been successfully implicated by Plasma Discharge technology today [57].

Recently, NTP technology design has shown promise for proper sanitation of surfaces for food processing, e.g., conveyor belts thoroughly moving the cycles of intermittent disinfection. Similarly, the technology can also be employed for disinfecting containers of food before the filling with the product. Given the point that plasma is broadly applied for modification of food surface area and packaging; and in other industries the transferring to the food phase for such requirements is correlated and such surface areas are basically smoother to allow rapid treatment, efficacy of sanitation and validation of process. Recently, active packaging has been prepared by coating up surface of antimicrobial elements into packing of polymers by use of a plasma discharge processing and it has resulted in reducing the viable microbial cell count loads for beef products with significant increase in the storage life of the product [16].

In the case of meat processing and packaging, modified gas atmosphere may easily be employed inside the packaging for increasing the quality of meat retention and lowering the load of spoilage microbes [40, 43].

Wang et al. [66] employed on-thermal plasma process to the meat products and they reported that the storage life of chicken was extended up to 14 days because the cold plasma atmosphere with slight modifications (66% O_2 + 31% CO_2 + 6% N_2) prevented growth of microbes (>4 log10) than that present in the air. Further, insignificant alterations were observed in the color of chicken treated with NTP. Henceforth in-packaging employing NTP, technology can be further improved for continuous decontamination of food on a commercial scale in a seal-packed container amongst the electrodes of a DBD on a conveyer belt with continuous motion [44, 49].

9.4.6 IMPACT OF PLASMA TECHNOLOGY ON PHENOLIC AND ANTIOXIDANT COMPOUNDS

Antioxidants inhibit the toxic dosages of ROS (namely, radical of singlet oxygen, superoxide radical or peroxyl radicals, etc.) [3]. Although, there exists only limited data regarding the impact of applying plasma on phenolics, yet few species of plant were tolerable in nature to UV irradiations and accumulation of metabolites of flavonoids in cells (epidermal) [12]. During the liberation of plasma, the UV irradiations that were developed might be considered for the accumulation of phenolics that are obtained from upper leaf epidermis's cells. The application of cold plasma method along with pasteurization for treating of Morello cherry (sour) juice revealed a high amount of phenolic acid and anthocyanin factor 21].

9.5 FUTURE PERSPECTIVES

Food processing industries aim to satisfy the demands of consumers with a better quality of foods with high shelf-life for increasing economic standards, net profits, and these industries for the past few years have been constantly striving for consumption due to reasons mentioned in Figure 9.4. This can only be accomplished by employing non-conventional technologies (such as thermal preservation techniques requiring high levels of water with additional cost for managing wastewater) [2]. There exist various prototypes of standards and regulations based on hygiene. The demand of advanced food processing instruments and specific tool for analyzing new technologies based on NTP is the need of the hour [8].

In near future, research-based on novel and conventional techniques is subjected to two facts: Improvement of the non-thermal technologies for proving a fruitful environmental impact; and comparison of non-thermal against conventional techniques (for example, the final product quality, shelf life, the entire cost of investment) [18]. One of a promising factor of this non-thermal technology can be justified by a hypothesis via connecting this with related methods engaged in decontaminating processes as photocatalysis and this kind of bridging are considered as promising since the intake of energy is very less with a higher degree of decontamination [28].

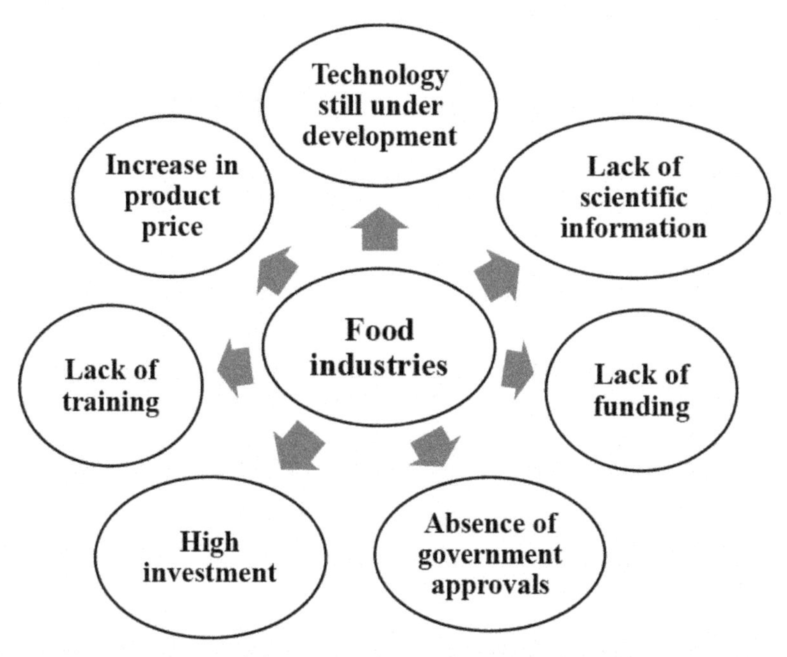

FIGURE 9.4 Parameters limiting food industries from employing non-thermal food technologies.

Today, consumers are not only focused about the nutrition content, a source with potential health benefits, and food safety but also on the detailed protocols and various technologies that are applied to the food processing and production chain simultaneously [17]. Irrespective of any of the selective processed technology for decontamination implicated, food products might be exposed to pathogens and turn sensitive just before or during the end of their respective sensory quality and storage period, when the temperature is fluctuated and packaged improperly.

Hence, this has become a primary important factor to safeguard the packing techniques along with the supply chain in such a way that the growth of spoilage microbes halts after treating with NTP. Therefore, the material used in packaging is considered a major parameter of the decontaminating phase for inactivation of microorganisms after being projected for discharging of plasma discharge and this factor represents itself to be a platform and a need from the demand of hygienic packaging materials of food-grade nature that has the potential of withstanding the treatment without disturbing the motion of packaging elements and concluding to pinholes. The movement beyond the regulatory ranges is considered or taken into account when employing packaging based on biopolymers (say, films of polylactic acid) [50]. The opportunity arises due to the possible nature of incorporating natural antimicrobial compounds like bacteriocins or essential oils (EOs) into the packaging so that plasma exposing to plasma will conclude with etching and final generation of proactive elements [51, 52].

9.6 SUMMARY

The emerging technology based on NTP has the potential of increasing the storage life and preservation of the food by disrupting the cell organelles of the spoilage microorganisms along with biodegradation of pesticides. Considering, the studies and reported interpretations in this chapter by employing NTP for further exposure and circulation, it is hence concluded that the efficient irradiation of plasma is enhanced marginally by the confined environmental set up by the elimination of continuous airflow. This system can be applied broadly for storing agricultural-based products aseptically and preventing of harmful materials on the products. Not enough research reports have been recorded on the efficient degradation of pesticide compounds along with the inactivation of microorganisms by NTP processes in India.

Therefore, there is an urgent need for standardizing non-thermal methods and technologies that are promising enough for the generation of effective NTP with potential for the degrading residues of pesticide, disruption of microbes residing on the surface of fruits, vegetables, and spices. The transformation of plasma technology from laboratory conditions to the food preservation premier industries is surrounded by questionable challenging objectives with reasonable opportunities that need to accomplish the health and hygiene of the consumers across the globe. Novel designs and technologies for delivery of specific attempted species and treating/modifying of road areas of food types by this technology are raising rapidly that generally includes: discharges for atmospheric air, plasma infused activated water, sprays, and several packaging technologies.

KEYWORDS

- **atmospheric pressure plasma**
- **Corona discharges**
- **food preservation**
- **food-borne pathogens**
- **microbial decontamination**
- **non-thermal plasma**

REFERENCES

1. Abd, E. A., Mahmoud, M. F., & Elaragi, E. A., (2014). Non-thermal plasma for control of the Indian meal moth, *Plodia interpunctella* (Lepidoptera: Pyralidae). *Journal of Stored Products Research*, *59*, 215–221.
2. Aarnisalo, K., Tallavaara, K., Wirtanen, G., Maijala, R., & Raaska, L., (2006). The hygienic working practices of maintenance personnel and equipment hygiene in the Finnish food industry. *Food Control*, *17*(12), 1001–1011.
3. Andrade, C. T., Simao, R. A., Thire, R. M., & Achete, C. A., (2005). Surface modification of maize starch films by low-pressure glow 1-butene plasma. *Carbohydrate Polymers*, *61*(4), 407–413.
4. Banu, M. S., Sasikala, P., Dhanapal, A., Kavitha, V., Yazhini, G., & Rajamani, L., (2012). Cold plasma as a novel food processing technology. *International Journal of Emerging Trends in Engineering and Development*, *4*(2), 803–818.
5. Barbosa-Canovas, G. V., Pothakamury, U. R., Gongora-Nieto, M. M., & Swanson, B. G., (1999). Fundamentals of high-intensity pulsed electric fields (PEF). In: *Preservation of Foods with Pulsed Electric Fields* (pp. 1–19). New York: Elsevier.
6. Baptista, I., Rocha, S. M., Cunha, A., Saraiva, J. A., & Almeida, A., (2016). Inactivation of *Staphylococcus aureus* by high pressure processing: An overview. *Innovative Food Science and Emerging Technologies*, *36*, 128–149.
7. Becker, K. H., Schoenbach, K. H., & Eden, J. G., (2006). Micro plasmas and applications. *Journal of Physics D: Applied Physics*, *39*(3), R55–R70.
8. Betta, G., Barbanti, D., & Massini, R., (2011). Food Hygiene in aseptic processing and packaging system: A survey in the Italian food industry. *Trends in Food Science and Technology*, *22*(6), 327–334.
9. Blany, C., & Masson, P., (1993). Effects of high pressure on proteins. *Food Reviews International, 9*(4), 611–628.
10. Borneff-Lipp, M., Okpara, J., Bodendorf, M., & Sonntag, H. G., (1997). Validation of low-temperature-plasma (LPT) sterilization systems: Comparison of two technical versions, the Sterrad 100, 1.8 and the 100S. *Hygiene und Mikrobiologie*, *3*, 21–28.
11. Boettner, H., Waskoenig, J., O'Connell, D., Kim, T. L., Tchertchian, P. A., Winter, J., & Schulz-von, D. G. V., (2010). Excitation dynamics of micro-structured atmospheric pressure plasma arrays. *Journal of Physics D: Applied Physics*, *43*(12), 124–128.

12. Brandenburg, R., (2017). Dielectric barrier discharges: Progress on plasma sources and on the understanding of regimes and single filaments. *Plasma Sources Science and Technology*, *26*(5), 5, e-article ID 053001.
13. Brennan, J. G., & Grandison, A. S., (2011). *Food Processing Handbook* (2nd edn, Vol. I & II, p. 826). Weinheim: Germany: John Wiley VCH Verlag.
14. Chandrapala, J., Oliver, C., Kentish, S., & Ashokkumar, M., (2012). Ultrasonics in food processing-food quality assurance and food safety. *Trends in Food Science and Technology*, *26*(2), 88–98.
15. Chang, J. S., Lawless, P. A., & Yamamoto, T., (1991). Corona discharge processes. *IEEE Transactions on Plasma Science*, *19*(6), 1152–1166.
16. Clarke, D., Tyuftin, A. A., Cruz-Romero, M. C., Bolton, D., Fanning, S., Pankaj, S. K., & Kerry, J. P., (2017). Surface attachment of active antimicrobial coatings onto conventional plastic-based laminates and performance assessment of these materials on the storage life of vacuum packaged beef sub-primals. *Food Microbiology*, *62*, 196–201.
17. Cullen, P. J., Lalor, J., Scally, L., Boehm, D., Milosavljević, V., Bourke, P., & Keener, K., (2018). Translation of plasma technology from the lab to the food industry. *Plasma Processes and Polymers*, *15*(2), 1700085.
18. Djekic, I., Sanjuán, N., Clemente, G., Jambrak, A. R., Djukić-Vuković, A., Brodnjak, U. V., & Tonda, A., (2018). Review on environmental models in the food chain-current status and future perspectives. *Journal of Cleaner Production*, *176*, 1012–1025.
19. Dobrynin, D., Fridman, G., Friedman, G., & Fridman, A., (2009). Physical and biological mechanisms of direct plasma interaction with living tissue. *New Journal of Physics*, *11*(11), 115020.
20. Ehlermann, D. A., (2016). The early history of food irradiation. *Radiation Physics and Chemistry*, *129*, 10–12.
21. Elez, G. I., Režek, J. A., Milošević, S., Dragović-Uzelac, V., Zorić, Z., & Herceg, Z., (2015). The effect of gas phase plasma treatment on the anthocyanin and phenolic acid content of sour cherry Marasca (*Prunus cerasus* var. Marasca) juice. *LWT-Food Science and Technology*, *62*(1), 894–900.
22. Elez-Martínez, P., Aguiló-Aguayo, I., & Martín-Belloso, O., (2006). Inactivation of orange juice peroxidase by high-intensity pulsed electric fields as influenced by process parameters. *Journal of the Science of Food and Agriculture*, *86*(1), 71–81.
23. Fellows, P., (2009). *Food Processing Technologies: Principles and Practices* (3rd edn.). CRC Press, Boca Raton, FL.
24. Fridman, A., (2008). *Plasma Chemistry* (p. 887). Cambridge University Press, New York, USA.
25. Gao, G., Ren, P., Cao, X., Yan, B., Liao, X., Sun, Z., & Wang, Y., (2016). Comparing quality changes of cupped strawberry treated by high hydrostatic pressure and thermal processing during storage. *Food and Bioproducts Processing*, *100*, 221–229.
26. Griffiths, M., & Walking-Riberio, M., (2014). Pulsed electric field processing of liquid food and beverages. *Emerging Technologies of Food Processing*, 115–145.
27. Harder, M. N. C., Arthur, V., & Arthur, P. B., (2016). Irradiation of foods: Processing technology and effects on nutrients: Effect of ionizing radiation on food components. *Encyclopedia of Food and Health*, 476–481.
28. Hati, S., Mandal, S., Vij, S., Minz, P. S., Basu, S., Khetra, Y., Yadav, D., & Dahiya, M., (2012). Nonthermal plasma technology and its potential applications against food borne microorganisms. *Journal of Food Processing and Preservation, 36*(6). 518–524.

29. Hati, S., Das, S., & Mandal, S., (2019). Technological advancement of functional fermented dairy beverages. *Engineering Tools in the Beverage Industry*, pp. 101–136.
30. Hygreeva, D., & Pandey, M. C., (2016). Novel approaches in improving the quality and safety aspects of processed meat products through high pressure processing technology: A review. *Trends in Food Science and Technology*, *54*, 175–185.
31. Jabbar, S., Abid, M., Hu, B., Wu, T., Hashim, M. M., Lei, S., & Zeng, X., (2014). Quality of carrot juice as influenced by blanching and sonication treatments. *LWT-Food Science and Technology*, *55*(1), 16–21.
32. Khan, M. A., Ali, S., Abid, M., Cao, J., Jabbar, S., Tume, R. K., & Zhou, G., (2014). Improved duck meat quality by application of high pressure and heat: A study of water mobility and compartmentalization, protein denaturation and textural properties. *Food Research International*, *62*, 926–933.
33. Kim, H. J., Jayasena, D. D., Yong, H. I., & Jo, C., (2016). Quality of cold plasma treated foods of animal origin: Chapter 11. In: Misra, N. N., Schlüter, O., & Cullen, P. J., (eds.), *Cold Plasma in Food and Agriculture* (pp. 273–291). Apple Academic Press Inc., New Jersey-USA.
34. Knoor, D., Angerbach, A., Eshtiaghi, M., Heinz, V., & Lee, D. U., (2001). Processing concepts based on high intensity electric field pulses. *Trends in Food Science and Technology, 12*, 129–135.
35. Lavilla, M., Orcajo, J., Díaz-Perales, A., & Gamboa, P., (2016). Examining the effect of high pressure processing on the allergenic potential of the major allergen in peach. *Innovative Food Science and Emerging Technologies*, *38*, 334–341.
36. Lee, H., Kim, J. E., Chung, M. S., & Min, S. C., (2015). Cold plasma treatment for the microbiological safety of cabbage, lettuce, and dried figs. *Food Microbiology*, *51*, 74–80.
37. Leite, T. S., De Jesus, A. L. T., Schmiele, M., Tribst, A. A., & Cristianini, M., (2017). High pressure processing of pea starch: Effect on the gelatinization properties. *LWT-Food Science and Technology*, *76*, 361–369.
38. Manas, P., & Pagán, R., (2005). Microbial inactivation by new technologies of food preservation. *Journal of Applied Microbiology*, *98*(6), 1387–1399.
39. McClements, D. J., (1995). Advances in the application of ultrasound in food analysis and processing. *Trends in Food Science and Technology*, *6*(9), 293–299.
40. McMillin, K. W., (2008). Where is MAP going?-A review and future potential of modified atmosphere packaging for meat. *Meat Science*, *80*(1), 43–65.
41. Mendis, D. A., Rosenberg, M., & Azam, F., (2000). A note on the possible electrostatic disruption of bacteria. *IEEE Transactions on Plasma Science*, *28*(4), 1304–1306.
42. Misra, N. N., Schlüter, O., & Cullen, P. J., (2016). Plasma in food and agriculture: Chapter 1, In: Misra, N. N., Schlüter, O., & Cullen, P. J., (eds.), *Cold Plasma in Food and Agriculture: Fundamentals and Applications* (pp 1–13). Academic Press, Elsevier, Amsterdam, Netherlands.
43. Misra, N. N., Sullivan, C., Pankaj, S. K., Alvarez-Jubete, L., & Cullen, P. J., (2014). Enhancement of oil spreadability of biscuit surface by nonthermal barrier discharge plasma. *Innovative Food Science and Emerging Technologies*, *26*, 456–461.
44. Misra, N. N., Ziuzina, D., Cullen, P. J., & Keener, K. M., (2013). Characterization of a novel atmospheric air cold plasma system for treatment of packaged biomaterials. *Transactions of the ASABE, 56*, 1011–1016.

45. Nehra, V., Kumar, A., & Dwivedi, H., (2008). Atmospheric nonthermal plasma sources. *International Journal of Engineering*, *2*(1), 53–57.
46. Goldman, M., Goldman, A., & Sigmond, R. S., (1985). The corona discharge, its properties, and specific uses. *Pure and Applied Chemistry*, *57*(9), 1353–1362.
47. Niemira, B. A., (2012). Cold plasma decontamination of foods. *Annual Review of Food Science and Technology*, *3*, 125–142.
48. Novickij, V., Grainys, A., Lastauskienė, E., Kananavičiūtė, R., Pamedytytė, D., Kalėdienė, L., & Miklavčič, D., (2016). Pulsed electromagnetic field assisted in vitro electroporation: A pilot study. *Scientific Reports*, *6*(4), e-article ID 33537.
49. Pankaj, S. K., Bueno-Ferrer, C., Misra, N. N., Milosavljević, V., O'donnell, C. P., Bourke, P., & Cullen, P. J., (2014). Applications of cold plasma technology in food packaging. *Trends in Food Science and Technology*, *35*(1), 5–17.
50. Pankaj, S. K., Bueno-Ferrer, C., Misra, N. N., O'Neill, L., Jiménez, A., Bourke, P., & Cullen, P. J., (2014). Characterization of polylactic acid films for food packaging as affected by dielectric barrier discharge atmospheric plasma. *Innovative Food Science and Emerging Technologies, 21*, 107–113.
51. Pankaj, S. K., Bueno-Ferrer, C., Misra, N. N., O'Neill, L., Jiménez, A., Bourke, P., & Cullen, P. J., (2014). Surface, thermal, and antimicrobial release properties of plasma-treated Zeinfilms. *Journal of Renewable Materials, 2*, 77–84.
52. Pankaj, S. K., Bueno-Ferrer, C., Misra, N. N., O'Neill, L., Bourke, P., & Cullen, P. J., (2017). Effects of cold plasma on surface, thermal, and antimicrobial release properties of chitosan film. *Journal of Renewable Materials, 5*, 14–20.
53. Pankaj, S. K., Misra, N. N., & Cullen, P. J., (2013). Kinetics of tomato peroxidase inactivation by atmospheric pressure cold plasma based on dielectric barrier discharge. *Innovative Food Science and Emerging Technologies, 19*, 153–157.
54. Pasquali, F., Stratakos, A. C., Koidis, A., Berardinelli, A., Cevoli, C., Ragni, L., & Trevisani, M., (2016). Atmospheric cold plasma process for vegetable leaf decontamination: A feasibility study on radicchio (red chicory, *Cichorium intybus* L.). *Food Control*, *60*, 552–559.
55. Ragni, L., Berardinelli, A., Vannini, L., Montanari, C., Sirri, F., Guerzoni, M. E., & Guarnieri, A., (2010). Non-thermal atmospheric gas plasma device for surface decontamination of shell eggs. *Journal of Food Engineering*, *100*(1), 125–132.
56. Rastogi, N. K., (2003). Application of high-intensity pulsed electrical fields in food processing. *Food Reviews International*, *19*(3), 229–251.
57. Scholtz, V., Pazlarova, J., Souskova, H., Khun, J., & Julak, J., (2015). Nonthermal plasma: A tool for decontamination and disinfection. *Biotechnology Advances*, *33*(6), 1108–1119.
58. Surowsky, B., Fischer, A., Schlueter, O., & Knorr, D., (2013). Cold plasma effects on enzyme activity in a model food system. *Innovative Food Science and Emerging Technologies*, *19*, 146–152.
59. Tappi, S., Berardinelli, A., Ragni, L., Rosa, M. D., Guarnieri, A., & Rocculi, P., (2014). Atmospheric gas plasma treatment of fresh-cut apples. *Innovative Food Science and Emerging Technologies*, *21*, 114–122.
60. Terrier, O. E. B., Yver, M., Barthelemy, M., Bouscambert, Duchamp, M. K. P., & Van, M. D., (2009). Cold oxygen plasma technology efficiency against different airborne respiratory viruses. *Journal of Clinical Virology*, *45*(2), 119–124.

61. Thakur, B. R., & Nelson, P. E., (1998). High pressure processing and preservation of foods. *Food Reviews International, 14*(4), 427–447.
62. Tiwari, B. K., & Mason, T. J., (2012). Ultrasound processing of fluid foods. Chapter 6: In: Cullen, P. J., Valdramidis, V. P., & Tiwari, B. K., (eds.), *Novel Thermal and Non-Thermal Technologies for Fluid Foods* (pp. 135–165). Apple Academic Press Inc., New Jersey-USA.
63. Thirumdas, R., Sarangapani, C., & Annapure, U., (2014). Cold plasma: Novel non-thermal technology for food processing. *Food Biophysics, 10*(1), 1–11.
64. Toepfl, S., Mathys, A., Heinz, V., & Knorr, D., (2006). Potential of high hydrostatic pressure and pulsed electric fields for energy efficient and environmentally friendly food processing. *Food Reviews International, 22*(4), 405–423.
65. Vesel, A., & Mozetic, M., (2012). Surface modification and aging of PMMA polymer by oxygen plasma treatment. *Vacuum, 86*(6), 634–637.
66. Wang, J., Zhuang, H., Hinton, A., & Zhang, J., (2016). Influence of in-package cold plasma treatment on microbiological shelf life and appearance of fresh chicken breast fillets. *Food Microbiology, 60*, 142–146.
67. Wiseman, H., & Halliwell, B., (1996). Damage to DNA by reactive oxygen and nitrogen species: Role in inflammatory disease and progression to cancer. *Biochemical Journal, 313*(1), 17.
68. Yepez, X. V., & Keener, K. M., (2016). High-voltage atmospheric cold plasma (HVACP) hydrogenation of soybean oil without trans-fatty acids. *Innovative Food Science and Emerging Technologies, 38*, 169–174.
69. Yikmis, S., (2016). New approaches in non-thermal processes in the food industry. *International Journal of Nutrition and Food Science, 5*(5), 344–351.
70. Zou, J. J., Liu, C. J., & Eliasson, B., (2004). Modification of starch by glow discharge plasma. *Carbohydrate Polymers, 55*(1), 23–26.

CHAPTER 10

POTENTIAL OF HIGH HYDROSTATIC PRESSURE TECHNOLOGY IN FOOD PRESERVATION AND FOOD SAFETY

REKHA CHAWLA, VENUS BANSAL, S. SIVAKUMAR,
NARENDER K. CHANDLA, and SANTOSH K. MISHRA

ABSTRACT

Health consciousness consumers prefer 'healthy food products' devoid of contaminants and food-borne microorganisms and this has led to use of novel food processing methods, such as pulsed electric field (PEF), infrared (IR) processing, cold plasma, high hydrostatic pressure (HHP), ultrasound, radiation, and many more. Amongst these, being operative on the principle of pressure and independent of mass and time, HHP is becoming the choice of modern processors due to benefits, such as 'fresh-like' appearance of food products, decreased microbial load, enhanced desired attributes like digestibility, and extended shelf-life. Applications of HHP is not only limited to the food industry but it offers its advantages to dairy, meat, and seafoods. Inactivation of bacteria within a pressure ranging from 300–700 MPa is not only effectual in terms of long-term preservation but also helps in manipulating the textural characteristics of foods. Therefore, HPP is a potent processing method to obtain novel food products with improved nutritional properties and increased shelf-life. This chapter covers the basic information on HHP, mechanisms of action, process categorization, and applications of HHP in food and dairy industries.

10.1 INTRODUCTION

Increasing acceptance towards healthful options has created a boom in the food industry to practice alternative technologies for the processing of foods.

Not only products of milk and dairy industry, seafoods, meats, and meat products and fruits and vegetables, all are in demand owing to least thermally processed. Therefore, such commodities can preserve its maximum nutrition features, and are microbiologically safe with extended shelf-life. Among various novel technologies available, high hydrostatic pressure (HHP) processing is one of the promising technologies, fulfilling the criteria of acceptable food quality. Also, elimination of thermal treatment helps in maintaining maximum nutritional benefits of the commodity-based on two principles of HHP [51]:

- Le-Chatelier's Principle: If any change is imposed on a system that is in equilibrium then the system tends to adjust to a new equilibrium counteracting the change. Reduction in volume in a chemical reaction, phase transition, or change in molecular configuration can be elevated through pressure.
- Isostatic Principle: The pressure acting on a food material is consistent, equal, and prompt due to applied HPP (Figure 10.1).

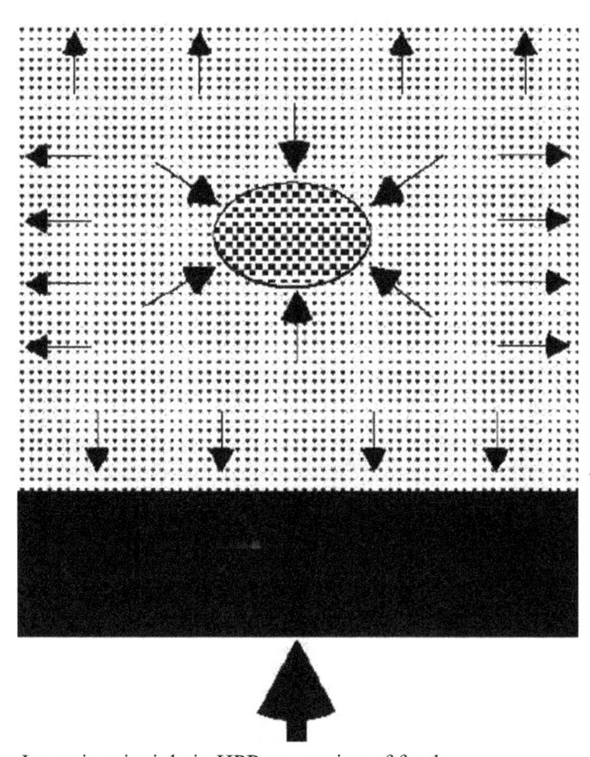

FIGURE 10.1 Isostatic principle in HPP processing of foods.

The working principle is affected by treatment parameters, such as pressure, time, and temperature and food matrix that endorse maximum and instant inactivation of microorganisms with least changes in color, texture, and flavor of the product under study, which is dependent on time and temperature only. Therefore, permission for three working parameters allows greater flexibility in the process design [28, 46]. Therefore, this advantage of HHP processing can be harnessed for heat-sensitive foods and HHP is sometimes also referred as 3rd dimension of food technology, next to time and temperature [25].

This chapter summarizes the effects of HPP technology on different food products and research studies undertaken on meat and meat produces, poultry, seafoods, and milk products. This chapter also includes basic information on HHP, mechanisms of action, process categorization, and applications of HHP in food and dairy industries.

10.2 HIGH HYDROSTATIC PRESSURE (HHP) PROCESSING OF PRODUCTS: METHOD OF OPERATION

Operational methods of HHP processing of products is quite simple and usually requires a flexible pouch and container for the same. Package or food to be treated is placed into a vessel, packed with fluid to transmit high pressure, and usually water is a medium for the same. The fluid inside the vessel is subjected to high pressure using a pump and high pressure is transmitted to the package. Usually, 3 to 30 min is the time required for the processing of the object inside the vessel. Afterward, pressure is released and the product can be stored in a conventional manner. The advantage of HHP processing of food products is the uniform processing because pressure is transmitted equally in all the directions (Figure 10.1).

10.3 MECHANISM OF ACTION FOR HIGH HYDROSTATIC PRESSURE (HHP) PROCESSING OF FOODS

According to Le Chatelier, any food matrix where a change in volume is involved, HHP results in a decrease in volume [2]. The food processing through HHP can be divided into 3 major operational steps [29]:

- Come up Time: An initial time required to reach up to a pressure of treatment.

- Holding Time: The time required for necessary processing.
- Release Time: A very little time to release the pressure.

Typical HHP systems consist of four major parts:

- A vessel and its closure that can bear high pressure;
- A mechanism to generate high pressure;
- A device to control temperature; and
- And a system to handle food material.

As the process is effective with pressure only, only foods with enough quantity of water can be pressurized and can be treated using HHP, thus adding a limiting factor to this technology. However, HHP processing has various advantages, such as the killing of pathogenic and non-pathogenic microorganisms, enhanced storage life, foods without preservatives; and fresh like attributes, changes in the texture and improved desirable characteristics [13]. HHP processing of foods can be grouped into three broad classes [3] depending upon the automation and feed flow: batch, semi-continuous, and continuous.

- Batch Operation: Packed food is loaded manually into pressure container followed by sealing of the container, and removal of air through pressurized water. Afterward, the pressure is raised within the container to a predefined level and is held for a definite specific time. An elevated pressure maintained in the vessel is transmitted to package for a predefined time and thereafter pressure is released by opening pressure relief valve. When the water employed for high pressure comes to its normal atmospheric pressure, the packed food is taken out from the high-pressure container. The treated food now can be used for distribution to the consumers. The benefits of batch systems over semi-continuous systems are: simple apparatus and post-contamination prevention as the food is placed after packing.
- Semi-Continuous System: A piston is employed to compress the liquid within the vessel. Thereafter, pressurized water (300 to 600 MPa) is introduced at a temperature from –20°C to 80°C depending on food type and equipment used [3]. After a predefined treatment time at the desired pressure, the pressure of the system is released. Thereafter, the processed semiliquid food material is pumped to a pre-sterilized vat through a sterile discharge port followed by filling of the liquid food product under aseptic environment.

- Continuous Processing: The fluid food product is compressed continuously by means of hold tube or hold vessel [3].

10.4 EFFECTS OF HIGH HYDROSTATIC PRESSURE (HHP) PROCESSING ON MICROBIAL QUALITY OF FOODSTUFFS

Inactivation of various microbes is basically a fundamental approach towards kinetic aspects of specific products. Among various categories (Table 10.1), the vegetable cells are more prone to pressure treatment including yeast and mold, as these can be inactivated with a pressure from 300 to 600 MPa [36, 50], whereas spores on the other side are highly resistant and require high pressure up to 1200 MPa for their inactivation [3, 6, 38]. Basically, the use of pressure is strongly dependent on the inactivation of the inherent enzyme system of food products under consideration. Therefore, such a wide range from 100 to 1000 MPa can be observed [9, 42]. Also in a few cases, mild heat treatment is sometimes combined with HHP to inactivate the enzymes. Considering bacterial growth in foods, Gram-negative microflora is highly vulnerable towards HPP treatment compared to Grampositive cells.

TABLE 10.1 Outcome of HHP Inactivation on Different Microflorain Foods

Food Product	Target Organism	References
Apple juice	*Talaromyces avellaneus*	[62]
Apple-broccoli juice	*S. cerevisiae; A. flavus*	[30]
Carrot juice	*B. licheniformis*	[57]
Cheese	*P. Roqueforti*	[47]
Chicken	*C. botulinum*	[39]
Cold-smoked salmon	*S. enteritidis*	[44]
	E. coli O157:H7	
Fruit juice	*Aspergillus awamori*	
Mucor plumbeus	[7]	
Milk	*L. monocytogenes LSD 105–1*	[61]
	E. coli O157:H7 ATCC35150	
	S. enteritidis ATCC 13047	
Orange juice	*A. acidorterrestris*	[26]
Pineapple juice	*Byssochlamy snivea*	[20]
Tomato pulp	*B. coalulans*	[65]

Source: Bello et al. [24].

10.5 EFFECT OF HIGH HYDROSTATIC PRESSURE (HHP) PROCESSING ON MILK CONSTITUENTS AND MILK PROPERTIES

Milk is an ideal food, but the sensitive nature of milk requires immediate processing. The severe the temperature of processing, the maximum is the shelf-life that can be achieved. However, HHP can be employed for the extended shelf-life along with the least changes in chemical constituents. The non-covalent bonds are more vulnerable to HHP processing than covalent bonds. Also, high molecular weight (MW) compounds are more vulnerable towards HHP treatment than the compounds with low MW [12, 17].

The severity of the visible effect of HHP is dependent on the pressure applied. However, for improvement in quality, HHP has been verified as a potential tool for enhancement of rennet coagulation, curd formation, salting, and accelerated maturing, and enhancement in the microstructure of cheese [16]. Such cheese has increased moisture content, salt, and free amino acids [60]. Similarly, HHP also improves the texture of yogurt from such milk along with decreasing syneresis effect [32], apart from controlling spoilage microbiota.

Application of HHP treatment helps to reduce the size of casein micelles, which led to the decrease in turbidity and lightness value of the milk along with decreasing the viscosity of the milk. A detailed study of the HHP on milk and various milk products can be cited from the review article by authors [58].

10.6 EFFECT OF HIGH HYDROSTATIC PRESSURE (HHP) PROCESSING ON SEAFOODS

In general, chilling is employed to enhance the storage period and to maintain the quality of foodstuffs prepared from fish. However, the risk of foodborne illness has caused the researchers to explore novel technologies to obtain safe foods without deteriorating the quality of products. In recent years, the practicability of HHP to produce safe and extended shelf-life fish products has been studied by researchers. Most of the studies conducted on seafoods have revealed that pressure ranging from 200 to 700 MPa for time-varying from 3, 5, 10 to 20 minutes can be employed successfully to extend its shelf-life [58], wherein this high pressure can be achieved within 10 seconds.

Researchers have reported that HHP treatment along with refrigerated storage and good manufacturing practices can increase the storage stability of fish food products. Tuna treated with HHP treatment at 200 MPa and

220 MPa for 30 minutes resulted in extended storage stability of 18 days compared to 6 days of untreated sample stored at the refrigeration temperature [64]. Ramirez-Suarez and Morrissey [52] reported the storage life of >22 days at 4°C storage of albacore tuna treated with 275 MPa and 310 MPa for 2, 4, and 6 minutes. Similarly, Jo et al. [33] observed the increase in storage stability by 10 days of HHP treated (200 MPa for 5 minutes) abalone compared to 3 days of the untreated sample at refrigerated storage.

Yagiz et al. [63] reported the log reduction of 6.0 and 4.0 in *Oncorhynchus mykiss* (rainbow trout) and *Coryphaena hippurus* (mahi-mahi) when treated at a pressure of 3000 bars. At refrigerated storage of 5 or 10°C, the increase in shelf-life by 3 weeks for fish salad with mayonnaise has been reported when treated with 450 or 600 MPa for 5 min [53]. He et al. [27]observed effects of HHP (207 to 310 MPa) on storage life of Oyster and reported zero increase in microbial load when the product was kept at refrigeration temperature for 27 days. Extension of storage life to three weeks in HHP treated (300 MPa for 15 minutes) cold-smoked dolphin fish has been observed [22]. The author reported decrease in aerobic and LAB count lesser than the detection threshold throughout the storage period.

Though, the increase in storage stability of fish and seafood products have been reported in the literature, yet the HHP treatment also results in the undesirable sensory characteristics in fish products. HHP can cause accelerated lipid oxidation, increase in hardness and cooked appearance in the HHP treated products [49]. Cheah and Ledward [14] reported that increase in lipid oxidation might be due to the breakdown of metal ions from their compounds. Angsupanich and Ledward [1] concluded that increase rate of lipid oxidation is due to more concentration of polyunsaturated fats (PUFA) and oxidation prompted by HHP.

Torres et al. [59] reported the inhibition in lipid hydrolysis of frozen mackerel during three months of storage at –10°C when treated at 150, 300, and 450 MPa. The increase in hardness of fish is due to cross-linking of proteins, which occurs due to the formation of formaldehyde at an elevated pressure. Master et al. [43] observed the proliferation in hardness of frozen code at 200 and 400 MPa while no change in hardness was observed at 100 MPa. The denaturation of proteins at >300 MPa may result in cooked appearance especially in high protein fish products [34, 37]. The applications of HHP to produce extended shelf-life seafood products are magnificent; but the undesirable modifications in sensory and chemical characteristics of foodstuffs (viz., texture, color, and lipid oxidation) may hamper the exploitation of this technology for seafood industry.

10.7 EFFECT OF HIGH HYDROSTATIC PRESSURE (HHP) PROCESSING ON MEAT PRODUCTS

The application of HHP method has been explored in the recent past to produce high quality meat and meat products having fresh-like characteristics and at the same time safe for human consumption. Also, this technology has been approved by USDA-FSIS and Health Canada in recent years to inactivate *Listeria monocytogenes* in ready-to-eat food products, such as meat, poultry, and RTE pork. Moreover, the availability of HHP treated meat and meat products on shelves across the globe as natural minimally processed products proved the significant recognition of this technology in the present era.

The inactivation of pathogenic and non-pathogenic microorganisms employing HHP method to certify food safety and to enhance storage stability of meat and meat products has been extensively explored by several scientists. Aymerich et al. [4] studied the destruction of *L. monocytogenes* in cooked ham treated at 4000 bar for 10 min and they reported log reduction of 1.9 from an initial count of 2.6 when the product was stored at 6°C for 42 days. Similarly, a log reduction of >6.0 after 60 days at 4°C was observed in HHP with 600 MPa at 31°C for 6 minutes for prepared ham [22]. A log reduction of 10 for *Listeria* monocytogenes was obtained in HHP treated fresh pork sausage sat 400 MPa for 6 minutes [48].

Murano et al. [45] noticed the impact of HPP treatment (414 MPa for 6 minutes) on the shelf-life of ground pork patties and reported an increase in shelf-life to 28 days at 4°C than 5 days for the control sample. A log reduction of 4 was observed in marinated beef after HHP treatment at 600 MPa and it was less than the detection threshold throughout the storage period of 4 months at refrigeration temperature [21]. Similarly, Tanzi et al. [56] observed the microbial count below the threshold value in dry-cured Parma ham after HHP treatment at 6000 bar for 9 min. However, the authors reported a change in color and enhanced salty perception after the treatment.

Presently, the application of HHP to inactivate pathogenic and non-pathogenic microorganisms and to retain the nutritional grade of meat products has been exploited in the meat industry at par. Also, an enhanced tenderization of meat in curing has been observed in HHP treated meat products due to structural deformations of the myofibrils [8, 35, 41, 55]. However, the deteriorating outcomes of HHP on the characteristics of meat products limit the applicability of this technology to adventure its full potential in the meat industry. The denaturation of globular proteins and the conversion of ferrous to ferric myoglobin takes place above 400 MPa. Researchers also reported that HHP prompted denaturation of proteins is altogether different than heat-induced denaturation.

Okamoto et al. [48] concluded that denaturation of proteins at high pressure is due to a reduction in protein volume while heat denaturation is due to the break down of hydrogen and covalent bonds of protein molecules. The change in color has been reported by researchers for pressure >150 MPa [31]. However, change in color is appreciable only in fresh meat products while no change in color has been reported in cured meat products.

Cheftel and Culioli [15] stated that HHP treatment of fresh red meats resulted in harsh changes in the red color of the meat and were not acceptable by consumers. However, HHP processing of white meat did not result in any adverse change in the color. The impact of HHP on textural properties (viz., springiness, chewiness, and juiciness) has also been studied by many researchers. A pressure treatment of 200 MPa causes more hardness and chewiness in chicken batters compared to untreated counterparts [19]. Table 10.2 indicates results of Outcome of HPP treatment on shelf-life and log reduction of fish and meat products.

TABLE 10.2 Outcome of HPP Treatment on Shelf-Life and Log Reduction of Fish and Meat Products

Food Matrix	Experiment	Outcome	References
Seafoods			
Abalone	500 MPa	Shelf-life increased by >35 days	[9]
Herring	300 MPa at 2°C for 1 minute	Shelf-life increased by 12 days	[34]
Rainbow trout	220 MPa at 5°C for 15 minutes	Shelf-life increased by 4 days at refrigeration temperature	[24]
Red abalone	500 MPa at 20°C for 8 minutes	Shelf-life increased by 3 days	[9]
Sea bream	250 MPa for 3 minutes	Shelf-life increased by 3 days	[18]
Meat and Meat Products			
Cooked ham	600 MPa at 31°C for 6 minutes	1.12 log reduction in *S. aureus* count	[31]
Minced beef muscle	450 MPa at 20°C for 20 minutes	3–5 log reduction in total microflora	[12]
Minced chicken	500 MPa at 40°C for 15 minutes	1 log reduction in aerobic plate count	[40]
Pork slurry	400 MPa at 10°C for 25 minutes	6 log reduction in *Campylobacter jejuni* count	[54]

10.8 CONSIDERATIONS IN PACKAGING

A variety of shapes and materials can be employed for the HHP method in foodstuffs. However, the general guidelines stick to the rule that package must be flexible and impervious to force applied and should uphold its coherence under high pressure. The films [such as PE (polyethylene), BOPA (biaxially oriented nylon), BoPET (biaxially oriented PE terephthalate), PET (polyethylene terephthalate), PP (polypropylene or polypropene), EVOH (ethylene vinyl alcohol copolymer), and PA (Polyamide) and nylon] can be used for HHP treatments in packaging [23, 27].

10.9 PRESENT STATUS OF HIGH HYDROSTATIC PRESSURE (HHP) PROCESSING

Currently, 167 commercial HHP industrial plants in India are in operation with volume ranging between 55–420 liters, wherein reported annual production volume has increased from 0.2 million tons in 2009 [28] to 0.35 million tons in 2012 [49]. Though the majority of HHP product market lies in developed countries (such as Japan, USA, and Europe), yet Indian market is also emerging as a new place for HHP products with the basic advancements. Various categories of juice, salsa, smoothies, meat, and vegetable products are occupying shelves of supermarkets in the Indian context. Also, based on reports collected, the vegetable industry accounts for 28% of total products processed using HHP method, wherever meat and meat products account for 26%, and beverages/juices for 14% [14]. Thus, an increasing percentage of growing HHP products in the Indian market is an example of consumer awareness and desirability for quality products along with extended shelf-life.

10.10 SUMMARY

HHP processed products are an attractive choice of modern buyers sowing to the benefits associated with the process and increasing awareness. Increased shelf-life of endless food products decreased the number of microbiota (pathogenic and non-pathogenic), intact nutritional status, visual appearance, etc., are some of the key features of HHP-based products, leading to increasing acceptability of such products. Though cost is still a limiting factor for such products towards wider acceptability among consumers with middle-income group cadre, yet more number of commercial HHP-based products in developing countries can be a solution to this problem.

KEYWORDS

- **high hydrostatic pressure processing**
- **Le-Chatelier principle**
- **meat products**
- **microbial quality**
- **milk products**
- **seafood**

REFERENCES

1. Angsupanich, K., & Ledward, D. A., (1998). High pressure treatment effects on cod (*Gadusmorhua*) muscle. *Food Chemistry*, *63*(1), 39–50.
2. Anonymous, (2007). *Chemical Equilibrium Lab: Le Châtelier's Principle.* Saskschools.ca/curr_content/chem30_05/3_equilibrium/labs/le_chatelier (accessed on 25 May 2020).
3. Anonymous, (2007). *Technical Elements of New and Emerging Non-Thermal Food Technologies*.https://hortintl.cals.ncsu.edu/content/technical-elements-new-and-emerging-non-thermal-food-technologies (accessed on 25 May 2020).
4. Aymerich, M. T., Jofre, A., Garriga, M., & Hugas, M., (2005). Inhibition of *Listeria monocytogenes* and *Salmonella* by natural antimicrobials and high hydrostatic pressure in sliced cooked ham. *Journal of Food Protection*, *68*, 173–177.
5. Barbosa-Canovas, G. V., Gongora-Nieto, M. M., Pothakamury, U. R., & Swanson, B. G., (1999). *Preservation of Foods with Pulsed Electric Fields* (pp 1–9, 76–107, 108–155). Academic Press Ltd., London, UK.
6. Bello, E., Martinez, G., Ceberio, B., Rodrigo, D., & López, A., (2014). High pressure treatment in foods. *Foods*, *3*(3), 476–490. doi: 10.3390/FOODS3030476.
7. Black, E. P., Setlow, P., Hocking, A. D., Stewart, C. M., Kelly, A. L., & Hoover, D. G., (2007). Response of spores to HPP. *Comprehensive Reviews in Food Science and Food Safety*, *6*, 103–119.
8. Bouton, P. E., Ford, A. L., Harris, P. V., Macfarlane, J. J., & O'Shea, J. M., (1977). Pressure-heat treatment of post-rigor muscle: Effect on tenderness. *Journal of Food Science*, *42*, 132–135.
9. Briones, L. S., Rets, J. E., Tabilo-Munzaga, G. E., & Pérez-Won, M. O., (2010). Microbial shelf-life extension of chilled Coho salmon and abalone by HPP treatment. *Food Control*, *21*, 1530–1535.

10. Briones-Labarca, V., Perez-Won, M., Aguilera-Radic, J. M. Z., & Tabilo-Munizaga, G. T., (2012). Effects of high hydrostatic pressure on microstructure, texture, color and biochemical changes of red abalone during cold storage time. *Innovative Food Science and Emerging Technologies, 13*, 42–50.
11. Cano, M. P., Hernandez, A., & Ancos, D. E. B., (1997). High pressure and temperature effects on polyphenoloxidase from *La France* pear fruit and its pressure-activation. *Bioscience, Biotechnology and Biochemistry, 58*, 1486–1489.
12. Carlez, A., Rosec, J. P., Richard, N., & Cheftel, J. C., (1994). Bacterial growth during chilled storage of pressure-treated minced meat. *Lebensmittel-Wissenschaft und Technologies, 27*, 48–54.
13. Chawla, R., Patil, G. R., & Singh, A. K., (2011). HPP technology in dairy processing: A review. *Journal of Food Science and Technology, 48*(3), 260–268.
14. Cheah, P. B., & Ledward, D. A., (2006). Catalytic mechanism of lipid oxidation following HPP Treatment in pork fat and meat. *Food Science, 62*(6), 1135–1139.
15. Cheftel, J. C., & Culioli, J., (1997). Effects of high pressure on meat: A review. *Meat Science, 46*, 211–236.
16. Devi, A. F., Buckhow, R., Hemar, Y., & Kasapis, S., (2013). Structuring dairy systems through high pressure processing. *Journal of Food Engineering, 114*(1), 106–122.
17. Dzwolak, W., Kato, M., & Taniguchi, Y., (2002). Fourier transforms infrared spectroscopy in high-pressure studies on proteins. *Biophysica Acta, 1595*, 131–144.
18. Erken, N., Uretener, G., & Alpas, H., (2010). Effect of high pressure (HP) on the quality and shelf of red mullet. *Innovative Food Science and Emerging Technologies, 11*, 259–264.
19. Fernandez, P., Confrades, S., Solas, M. T., Carballo, J., & Jimenez-Colmenero, F., (1998). High pressure cooking of chicken meat batters with starch, egg white, and iota carrageenan. *Journal of Food Science, 63*, 261–271.
20. Ferreira, E. H. D. R., Rosenthal, A., Calado, A. V., Saraiva, J., & Mendo, S., (2009). *Byssochlamys nivea* inactivation in pineapple juice and nectar using high pressure cycles. *Journal of Food Engineering, 95*, 664–669.
21. Garriga, M., Costa, S., Monfort, J. M., & Hugas, M., (2002). Prevention of ropiness in cooked pork by bacteriocinogenic cultures. *International Dairy Journal, 12*, 239–246.
22. Garriga, M., Grebol, N., Aymerich, M. T., Monforta, J. M., & Hugas, M., (2004). Microbial inactivation after high pressure processing at 600MPa in commercial meat products over its shelf life. *Innovative Food Science and Emerging Technologies, 5*, 451–457.
23. Gómez-Estaca, J., Gómez-Guillén, M. C., & Montero, P., (2007). High pressure effects on the quality and preservation of cold-smoked dolphin fish (*Coryphaena hippurus*) fillets. *Food Chemistry, 102*(4), 1250–1259.
24. Günlü, A., Sipahioğlu, S., & Alpas, H., (2014). The effect of chitosan-based edible film and high hydrostatic pressure process on the microbiological and chemical quality of rainbow trout fillets during cold storage (4 ± 1°C). *High Pressure Research, 34*(1), 110–121.
25. Hafsteinsson, H., (2006). Application of ultrasonic waves to detect seal worms in fish tissue. *Journal of Food Science, 54*(2), 244–247.
26. Hartyáni, P., Dalmadi, I., & Knorr, D., (2013). Electronic nose investigation of *Alicyclobacillus acidoterrestris* inoculated apple and orange juice treated by high hydrostatic pressure. *Food Control, 32*, 262–269.

27. He, H., Adams, R. M., Farkas, D. F., & Morrissey, M. T., (2002). Use of high-pressure processing for oyster shucking and shelf-life extension. *Journal of Food Science*, *67*(2), 640–645. doi: 10.1111/j.1365-2621.2002.tb10652.x.
28. Heinz, V., & Buckow, R., (2009). Food preservation by high pressure. *Journal of Consumer Protection and Food Safety*, *5*(1), 73–81. doi: 10.1007/s00003-009-0311-x.
29. Hogan, E., Kelly, A. L., & Sun, D. W., (2005). High-pressure processing of foods: An Overview. In: *Emerging Technologies for Food Processing* (pp. 1–31). Elsevier Academic Press, San Diego-USA.
30. Houŝka, M., Strohalm, J., Kocurova, K., Totusek, J., Lefnerova, D., & Triska, J., (2006). High pressure and foods-fruit/vegetable juices. *Journal of Food Engineering*, *77*, 386–398.
31. Hugas, M., Garriga, M., & Monfort, J. M., (2002). New mild technologies in meat processing: High pressure as a model technology. *Meat Science*, *62*, 359–371.
32. Jankowska, A., Wisniewska, K., & Reps, A., (2005). Application of probiotic bacteria in production of yoghurt preserved under high pressure. *High Pressure Research*, *25*(1), 57–62. doi: 10.1080/08957950500062023.
33. Jo, Y. J., Jung, K. H., Lee, M. Y., Min, C. S. G., & Hong, G. P., (2014). Effects of high-pressure short-time processing on the physicochemical properties of abalone during refrigerated storage. *Innovative Food Science and Emerging Technologies, 23*, 33–38. doi: 10.10.1016/j.ifst.201.02.011.
34. Karim, N. U., Kennedy, T., Linton, M., Watson, S., & Gault, P. M. F., (2011). Effect of high pressure processing on the quality of herring and haddock stored on ice. *Food Control, 22, 76–484.*
35. Kennick, W. H., Elagism, E. A., Holmes, Z. A., & Meyer, P. F., (1980). The effect of pasteurization of pre-rigor muscle on post-rigor meat characteristics. *Meat Science*, *40*, 33–40.
36. Knorr, D., (1995). Hydrostatic pressure treatment of food: Microbiology. In: Gould, G. W., (ed.), *New Methods of Preservation* (pp. 159–175). New York: Blackie Academic and Professional.
37. Lakshmanan, R., Miskin, D., & Piggott, J. R., (2005). Quality of vacuum packed clod-smoked salmon during refrigerated storage as affected by high pressure processing. *Journal of the Science of Food and Agriculture*, *85*, 655–661.
38. Larson, W. P., Hartzell, T. B., & Diehl, H. S., (1918). The effect of high pressures on bacteria. *Journal of Infectious Diseases*, *22*, 271–279.
39. Linton, M., Connolly, M., Houston, L., & Patterson, M. F., (2014). The control of *Clostridium botulinum* during extended storage of pressure-treated, cooked chicken. *Food Control, 37*, 104–108.
40. Linton, M., McClements, J. M. J., & Patterson, M. F., (2004). Changes in microbiological quality of vacuum-packaged, minced chicken treated with high hydrostatic pressure. *Innovative Food Science and Emerging Technologies, 5*, 151–159.
41. MacFarlane, J. J., (1973). Pre-rigor pressurization of muscle: Effect on pH, shear value and taste panel assessment. *Journal of Food Science*, *38*, 294–298.
42. Macheboeuf, M. A., & Basset, J., (1934). The effects of high pressures on enzymes. *Ergebnisse Der Enzymforcung*, *3*, 303–308.
43. Master, A. M., Stegeman, D., Kals, J., & Bartels, P. V., (2000). Effects of high pressure on color and texture of fish. *High Pressure Research*, *19*, 109–115.

44. Montiel, R., Martin-Cabrejas, I., & Medina, M., (2016). Natural antimicrobials and high-pressure treatments on inactivation of *Salmonella enteritidis* and *Escherichia coli* O157:H7 in cold-smoked salmon. *Journal of the Science of Food and Agriculture, 96*(7), 2573–2578.
45. Murano, E. A., Murano, P. S., Brennan, R. E., Shenoy, K., & Moriera, R. G., (1999). Application of high hydrostatic pressure to eliminate *Listeria monocytogenes* from fresh pork sausage. *Journal of Food Protection, 62*, 480–483.
46. Naik, L., Sharma, R., Rajput, Y. S., & Manju, G., (2013). Application of high pressure processing technology for dairy food preservation-future perspective: A review. *Journal of Animal Production Advances, 3*(8), 232–241. doi: 10.5455/japa.20120512104313.
47. O'Reilly, C. E., O'Connor, P. M., Kelly, A. L., Beresford, T. P., & Murphy, P. M., (2000). Use of hydrostatic pressure for inactivation of microbial contaminants in cheese. *Applied Environmental Microbiology, 66*, 4890–4896.
48. Okamoto, M., Kawamura, Y., & Hayashi, R., (1990). Application of high pressure to food processing: Textural comparison of pressure and heat-induced gels of food proteins. *Agricultural and Biological Chemistry, 54*, 183–189.
49. Olatunde, O. O., & Benjakul, S., (2018). Nonthermal processes for shelf-life extension of seafood's: Revisited. *Comprehensive Reviews in Food Science and Food Safety, 17*(4), 892–904.
50. Patterson, M. F., Quinn, M., Simpson, R., & Gilmour, A., (1995). Effects of high pressure on vegetative pathogens. In: Ledward, D. A., Johnston, D. E., Earnshaw, R. G., & Hasting, A. P. M., (eds.), *High Pressure Processing of Foods* (pp. 47–63). Notting-ham, University Press.
51. Ramaswamy, H. S., Chen, C., & Marcotte, M., (1999). Novel processing technologies in food preservation. In: Barrett, D. M., Somogyi, L. P., & Ramaswamy, H. S., (eds.), *Processing Fruits Science and Technology* (2nd edn., pp. 201–220). CRC Press: Boca Raton.
52. Ramirez-Suarez, J. C., & Morrissey, T. M., (2006). Effect of high pressure processing (HPP) on shelf-life of albacore tuna minced muscle. *Innovative Food Science and Emerging Technologies, 1, 2*, 19–27.
53. Salamon, B., Toth, A., Palotas, P., Sudi, G., Csehi, B., Nemeth, C. S., & Friedrich, L., (2016). Effects of high hydrostatic pressure (HHP) processing on organoleptic properties and shelf-life salad with mayonnaise. *Acta Alimentaria an International Journal of Food Science, 45*, 558–564. doi: 10.1556/066.2016.45.4.4.13.
54. Shigehisa, T., Ohmori, T., Saito, A., & Taji, S., (1991). Effects of high pressure on the characteristics of pork slurries and the inactivation of micro-organisms associated with meat and meat products. *International Journal of Food Microbiology, 12*, 207–216.
55. Suzuki, A., Watanabe, M., Iwamura, K., Ikeuchi, Y., & Saito, M., (1990). Effects of high pressure treatment on ultrastructure and myofibrillar protein of beef skeletal muscle. *Agricultural and Biological Chemistry, 54*, 3085–3091.
56. Tanzi, E., Saccani, G., Barbuti, S., Grisenti, M. S., Lori, D., Bolzoni, S., & Parolari, G., (2004). High pressure treatment of raw ham: Sanitation and impact on quality. *Industria Conserve, 79*, 37–50.
57. Tola, Y. B., & Ramaswamy, H. S., (2014). Combined effects of high pressure, moderate heat and pH on the inactivation kinetics of *Bacillus licheniformis* spores in carrot juice. *Food Research International, 62*, 50–58.

58. Torres, J. A., & Velazquez, G., (2005). Commercial opportunities and research challenges in high pressure processing foods. *Journal of Food Engineering, 7*(1–2), 95–112.
59. Torres, J. A., Vazquez, M., Saraiva, J. A., Gallardo, J. M., & Aubourg, S. P., (2013). Lipid damage inhibition by previous HPP in white muscle of frozen horse mackerel. *European Journal of Lipid Science and Technology, 115*(12), 1454–1461.
60. Trujillo, A. J., Capellas, M., Saldo, J., Gervilla, R., & Guamis, B., (2002). Applications of high-hydrostatic pressure on milk and dairy products: A review. *Innovative Food Science and Emerging Technologies, 3*(4), 295–307. doi: 10.1016/S1466-8564(02)00049-8.
61. Vachon, J. F., Kheadr, E. E., Giasson, J., Paquin, P., & Fliss, I., (2002). Inactivation of food-borne pathogens in milk using dynamic high pressure. *Journal of Food Protection, 65*, 345–352.
62. Voldrich, M., Dobiáš, J., Tichá, L., Cerovský, M., & Krátká, J., (2004). Resistance of vegetative cells and ascospores of heat resistant mold *Talaromyces avellaneus* to the HPP in apple juice. *Journal of Food Engineering, 61*, 541–543.
63. Yagiz, Y., Kristinsson, H. G., Balaban, M. O., & Marshall, M. R., (2007). Effect of high pressure treatment on the quality of rainbow trout and mahi-mahi. *Journal of Food Science, 72*(9), 509–515. doi: 10.1111/j.1750-3841.2007.00560.x.
64. Zare, Z., (2004). High pressure processing of fresh tuna fish and its effects on shelf life. *Master's Degree Thesis* (p. 117). Faculty of Graduate Studies and Research, McGill University, Montreal, Quebec, Canada.
65. Zimmermann, M., Schaffner, D. W., & Aragão, G. M. F., (2013). Modeling the inactivation kinetics of *Bacillus coagulans* spores in tomato pulp from the combined effect of high pressure and moderate temperature. *LWT Food Science and Technology, 53*, 107–112.

CHAPTER 11

APPLICATION OF PULSED LIGHT TECHNOLOGY IN MICROBIAL SAFETY AND FOOD PRESERVATION

VENUS BANSAL, NARENDER K. CHANDLA, REKHA CHAWLA, VEENA NAGARAJAPPA, and SANTOSH K. MISHRA

ABSTRACT

The processing of foods to mend the storage stability and to conserve the food quality can be traced since human civilization. The preservation technologies have evolved from the application of salt, sugar, preservatives, and thermal inactivation to novel processing techniques such as membrane processing, high hydrostatic pressure (HHP), thermo-sonication, cold plasma, pulsed electric field, ultrasound, pulsed light technology (PLT), ozone, and active packaging, etc. Conventionally, inactivation of microbes through heating is one of the most widely used tools for preservation and microbial safety.

To address the issues in thermal processing, researchers are exploring different non-thermal technologies to preserve the foods without loss in the nutritional quality and sensory properties. PLT is such a technology, which is recognized by the FDA for the sanitization and microbial safety of foods. The PLT is grounded on the discharge of minute duration of high-frequency pulses of light on surfaces and into the food products to obtain foods safe for consumption. PLT is an eco-friendly technology that is emerging in this era owing to its wide applications from the decontaminations of foods, surfaces, environments, devices, and packages to reduce the allergen potent of some naturally occurring foods. Therefore, with growing demand in minimally processed foods and eco-friendly technologies, PLT can be an auspicious technique to conserve the quality of food products having fresh like characteristics throughout their storage period.

11.1 INTRODUCTION

Food processing has played a pivotal role to reduce the risk of food poisoning together with enhanced storage stability of food products. As human evolution progressed, methods to preserve foods evolved from preservation by smoking, salt, sugar to heat preservation, which is presently the universally used media for the preservation of foods. Heat preservation methods like pasteurization, appertization, sterilization, UHT, etc. involve the use of high temperatures to eliminate pathogenic and non-pathogenic microorganisms to obtain foods safe for consumption with prolonged shelf-life. However, the detrimental effects of heat on the heat-sensitive nutrients which include minerals, vitamins, pigments, antioxidants, and bioactive compounds have motivated researchers to develop non-thermal processing methods that not merely retain the nutritional quality but also are environmentally benign, energy-efficient and cost-effective.

During the last 10 years, the rate of technological advancement has shown a spectacular inversion from thermal processing to non-thermal technologies to inactivate or remove the microorganisms from food and to reduce the cost of production, manufacturing time, and to produce uniformly consistent and better-quality products. Of these non-thermic processing techniques, pascalization (HPP), pulsed electric field (PEF), microfiltration (MF), and pulsed white light are most popular in the food industry to augment the shelf-life without deteriorating the quality of food products. Though, PLT is not sternly a non-thermal process, yet is still grouped under non-thermal technologies by food scientists owing to minimum deteriorating effects on nutrient content and physical properties of food products. PLT is an unconventional non-thermal processing method that operates on the principle of ejection of intense light pulses (ILP) of short duration for the decontamination of food products.

Since PLT involves the utilization of a wide wavelength range of 200–1100 nm, therefore this technology is also named as:

- High intensity broad spectrum PL [61];
- Intense light pulses [26];
- Intense pulsed light (IPL) [13];
- Pulsed ultraviolet (UV) light [67]; or
- Pulsed white light [47].

The utilization of UV rays to inactivate microorganisms can be flashed back to 1928 [23], but the scientific technology to inactivate microorganisms using UV pulses triggered from gas discharge flash-lamps was developed during 70s to 80s. Hiramoto patented the method of sterilization utilizing PL in 1984 [32]. Though, a steep development in PLT for the processing of foods occurred when FDA [20] approved this technology in 1996 for processing and preservation of food products; and since then different types of devices and equipments have been patented during the subsequent years. The recommended conditions by FDA for the use of PLT in food processing involve overall aggregate treatment energy not exceeding 12 J/cm^2 [20].

In the last two decades, the scientific studies on PLT have successfully demonstrated the efficacy of ILP for the elimination of pathogenic and non-pathogenic microorganisms. Commercially, PLT can be exploited not only for the decontamination of foods but also for the sterilization of utilities involved in the food industry like water, air, packaging materials as well as surfaces and environments, etc.

This chapter discusses principles of PLT, mechanism of bacterial destruction, factors affecting microbial destruction by PLT, the kinetics of microbial destruction, and application of PLT in the food preservation.

11.2 PRINCIPLE OF PULSED LIGHT TECHNOLOGY (PLT)

Pulsed light technology (PLT) involves the generation of broad-spectrum intense pulses of light from xenon lamp, which includes UV rays, visible light, and infrared (IR) rays of wavelength 180–400, 400 to 700 and 700–1100 nm, respectively. The strength of light is 20,000 Fluence compared to sunbeam at the surface of earth [17, 45]. PLT is described using the following terms that are consistently employed in scientific literature to better understand the technology [27, 42, 55]:

- Exposure Time: It is the total time of the treatment.
- Fluence/Dose Ratio (J/m^2): It can be described as the energy incident on the matrix from the light source divided by area of that sample during the treatment time.
- Fluence Rate (W/m^2): It can be described as the energy received by the treatment matrix from the pulses pointing from all directions by the lamp per unit area of that sample.
- Frequency (Hz or kHz): It depicts number of pulses per second. It is also called pulse repetition rate.
- Peak Power (W): It can be described as energy of the pulse divided by duration of the pulse during which that energy acts.

- Pulse Width (s): It can be described as the time taken by the light source to deliver one pulse (fractions of seconds).

PL equipment consists of number of components to dissipate high intensity pulses of light on the target. High voltage power supply converts input electrical energy into high voltage DC electrical power. The function of capacitor is to store high voltage electrical energy from power supply and to deliver the same to the switches. Gas discharge flash lamp converts this high voltage DC power into high intensity light pulses. A trigger signal is used to deliver the obtained energy to the target.

11.3 PULSED LIGHT TECHNOLOGY (PLT): MECHANISM OF MICROBIAL INACTIVATION

Researchers have proposed few mechanisms to explain the lethal influence of PLT on bacteria and is largely anticipated that the UV component is mainly responsible for bactericidal effects of PLT [27]. However, mechanism that explains lethality of microbes by PL can be simplified with three different mechanisms, i.e., photo-chemical mechanism, photo-thermal mechanism and photo-physical mechanism. Several authors have reported that irreversible inactivation of microbes by ILP is due to inter-related effect of photochemical, photothermal, and photophysical mechanisms [12, 19, 38, 40, 69, 74]. Though, the detail of mechanism by which PL inactivates microorganisms is still an area that needs to be explored more extensively for better understanding of the process.

11.3.1 PHOTOCHEMICAL MECHANISM

The photochemical inactivation (Figure 11.1) of microorganism postulates that the UV-C of wavelength 25–27 A° is predominantly liable for bactericidal effect of ILP and no antimicrobial effect is recognized to wavelength 400–1100 nm [1, 56, 60].

The mutation of DNA and RNA owing to UV light absorbed through conjugated C-C double bonds of protein molecules and nucleic acids (NAs) is the basic principle of inactivation of microbes [12, 49, 59, 70, 73]. This results in the formation of cyclobutane thiamine dimers [5, 66], which result in the chologenic death of affected microorganism due to inhibition of DNA synthesis during the DNA replication [6]. The peak destruction of *E. coli* by UV-C was reported at wavelength of 260 nm, as this wavelength is highly absorbed by DNA [73].

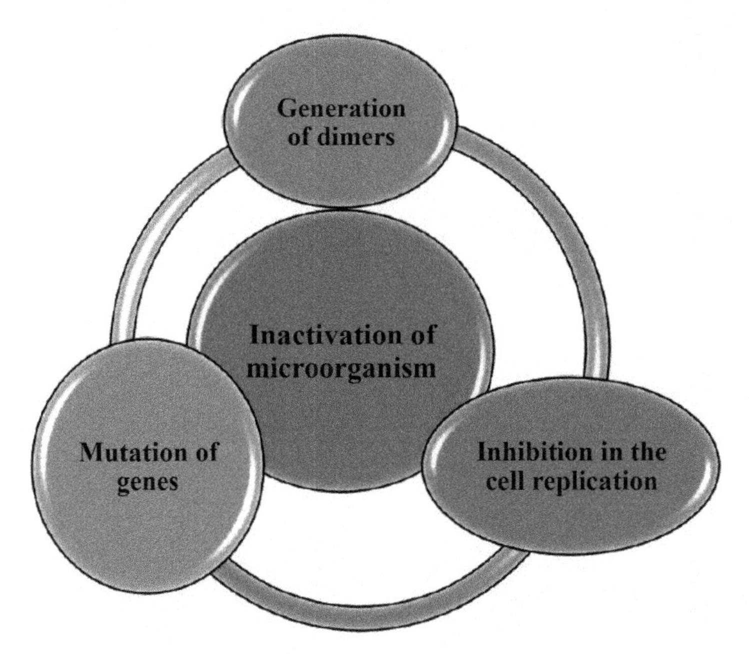

FIGURE 11.1 Microbial inactivation with UV light.

11.3.2 PHOTOTHERMAL MECHANISM

Photothermal hypothesis of inactivation is based on localized heating of microbial cells due to dissipation of energy by light pulses into heat, which causes the rupture of microbial cell membrane. The disruption of cell membrane occurs owing to vaporization of water and generation of small steam flow [69]. The photochemical inactivation of microbes is mainly because of UV-C spectrum while the photothermal effect is manifested to spectrum consisting of UV light to IR (Infra-red, 180–1100 nm wavelength).

The surface temperature depends on the dose of the light pulses; and at higher doses, a temperature of up to 130°C can be observed [64]. Photothermal effect of inactivation is observed when the fluence exceeds 0.5 J/cm^2; and at low fluence, the antimicrobial effect is predominantly acknowledged to the photochemical effect of UV-C [74]. In another experiment conducted by Wekhof et al. [75] on *Aspergillus niger* spores, the inactivation of microorganisms was mainly because of photochemical effect at low fluence of 10–30 kJ/cm^2 while at 50–60 kJ/cm^2, the inactivation was due to severe deformation and rupture of cells.

11.3.3 PHOTOPHYSICAL MECHANISM

Physical mechanism of microbial destruction can be attributed to depolarization of bacterial membrane, change in cell permeability and change in ion flow [51]. Other structural damages like enlargement of vacuoles and collapse of cell structure may also be involved in the inactivation of microbes [48].

11.4 PULSED LIGHT TECHNOLOGY (PLT): FACTORS AFFECTING MICROBIAL INACTIVATION

It is important to know the factors that influence the efficiency of microbial inactivation to successfully implement the food preservation process. The knowledge of resistance factors helps to envisage microbial inactivation in a complex system and to develop mathematical models of microbial inactivation. An adequate knowledge of microbial characteristics, interaction between microbes and food matrix, interaction between the light and microbial cell, target microorganism, and design of the PL arrangement is essential for the optimization of PLT process to diminish the risk of food poisoning and to retard the physiological, microbiological, and deteriorative processes occurring in the product during storage.

- **Microbial Characteristics:** The optical characteristics of microbes (viz. degree of scattering and absorption of light) have great influence on the efficiency of microbial destruction through PL processing [1]. The microorganisms that have proficient DNA repair mechanism are insensitive to PL [46, 58, 72]. Gomez et al. [26] experimented the sensitivity of microorganisms on agar media flashed 50 times with pulse width of 30 μs and 7 J pulse intensity. Authors reported no clear pattern regarding the inactivation sensitivity of various groups of microbes, though *Listeria monocytogenes* was found to have the maximum resistant to ILP. Turtoi and Nicolau [70] reported that fluence required to inactivate different spores depends on their color as it affects the resistance to pulsed light (PL).
- **Treated Matrix:** The nature of food material and surface integrity (rough or smooth) strongly affects the efficacy of microbial inactivation because it affects the absorption, reflection, refraction, and scattering of light. Effectiveness of inactivation is affected by

opaqueness, color, and viscosity of the material [57] as it affects absorption and transmission coefficients of the matrix [55]. The proportion of fat, carbohydrate, proteins, and water also impairs the microbial inactivation where fat and protein have negative effect on decontamination; and no detrimental effect of carbohydrates has been reported [26, 58]. Scientific studies conducted by researchers show that inactivation can also be influenced by the color of food products. Dunn et al. [14] suggested that color of carotenes, lime green, black cherry and cooking oil was affected by the PL processing of food products. Abida et al. [1] reported that refraction is prevailing in transparent and colored food materials while reflection for opaque food materials is pertinent phenomena. Interaction of light with internal structure tends to moderate the efficacy of PL processing owing to multiple internal reflections and redirections. Solid matrices with rough surface and minute grooves can reduce the efficacy of microbial inactivation, as it shadows some microorganisms from light [41].

- **Vertical Distance from the Strobe:** The vertical distance between the PL and the surface has a negative relationship on the efficiency of microbial destruction. This is owing to reduction in absorption and scattering of light due to decrease in intensity of light, as it travels through the substrate [1]. The term optical penetration depth has been devised to describe the dissemination of pulse inside a substrate and is described as the distance when light decreases the fluence rate by 37% of its primary value. Gomez-Lopez et al. [26] testified the negative impact of distance on the microbial destruction by PL processing.
- **Design of Pulsed Light Process:** The proficiency of ILP technology is ascertained by the conversion of electrical energy to radiant energy. In PL processing, gas-discharge lamp is an important component as it transforms about 45% of the given electric power to ILP [76]. The design of the PL apparatus varies from company to company but basic components (Figure 11.1) are common among all designs.

11.5 PULSED LIGHT TECHNOLOGY (PLT): REACTION KINETICS OF MICROBIAL INACTIVATION

Kinetic parameters and models, that accurately describe the inactivation rate, are fundamental for process optimization of PL technology. Lack of knowledge for the exact mechanism of microbial inactivation has limited the development of models that describe the inactivation rate by PL. However, authors have reported the use of different non-linear models to delineate the kinetics of microbial destruction by PLT. However, some scientists have used a simple first-order reaction to explain the kinetics of microbial destruction by PL. Rowan et al. [63] extensively reviewed the microbial inactivation models to describe the PL inactivation kinetics.

- **Log-Linear Model:** The simplest model and is based on the statement that cell count and their lethal inactivation have negative and linear correlation between each other. This model has been used by many researchers in different media to explain the inactivation reaction kinetics by PL. This model is based on simple systems (like laboratory media) and for a single strain population [63]; and for food matrixes, this model has not been identified yet. The Log-linear model has been used for describing elimination kinetics of *E. coli* [18] and *Listeria monocytogenes* [7] on agar and *ZygoSaccharomyces bailii, Zygosachromyces rouxii,* and *Sachromyces cerevisiae* in glucose solution [29].

$$\log_{10}\left(N_f\right) = \log_{10}\left(N_o\right) - k_{\max} \times \frac{F}{\ln(10)} \tag{11.1}$$

where, F (fluence, J/cm^2) is introduced instead of treatment time; N_f stands for final count; N_o for the initial population; and k_{max} for inactivation rate (cm^2/J).

- **Biphasic Model:** It is applied when two species or strains have different resistances to the treatment and it was originally proposed by Cerf in 1977 [9]. Ferrario et al. [21] applied following biphasic model to explicate microbial inactivation by PL in juices.

$$\log_{10}\left(N_f\right) = \log_{10}\left(N_o\right) + \log_{10}[f.\exp\left(-k_{\max 1} \times t\right) + \left(1 - f\right).\exp\left(-k_{\max 2} \times t\right)] \tag{11.2}$$

where, f, and (1-f) is the division of initial count conforming to species of more sensitive and resistant to PL, respectively; k_{max1} and k_{max2} stand for their specific inactivation rate, respectively.

- **Sigmoidal Model:** Many researchers have suggested that inactivation kinetics can be explained through a sigmoid consisting of three different phases: a shoulder (no decrease in cell count), a log-linear destruction period, and a tail. However, for PL microbial inactivation, authors used the reduced form of the sigmoidal model in which tailing was present but not shoulder. Izquier and Gomez-Lopez [34] applied the following equation to elucidate the kinetics of microbes inactivation by PL:

$$\log_{10}\left(N_f\right) = \log_{10}\left(10^{\log_{10}(N_o)} - 10^{\log_{10}(N_{res})}\right) + \log_{10}\left(\exp\left(-K_{max}.\,F\right) + 10^{\log_{10}(N_{res})}\right) \quad (11.3)$$

  where, N_{res} is the residual count (cfu/ml); and k_{max} stands for specific inactivation rate (1/min).

- **Weibull Model:** It has been used extensively for describing microbial inactivation by PL *in vitro* and on food contact surfaces. It can be represented as follows:

$$\log_{10}\left(N_f\right) = \log_{10}\left(N_o\right) - \left(F/\delta\right)^p \quad (11.4)$$

  where, δ (J/cm^2) stands for fluence that results in first one log reduction; and p is dimensionless that describes upward or downward concavity of the curve.

11.6 EFFICACY OF PULSED LIGHT TECHNOLOGY (PLT)

In the past two decades, many scientific studies have documented the microbial inactivation by PL *in vitro* as well as in different food products. However, the difference in experimental design renders it difficult to compare the studies. The results in Table 11.1 illuminate that PL processing has a high potential for the microbial inactivation of vegetative cells, yeasts, fungi together with spores. *In vitro* tests spotlight that *fungi* are the most heat resistant while vegetative cells the most sensitive to IPL [2, 26, 62]. However, no perceptible pattern can be drawn concerning the resistance of a different group of microbes. It has been reported that Gram-negative microbial cells are more vulnerable to PL compared with Gram-positive bacteria [2, 45, 62].

TABLE 11.1 Scientific Information for in vitro Microbial Decontamination on Agar by PLT

Targeted Microorganism	Fluence	Pulse Width	Number of Pulses	Initial Number	Log Reduction	References
	J	μs	—	log	—	
Vegetative Cells						
Enterobacer aerogenes	7	30	50	5.4	2.4 ± 0.5	[26]
Escherichia coli	7	30	50	5.3	4.7 ± 1.3	
Leuconostoc mesentroids	7	30	50	5.0	4.0 ± 0.8	
Listeria monocytogenes	7	30	50	5.0	2.8 ± 0.4	
Listeria monocytogenes	3	—	512	8.38	6.25	[45]
Photobacterium phosphoreum	7	30	50	4.8	4.4	[26]
Pseudomonas fluoroscens	7	30	50	4.8	>4.4	
Salmonella typhimurium	7	30	50	5.4	3.2 ± 0.7	
Shawnella putretaciens	7	30	50	5.1	3.9 ± 0.8	
Staphylococcus aureus	7	30	50	5.5	>5.1	
Bacterial Spores						
Alicyclobacillus acidoterrestris	7	30	50	3.3	2.5 ± 0.4	[26]
Bacillus cereus	7	30	50	6.3	>5.9	
Bacillus circulans	7	30	50	5.7	3.7 ± 0.3	
Yeasts						
Candida lambica	7	30	50	3.4	2.8 ± 0.4	[26]
Saccharomyces cerevisiae	3	—	100	8.4	3.7	[62]
Fungi						
Aspergillus flavus	7	30	50	5.2	2.2 ± 0.1	[26]
Botrytis cinerea	7	30	50	4.1	1.2 ± 0.1	

Although, *in vitro* studies show the bacterial reductions of up to 6.82 log on agar, yet it has been testified that complex matrices impair the microbial elimination by PL. Table 11.2 represents the scientific data on the destruction of microbes in liquid food products, powders, seeds, fresh produce, fruits, fish and seafoods, dairy products, and poultry and meat products. As discussed earlier, the properties of the food products (i.e., color, viscosity, opaqueness, etc.) have a major impact on the efficacy of microflora elimination by high intensity light pulses.

TABLE 11.2 Results for Bacterial Elimination in Foods using Pulsed Light Technology (PLT)

Targeted Microorganism	Treated Matrix	Experimental Conditions	Log Reduction	References
Liquid Food Products				
Clostridium sporogenes	Honey	5.6 J/cm^2; 135 pulses	0.89 to 5.46	[31]
Escherichia coli ATCC 25922	Apple juice	12.6 J/cm^2; Pulse width: 360; 3 pulses/s	2.66	[65]
Escherichia coli DH5-α	Apple juice	4 J/cm^2; Pulse width 360; 3 pulses/s	4	[57]
Escherichia coli DH5-α	Orange juice	4 J/cm^2; Pulse width 360;	2.90	
Listeria innocua 11288	Apple juice		2.98	
Listeria innocua 11288	Orange juice		0.93	
Powders				
Aspergillus niger	Corn meal	5.6 J/cm^2; 3 pulses/s	1.35–4.95	[35]
Listeria monocytogenes	Infant Food	Pulse width: 1.5	3	[13]
Saccharomyces cerevisiae	Wheat flour	31.12 J/cm^2; 64 pulses	0.7	[22]
Saccharomyces cerevisiae	Black pepper		2.93	
Seeds				
Escherichia coli	Alfalfa seeds	5.6 J/cm^2; 135 pulses	0.94–1.82	[67]
Fresh Produce				
Aerobic mesophiles	Iceberg lettuce	7 J; 675 pulses	1.24	[25]
Escherichia coli	Mushrooms	12 J/cm^2	3.03	[59]

TABLE 11.2 *(Continued)*

Targeted Microorganism	Treated Matrix	Experimental Conditions	Log Reduction	References
Fruits				
Aerobic plate count	Blueberries	3 pulses/s; time 90s; distance from lamp 13 cm	1.97	[28]
Botrytis cinerea	Strawberries	7 J; 3750 pulses	<1	[47]
Yeasts and mold	Blueberries	3 pulses/s; time 90s; distance from lamp 13 cm	1.27	[28]
Fish and Meat Products				
Escherichia coli	Raw salmon fillets	5.6 J/cm^2;135 pulses	0.24–0.91	[53]
Listeria monocytogenes	Raw shrimp	6.3 to 12.1 J/cm^2; pulse width 1.5 μs	2.2–2.4	[11]
Listeria monocytogenes	Raw salmon fillets	5.6 J/cm^2; 135 pulses	0.72–0.8	[53]
Pseudomonas spp.	Cottage cheese	16 J/cm^2; 2 pulses; Pulse duration 0.5 ms;	1.5	[15]
Milk				
Aerobic mesophiles	Bulk tank milk	25 J/cm^2; 110 pulses	>2	[68]
E. coli	Milk	7–28 J/cm^2; Pulse width 360	0.61–1.06	[54]
L. innocua		7–28 J/cm^2; Pulse width 360	0.51–0.84	
S. aureus		3 pulses/s; distance from the lamp 5–11 cm	0.55–7.23	[39]
S. thyphimurium		7–28 J/cm^2; Pulse width 360	0.51–1.73	[54]
Serratia marcescens	Bulk tank milk	25 J/cm^2; 110 pulses	>4	[68]
Poultry Products				
Salmonella	Egg shell	4 J/cm^2	>7.9	[16]
		4 J/cm^2	1	[30]
		2 J/cm^2	0.14	
		1.2–35.3 J/cm^2; 90 pulses	2–7.7	[36]

Gomez-Lopez et al. [25] reported log decrease ranging from 0.56 to 2.04 of mesophile and aerobic microorganisms in different vegetables. They reported that the difference in a log reduction of microbes among samples was due to differences in the sensitivity of inhabitant microbial population, protective substances in specific vegetables and due to shadow effect, i.e., site of microorganisms on and in the sample. Similarly, Sharma and Demirci [67] observed that the thickness of alfalfa seeds (solid food products) has a negative impact on the efficacy of microbial destruction by high intensity light pulses.

Surface topology (i.e., shape, and irregularities on the surface) also affect the efficacy of PL because bacterial cells can shield on crooked surfaces and consequently impair the microbial inactivation [65]. Huang et al. [33], Bialka, and Demirci [4] reported lesser decontamination in strawberry than blueberry and raspberry. Nicorescu et al. [50], Huang et al. [33] suggested that rough surface of the strawberry offers shielding/shadow effect to microbes from highly directional coherent PL leading to lesser decontamination compared to blueberry and raspberry. Similarly, Luksiene et al. [44] reported lesser decontamination in cauliflower than other fruits and vegetables. Koh et al. [37] experimented the impact of shape on microbial inactivation in cut cantaloupe. They observed 50% more reduction of the microbes in sphere than cuboid and triangular-prism shaped samples. A higher reduction in microbes can be attributed to higher area/volume ratio of sphere and due to less scattering of pulse light near the edges [37].

11.7 EFFECT OF PULSED LIGHT TECHNOLOGY (PLT) ON FOOD PROPERTIES

PLT is gaining the interest of researchers as it offers very low or almost nil changes in the nutritional as well as sensorial properties of the products. The scientific literature of PL on food properties is scanty and therefore no conclusive remarks can be extracted from the available data. However, the influence of PL on nutritional and physical properties depends on the experimental design (i.e., fluence, number of pulses and treatment time, etc.). Dunn et al. [17] showed no difference in nutritional properties of frankfurters exposed up to 300 kJ/m^2. Charles et al. [10] observed the effect of ILP (total fluence 8 J/cm^2) on fresh-cut mangoes and reported no change in phenol and total ascorbic acid (AC) concentration in the treated samples compared to control.

Lasagabaster et al. [43] demonstrated no devastating effects on the physical and functional characteristics of eggs after treating with high-intensity light pulses of fluence 2.1 J/cm^2. Oms-Oliu et al. [52] reported the decrease in phenolic compound and vitamin C content in fresh slices of mushrooms treated with high intensity light pulses at 12 and 28 J/cm^2. The negative effect on texture as well as color was also reported owing to thermal damage and increase in polyphenoloxidase activity.

Physical analysis was conducted by Gomez-Lopez et al. [25] on PL exposed vegetables and they reported that the overall impression depends upon the kind of vegetable being treated. A transient "plastic-off" odor was observed in PL treated white cabbage; however, no significant difference was reported for sensory scores of PL treated white cabbage from untreated counterparts. Correspondingly, PL exposed iceberg lettuce resulted in better sensory scores in relation to odor and overall visual quality compared to their unprocessed samples.

Dunn et al. [14] studied effect of ILP on sensory characteristics of various samples of summer flounder and reported lesser sensory scores after one week of refrigerated storage. While after two weeks of storage, the PL exposed sample resulted in better sensory scores compared to untreated counterparts.

11.8 COMBINATION PROCESSING WITH PULSED LIGHT TECHNOLOGY (PLT)

In the era of guaranteeing microbial safety and quality, researchers are working intensively to obtain food products with minimal changes in their nutritional and sensory characteristics. The limitations of PLT (viz., browning, uneven exposure, shadowing, and sample heating [3]) make it render to use in combination with other technologies to obtain minimally processed foods. In this regard, PLT have been studied with other preservation methods, such as ultraviolet light + ILP; PEF + PL; thermosonication + ILP; ultrasonication + ILP; AC + pulse white light; nisin + PL; H_2O_2 + ILP; malic acid + PL, etc.

The use of anti-browning elements (like AC and calcium chloride) have been studied with PL to reduce the browning of apples and mushrooms. AC (1%) was used on sliced mushroom prior to pulse light treatment and authors reported a significant reduction in browning during storage [52]. Gomez et al. [24] used AC/calcium chloride solution to reduce the change in the color of apples. However, they reported greater microbial growth in the combined treatment sample after 7-day refrigerated storage. The authors reported that it might be owing to tissue impairment and the antioxidant ability of AC.

Similarly, the pooled effect of PL + nisin resulted in a considerably greater reduction of *Listeriainnocua* in ready-made sausages compared to individual treatment [71].

The use of PL processing along with other non-thermal preservation methods have been experimented in different food products to study the influence on sensory characteristics and food safety. Combination of PL (3.3 J/cm^2) with other non-thermal technologies (such as UV (5.3 J/cm^2), PEF (34 KV/cm, 18 Hz, 93 ms) and manothermosonication (5 bar, 43°C, 750 W, 20 kHz)) was evaluated on mixture of cranberry and apple juice (10:90 v/v) [8]. In another study, Munoz et al. [49] specified that PL + thermosonication had an additive effect on *Escherichia coli* decontamination in orange juices compared to individual treatment.

The application of PL along with other non-thermal preservation methods and antimicrobial substances can be exploited in future to ensure the food safety characterizing fresh like products.

11.9 ADVANTAGES AND DISADVANTAGES OF PULSED LIGHT TECHNOLOGY (PLT)

PL is eco-friendlier compared to UV treatment owing to use of xenon lamp than mercury vapor lamp in PL system [27]. The PLT also have limitations that restrict its applications in the food industry. Shadow/shielding hampers the use of this decontamination process on food products having crevices on the effect of the surface. Moreover, the resistance of certain strains of microorganisms to the PLT may limit its applications [8].

11.10 SUMMARY

This chapter focuses on: the progress of different preservation methods in the food industry with main focus on non-thermal technologies; principle and basic components of PLT; different mechanisms of microbial inactivation and factors affecting microbial destruction by PLT; reaction kinetics of microbial inactivation; the efficacy of PLT to eliminate microbes *in vitro* as well as in different food products; effect of PLT on food properties; application of PLT as hurdle technology; and advantages and disadvantages of PLT. This chapter is of vital significance to fully adventure this technology in the food processing industry for future submissions.

KEYWORDS

- **ascorbic acid**
- **bacterial inactivation**
- **hurdle technology**
- **photothermal mechanism**
- **pulsed light technology**
- **reaction kinetics**

REFERENCES

1. Abida, J., Rayees, B., & Masoodi, F. A., (2014). Pulsed light technology: A novel method for food preservation. *International Food Research Journal*, *21*(3), 839–848.
2. Anderson, J. G., Rowan, N. J., MacGregor, S. J., Fouracre, R. A., & Farish, O., (2000). Inactivation of food-borne enteropathogenic bacteria and spoilage fungi using pulsed-light. *IEEE Transactions on Plasma Science*, *28*(1), 83–88.
3. Bhavya, M. L., & Umesh, H. H., (2017). Pulsed light processing of foods for microbial safety. *Food Quality and Safety*, *1*(3), 187–202.
4. Bialka, K. L., & Demirci, A., (2008). Efficacy of pulsed UV-light for the decontamination of *Escherichia coli* O157: H7 and *Salmonella* spp. on raspberries and strawberries. *Journal of Food Science*, *73*(5), M201–M207.
5. Bohrerova, Z., Shemer, H., Lantis, R., Impellitteri, C. A., & Linden, K. G., (2008). Comparative disinfection efficiency of pulsed and continuous-wave UV irradiation technologies. *Water Research*, *42*(12), 2975–2982.
6. Bolton, J. R., & Linden, K. G., (2003). Standardization of methods for fluence (UV dose) determination in bench-scale UV experiments. *Journal of Environmental Engineering*, *129*(3), 209–215.
7. Bradley, D., McNeil, B., Laffey, J. G., & Rowan, N. J., (2012). Studies on the pathogenesis and survival of different culture forms of *Listeria monocytogenes* to pulsed UV-light irradiation after exposure to mild-food processing stresses. *Food Microbiology*, *30*(2), 330–339.
8. Caminiti, I. M., Noci, F., Muñoz, A., Whyte, P., Morgan, D. J., Cronin, D. A., & Lyng, J. G., (2011). Impact of selected combinations of non-thermal processing technologies on the quality of an apple and cranberry juice blend. *Food Chemistry*, *124*(4), 1387–1392.
9. Cerf, O., (1977). A review tailing of survival curves of bacterial spores. *Journal of Applied Bacteriology*, *42*(1), 1–19.
10. Charles, F., Vidal, V., Olive, F., Filgueiras, H., & Sallanon, H., (2013). Pulsed light treatment as new method to maintain physical and nutritional quality of fresh-cut mangoes. *Innovative Food Science and Emerging Technologies*, *18*, 190–195.

11. Cheigh, C. I., Hwang, H. J., & Chung, M. S., (2013). Intense pulsed light (IPL) and UV-C treatments for inactivating *Listeria monocytogenes* on solid medium and sea foods. *Food Research International*, *54*(1), 745–752.
12. Cheigh, C. I., Park, M. H., Chung, M. S., Shin, J. K., & Park, Y. S., (2012). Comparison of intense pulsed light-and ultraviolet (UVC)-induced cell damage in *Listeria monocytogenes* and *Escherichia* coli O157: H7. *Food Control*, *25*(2), 654–659.
13. Choi, M. S., Cheigh, C. I., Jeong, E. A., Shin, J. K., & Chung, M. S., (2010). Nonthermal sterilization of *Listeria monocytogenes* in infant foods by intense pulsed-light treatment. *Journal of Food Engineering*, *97*(4), 504–509.
14. Dunn, J. E., Clark, R. W., Asmus, J. F., Pearlman, J. S., Boyer, K., Painchaud, F., & Hofmann, G. A., (1989). *U.S. Patent No. 4,871,559*. Washington, DC: U.S., Patent and Trademark Office.
15. Dunn, J. E., Clark, R. W., Asmus, J. F., Pearlman, J. S., Boyer, K., Painchaud, F., & Hofmann, G. A., (1991). *U.S. Patent No. 5,034,235*. Washington, DC: U.S. Patent and Trademark Office.
16. Dunn, J. E., (1996). Pulsed light and pulsed electric field for foods and eggs. *Poultry Science*, *75*(9), 1133–1136.
17. Dunn, J. E., (1995). Pulsed-light treatment of food and packaging. *Food Technology*, *49*(9), 95–98.
18. Farrell, H., Garvey, M., & Rowan, N., (2009). Studies on the inactivation of medically important Candida species on agar surfaces using pulsed light. *FEMS Yeast Research*, *9*(6), 956–966.
19. Farrell, H., Hayes, J., Laffey, J., & Rowan, N., (2011). Studies on the relationship between pulsed UV light irradiation and the simultaneous occurrence of molecular and cellular damage in clinically-relevant *Candida albicans*. *Journal of Microbiological Methods*, *84*(2), 317–326.
20. FDA (Food and Drug Administration), (1996). *Irradiation in the Production, Processing and Handling of Food* (p. 23). Code of Federal, regulations (CFR), Title 21, Part-179: FDA 21CFR179.41. Office of the Federal Register, US Government Printing Office, Washington DC.
21. Ferrario, M., Alzamora, S. M., & Guerrero, S., (2013). Inactivation kinetics of some microorganisms in apple, melon, orange and strawberry juices by high intensity light pulses. *Journal of Food Engineering*, *118*(3), 302–311.
22. Fine, F., & Gervais, P., (2004). Efficiency of pulsed UV light for microbial decontamination of food powders. *Journal of Food Protection*, *67*(4), 787–792.
23. Gates, F. L., (1928). On nuclear derivatives and the lethal action of ultra-violet light. *Science*, *68*(1768), 479–480.
24. Gómez, P. L., Salvatori, D. M., García-Loredo, A., & Alzamora, S. M., (2012). Pulsed light treatment of cut apple: Dose effect on color, structure, and microbiological stability. *Food and Bioprocess Technology*, *5*(6), 2311–2322.
25. Gomez-Lopez, V. M., Devlieghere, F., Bonduelle, V., & Debevere, J., (2005). Intense light pulses decontamination of minimally processed vegetables and their shelf-life. *International Journal of Food Microbiology*, *103*(1), 79–89.
26. Gómez-López, V. M., Devlieghere, F., Bonduelle, V., & Debevere, J., (2005). Factors affecting the inactivation of micro-organisms by intense light pulses. *Journal of Applied Microbiology*, *99*(3), 460–470.

27. Gomez-Lopez, V. M., Ragaert, P., Debevere, J., & Devlieghere, F., (2007). Pulsed light for food decontamination: A review. *Trends in Food Science and Technology*, *18*(9), 464–473.
28. Guner, S., & Topalcengiz, Z., (2018). Effect of pulsed ultraviolet light on natural microbial load and antioxidant properties of fresh blueberries. *Turkish Journal of Agriculture-Food Science and Technology*, *6*(6), 733–739.
29. Hayes, J. C., Laffey, J. G., McNeil, B., & Rowan, N. J., (2012). Relationship between growth of food-spoilage yeast in high-sugar environments and sensitivity to high-intensity pulsed UV light irradiation. *International Journal of Food Science and Technology*, *47*(9), 1925–1934.
30. Hierro, E., Manzano, S., Ordóñez, J. A., De La Hoz, L., & Fernández, M., (2009). Inactivation of *Salmonella enteric* serovar enteritidis on shell eggs by pulsed light technology. *International Journal of Food Microbiology*, *135*(2), 125–130.
31. Hillegas, S. L., & Demirci, A., (2003). *Inactivation of Clostridium Sporogenes in Clover Honey by Pulsed UV-Light Treatment* (p. 8). Presented at 2003 ASAE annual meeting, American Society of Agricultural and Biological Engineers, St. Joseph-MI.
32. Hiramoto, T., (1984). *U.S. Patent No. 4,464,336* (p. 10). Washington, DC: U.S. Patent and Trademark Office.
33. Huang, Y., Ye, M., Cao, X., & Chen, H., (2017). Pulsed light inactivation of murine norovirus, Tulane virus, *Escherichia coli* O157: H7 and *Salmonella* in suspension and on berry surfaces. *Food Microbiology*, *61*, 1–4.
34. Izquier, A., & Gómez-López, V. M., (2011). Modeling the pulsed light inactivation of microorganisms naturally occurring on vegetable substrates. *Food Microbiology*, *28*(6), 1170–1174.
35. Jun, S., Irudayaraj, J., Demirci, A., & Geiser, D., (2003). Pulsed UV-light treatment of corn meal for inactivation of *Aspergillus niger* spores. *International Journal of Food Science and Technology*, *38*(8), 883–888.
36. Keklik, N. M., Demirci, A., Puri, V. M., & Heinemann, P. H., (2012). Modeling the inactivation of *Salmonella Typhimurium, Listeria monocytogenes*, and *Salmonella enteritidis* on poultry products exposed to pulsed UV light. *Journal of Food Protection*, *75*(2), 281–288.
37. Koh, P. C., Noranizan, M. A., Karim, R., & Hanani, Z. A. N., (2016). Microbiological stability and quality of pulsed light treated cantaloupe (*Cucumis melo* L. reticulatus cv. Glamour) based on cut type and light fluence. *Journal of Food Science and Technology*, *53*(4), 1798–1810.
38. Kramer, B., & Muranyi, P., (2014). Effect of pulsed light on structural and physiological properties of *Listeria innocua* and *Escherichia coli*. *Journal of Applied Microbiology*, *116*(3), 596–611.
39. Krishnamurthy, K., Demirci, A., & Irudayaraj, J. M., (2007). Inactivation of *Staphylococcus aureus* in milk using flow-through pulsed UV-light treatment system. *Journal of Food Science*, *72*(7), M233–M239.
40. Krishnamurthy, K., Tewari, J. C., Irudayaraj, J., & Demirci, A., (2010). Microscopic and spectroscopic evaluation of inactivation of *Staphylococcus aureus* by pulsed UV light and infrared heating. *Food and Bioprocess Technology*, *3*(1), 93.
41. Kundwal, M. E., Lani, M. N., & Tamuri, A. R., (2015). Microbial inactivation using pulsed light: A review. *Jurnal Teknologi*, *77*(13), 99–109.

42. Lagunas-Solar, M. C., & Gómez-López, V. M., (2015). Cost sub-committee report on UV units, *UV Sources Specifications and Experimental Procedures, 2015*, 1–9.
43. Lasagabaster, A., Arboleya, J. C., & De Maranon, I. M., (2011). Pulsed light technology for surface decontamination of eggs: Impact on Salmonella inactivation and egg quality. *Innovative Food Science and Emerging Technologies*, *12*(2), 124–128.
44. Luksiene, Z., Buchovec, I., Kairyte, K., Paskeviciute, E., & Viskelis, P., (2012). High-power pulsed light for microbial decontamination of some fruits and vegetables with different surfaces. *Journal of Food Agriculture and Environment*, *10*(3–4), 162–167.
45. MacGregor, S. J., Rowan, N. J., McIlvaney, L., Anderson, J. G., Fouracre, R. A., & Farish, O., (1998). Light inactivation of food-related pathogenic bacteria using a pulsed power source. *Letters in Applied Microbiology*, *27*(2), 67–70.
46. Manzocco, L., Da Pieve, S., & Maifreni, M., (2011). Impact of UV-C light on safety and quality of fresh-cut melon. *Innovative Food Science and Emerging Technologies*, *12*(1), 13–17.
47. Marquenie, D., Michiels, C. W., Van, I. J. F., Schrevens, E., & Nicolaï, B. N., (2003). Pulsed white light in combination with UV-C and heat to reduce storage rot of strawberry. *Postharvest Biology and Technology*, *28*(3), 455–461.
48. Moraru, C. I., (2011). High-intensity pulsed light food processing: Chapter 9. In: Proctor, A., (ed.), *Alternatives to Conventional Food Processing* (pp. 367–386). Royal Society of Chemistry, Arkansas-USA.
49. Muñoz, A., Caminiti, I. M., Palgan, I., Pataro, G., Noci, F., Morgan, D. J., & Lyng, J. G., (2012). Effects on *Escherichia coli* inactivation and quality attributes in apple juice treated by combinations of pulsed light and thermosonication. *Food Research International*, *45*(1), 299–305.
50. Nicorescu, I., Nguyen, B., Moreau-Ferret, M., Agoulon, A., Chevalier, S., & Orange, N., (2013). Pulsed light inactivation of *Bacillus subtilis* vegetative cells in suspensions and spices. *Food Control*, *31*(1), 151–157.
51. Ohlsson, T., & Bengtsson, N., (2002). Minimal processing of foods with non-thermal methods: Chapter 3. In: Ohlsson, T., & Bengtsson, N., (eds.), *Minimal Processing Technologies in the Food Industries* (pp. 34–60). Woodhead Publishing Limited, Cambridge-England.
52. Oms-Oliu, G., Aguiló-Aguayo, I., Martín-Belloso, O., & Soliva-Fortuny, R., (2010). Effects of pulsed light treatments on quality and antioxidant properties of fresh-cut mushrooms (*Agaricus bisporus*). *Postharvest Biology and Technology*, *56*(3), 216–222.
53. Ozer, N. P., & Demirci, A., (2006). Inactivation of *Escherichia coli* O157: H7 and *Listeria monocytogenes* inoculated on raw salmon fillets by pulsed UV-light treatment. *International Journal of Food Science and Technology*, *41*(4), 354–360.
54. Palgan, I., Caminiti, I. M., Muñoz, A., Noci, F., Whyte, P., Morgan, D. J., & Lyng, J. G., (2011). Effectiveness of high intensity light pulses (HILP) treatments for the control of *Escherichia coli* and *Listeria innocua* in apple juice, orange juice and milk. *Food Microbiology*, *28*(1), 14–20.
55. Palmieri, L., & Cacace, D., (2005). High intensity pulsed light technology: Chapter 11. In: Da-Wen, S., (ed.), *Emerging Technologies for Food Processing* (pp. 279–306). Elsevier Academic Press, London-UK.
56. Paskeviciute, E., & Luksiene, Z., (2009). *Photosensitization as a Novel Approach to Decontaminate Fruits and Vegetables* (p. 110). PTEP, Serbia-Russia.

57. Pataro, G., Muñoz, A., Palgan, I., Noci, F., Ferrari, G., & Lyng, J. G., (2011). Bacterial inactivation in fruit juices using a continuous flow pulsed light (PL) system. *Food Research International*, *44*(6), 1642–1648.
58. Rajkovic, A., Tomasevic, I., Smigic, N., Uyttendaele, M., Radovanovic, R., & Devlieghere, F., (2010). Pulsed UV light as an intervention strategy against *Listeria monocytogenes* and *Escherichia coli* O157: H7 on the surface of a meat slicing knife. *Journal of Food Engineering*, *100*(3), 446–451.
59. Ramos-Villarroel, A. Y., Aron-Maftei, N., Martín-Belloso, O., & Soliva-Fortuny, R., (2012). The role of pulsed light spectral distribution in the inactivation of *Escherichia coli* and *Listeria innocua* on fresh-cut mushrooms. *Food Control*, *24*(1–2), 206–213.
60. Ramos-Villarroel, A., Aron-Maftei, N., Martín-Belloso, O., & Soliva-Fortuny, R., (2014). Bacterial inactivation and quality changes of fresh-cut avocados as affected by intense light pulses of specific spectra. *International Journal of Food Science and Technology*, *49*(1), 128–136.
61. Roberts, P., & Hope, A., (2003). Virus inactivation by high intensity broad spectrum pulsed light. *Journal of Virological Methods*, *110*(1), 61–65.
62. Rowan, N. J., MacGregor, S. J., Anderson, J. G., Fouracre, R. A., McIlvaney, L., & Farish, O., (1999). Pulsed-light inactivation of food-related microorganisms. *Applied and Environmental Microbiology*, *65*(3), 1312–1315.
63. Rowan, N. J., Valdramidis, V. P., & Gómez-López, V. M., (2015). Review of quantitative methods to describe efficacy of pulsed light generated inactivation data that embraces the occurrence of viable but non culturable state microorganisms. *Trends in Food Science and Technology*, *44*(1), 79–92.
64. Saikiran, K. C. H. S., Mn, L., & Venkatachalapathy, N., (2016). Different pulsed light systems and their application in foods: A review. *International Journal of Science, Environment, and Technology*, *5*(3), 1463–1476.
65. Sauer, A., & Moraru, C. I., (2009). Inactivation of *Escherichia coli* ATCC 25922 and *Escherichia coli* O157: H7 in apple juice and apple cider, using pulsed light treatment. *Journal of Food Protection*, *72*(5), 937–944.
66. Setlow, R. B., & Carrier, W. L., (1966). Pyrimidine dimmers in ultraviolet-irradiated DNAs. *Journal of Molecular Biology*, *17*, 237–254.
67. Sharma, R. R., & Demirci, A., (2003). Inactivation of *Escherichia coli* O157: H7 on inoculated alfalfa seeds with pulsed ultraviolet light and response surface modeling. *Journal of Food Science*, *68*(4), 1448–1453.
68. Smith, W. L., Lagunas-Solar, M. C., & Cullor, J. S., (2002). Use of pulsed ultraviolet laser light for the cold pasteurization of bovine milk. *Journal of Food Protection*, *65*(9), 1480–1482.
69. Takeshita, K., Shibato, J., Sameshima, T., Fukunaga, S., Isobe, S., Arihara, K., & Itoh, M., (2003). Damage of yeast cells induced by pulsed light irradiation. *International Journal of Food Microbiology*, *85*(1–2), 151–158.
70. Turtoi, M., & Nicolau, A., (2007). Intense light pulse treatment as alternative method for mold spores destruction on paper-polyethylene packaging material. *Journal of Food Engineering*, *83*(1), 47–53.
71. Uesugi, A. R., & Moraru, C. I., (2009). Reduction of *Listeria* on ready-to-eat sausages after exposure to a combination of pulsed light and nisin. *Journal of Food Protection*, *72*(2), 347–353.

72. Uysal, P. C., & Kirca, T. A., (2011). Effect of ultraviolet-C light on anthocyanin content and other quality parameters of pomegranate juice. *Journal of Food Composition and Analysis*, *24*(6), 790–795.
73. Wang, T., MacGregor, S. J., Anderson, J. G., & Woolsey, G. A., (2005). Pulsed ultra-violet inactivation spectrum of *Escherichia coli*. *Water Research*, *39*(13), 2921–2925.
74. Wekhof, A., (2000). Disinfection with flash lamps. *PDA Journal of Pharmaceutical Science and Technology*, *54*(3), 264–276.
75. Wekhof, A., Trompeter, F. J., & Franken, O., (2001). Pulsed UV disintegration (PUVD): A new sterilization mechanism for packaging and broad medical-hospital applications. In: *The First International Conference on Ultraviolet Technologies* (pp. 1–15).

CHAPTER 12

NOVEL PACKAGING SYSTEMS FOR FOOD PRESERVATION

NARENDER K. CHANDLA, VENUS BANSAL, GOPIKA TALWAR, SANTOSH K. MISHRA, and SUNIL K. KHATKAR

ABSTRACT

The packaging is an essential and integral part of our daily life and its demand has augmented due to revolution in industrial development. Traditionally, packaging helps in the containment of food products and it now continues to be the most innovative component of processed foods for accelerating demand and awareness of consumers towards packaging. Till date, novel food packaging systems (such as intelligent, smart/active, antimicrobial, and biodegradable/edible packaging) are the most recent developments to meet the preferences of consumers. Novel packaging systems not only serve the purpose of packaging but also offer specific functions in processing, transportation, and storage of foods. Novel packaging system maximizes shelf-life, and at the same time assesses the food quality while in pack, till the product reaches to the end-consumer. In the modern packaging system, internal changes in product's environmental conditions are controlled/sensed/detected/recorded or altered for the preservation action. Bioactive packaging, the release of antimicrobial agents from packaging material itself requires skillful efforts, to increase the shelf-life by preventing microbial growth and food spoilage. On the other hand, intelligent packaging (time-based temperature indicators (TTIs)), fresh-check indicator, gas-based indicators, thermochromic inks, and biochip-based sensors display the real-time information on the outside of the package. Hence, novel food packaging systems offer new and innovative food packaging solutions to achieve the best quality and safety of foods and convenience to consumers.

12.1 INTRODUCTION

The packaging is an integral part of food processing and the entire supply chain, where food losses generally occur due to spoilage of the food products. The packaging is a system of preparing products for transport, distribution, and storage. Packaging plays a substantial role in supply chain management right from the raw material conversion into processed foods, to the kitchen. Defective primary packaging leads to rejection of the product by the consumer at first sight.

Packaging protects the food from external environmental detrimental abuse and contains the whole, semi-finished, and final processed ready-to-eat food during transportation and storage. Preservation of food occurs due to packaging material as it serves as an interface between the environment and food products. To prevent the physical and chemical spoilage, packaging offers barrier properties between food and environmental conditions. However, food spoils itself after a certain time due to microbial incursion. Food packages protect food from mechanical/physical stress, light, oxygen, moisture, microbes, and dust.

Processed food provides quality, extended product life and convenience to use, and is increasing across the world [19]. Consumers are expecting better attributes in the novel and innovative packaging solution to the food and dairy sector. Prepared foods that are ready-to-use and ready-to-cook are the driven thrust behind the development of novel techniques to pack the food products. Engineering innovations have led to change the scenario of packaging in an earlier time as automatic packaging machinery to aseptically pack the food products in flexible packing and in aluminum; foils with printing facility during processing boosted the packing segment. Food spoilage and foodborne microbial outbreaks had given rise to develop novel food packaging methods with antimicrobial properties [2].

The packaging is required to contain food, to protect from environmental abuse and to display information to end-consumers. Development in packaging techniques in 20th century appeared as smart/intelligent, active, antimicrobial packaging and fresh-check indicators. Interventions in designing packaging material and to create inside the condition of the packages favorable to prevent the microbial spoilage, novel food packaging systems were developed and applied for the extended life of food products [9].

Due to increased demand of safe food, extension of shel-life and consumer accessibility/traceability, research is accelerating to develop novel packaging systems for sustainable quality of food product after processing.

To increase the efficiency of the packaging system, innovative intelligent-tailored packaging systems are advanced to ensure food safety. Active packaging control system maintains the internal conditions, however, monitoring and any significant change are depicted by intelligent packaging. Time temperature indicators (TTIs) track the overall temperature abuse and details about the history of products [6, 37, 53]. Intelligent and active packaging helps to maintain and regulate critical controls measures of the packaging system throughout the food chain. On the other hand, fresh check indicators help in real-time indication and to check the status of food products being spoiled in terms of the degree of deterioration within the package [23]. These fresh check indicators help us to decide the food condition so that the utilization of food products can be managed before its spoilage.

Novel food packaging systems help in designing, packaging material, and standardizing packages, which satisfy preservation action, functional, legislative, and marketing requirements effectively [15]. Novel food packaging systems offer new trends in food packaging and open new horizons and possibilities to retain the product quality as fresh produce. New horizons in packaging will minimize the losses, and enhance product quality to a substantial level. In return, a new packaging system will boost the food safety of processed food [3, 62]. Furthermore, novel packaging systems better satisfy the needs of today's customer by providing product real-time detailing of food to be spoiled, and ensure food safety without deteriorating the quality of food during storage.

This chapter discusses the novel packaging techniques, to acmes innovations in packaging technology, the development of new packaging material, and microbial inactivation. The principle and mechanism of action of intelligent, smart/active, antimicrobial/biodegradable/edible packaging would also be discussed. Recent methods of fresh-check, sensory, and intensity of microbial spoilage are also to be addressed. Real-time display of product information by novel packaging systems comprising sensors/biosensors and fresh check indicators are highlighted. Finally, it also includes the application of edible active packaging as a novel solution to the food industry.

12.1.1 PACKAGING AND SELECTION CRITERION

Choosing the right packaging for a particular product is a very crucial factor in the selection of a packaging system. The packaging is a science that needs the expertise of packaging technologists or engineers. Proper packaging material helps in reducing product spoilage, breakage, tempering, and in return, less product return will occur after proper packaging. Some benefits are durability, temper toughness, holding, cost-effective, convenience to use the packaging, and eco-friendly to the surroundings [44]. Scientific information helps in the selection of packaging in terms of their specifications conforming to standards and regulations of agencies. Techniques to select the packaging system depend on types of packaging: primary, secondary, and tertiary. Before selecting the material, the full detail of the product to be packaged is required.

Along with shielding the produce till it reaches to the end-consumer, primary packaging has a very significant role. Primary packaging becomes a key place to place the logo and nutritional and other information that will help to trace and differentiate the brand from others [15]. Primary packaging comes in physical contact to food material. During packaging material, evaluation, tensile strength, water vapor permeation, solubility, and transmission rates (gas and solvent), elongation at break, puncture test, sealability are the most important deciding factors to design a package before its commercialization [52].

Secondary packaging holds together every single unit. It is designed to transport mass capacities of the product to the point of sale or up to the retailer. Therefore, secondary packaging does not affect the quality of the food product packed, e.g., cartons, cardboard boxes, and plastic rings that hold a case of cans together.

In the case of tertiary packaging, pallets, and shippers are used to hold the secondary packs as a consignment to transport for long distances.

To design the tertiary packaging systems, details information include: where the food product is to market; the route and conditions under the product is to be transported. Nowadays biodegradable macromolecule-based polymer materials are gaining popularity and are used as a primary pack as biodegradable plastics and edible coatings for whole and processed food products. Biodegradable plastics are evaluated for their biodegradation rates as a critical quality parameter during their selection. In addition to these parameters, polymers are also assessed for their conversion after use, as recycled plastics to form other packaging materials for secondary packaging. Earlier packaging is to ensure the containment of the product, help, and prevent it from external damage but novel packaging technology has an important role in the attaining food safety due to preservation.

12.2 TYPES OF NOVEL PACKAGING SYSTEMS

Novel food packaging systems are active packaging, intelligent packaging, antimicrobial packaging, intelligent packaging integrated either with TTIs or biosensors, modified packaging, and bio-plastics (edible films/coatings) to ensure the safety and to improve the shelf-life of the whole and processed foods. These novel packaging systems are indicated in Figure 12.1.

12.2.1 ACTIVE PACKAGING

Active packaging is a distinct packaging system in which ingredients have been intentionally altered inside the package [59]. Active packaging alters the internal packed food conditions, to improve the shelf-life and maintain quality attributes of food. This packaging involves the fusion of active components within the plastic and changing the intrinsic attributes of package used for containment [17]. This packaging provides to the consumer, safe food as added advantages of recyclable and edible/degradable packaging system with less environment effect [12, 33, 41]. Active pack extends the storage period and enhances the food safety by altering the internal conditions of the processed food [36]. To preserve the food products, polymeric matrices release active agents (antioxidants and antimicrobials), and absorb undesirable food components (ethylene, oxygen, and water) [21].

Active packaging is basically an in-pack preservation technique to substitute customary food processing techniques (brining, acidification, dehydration, heat treatments, and chemical preservation) [40]. Controlled release and in-time delivery of ingredients into the food via packaging films is most important factor to achieve preservation action for extended shelf-life of product packaged [54]. Active packaging involves techniques that play a role in preservation, such as respiration (fruit and vegetables), lipid oxidation, microbial spoilage, and infestation (insects). In active packaging technique, food deterioration can be reduced to a significant level. Active packaging techniques are classified as: absorbers or scavengers and releasing systems. Absorbing system scavenges undesirable components like oxygen, ethylene, water, and specific compounds of interest. Discharging system such as antioxidants, carbon oxide and specific preservatives of antimicrobial properties either adds/emits active compounds directly into packaged food or placed in head space of food container/package [38, 72]. Self-heating/cooling and preservation by temperature sensitive films, fresh pads, and surface treated food packaging materials are other new active packaging techniques.

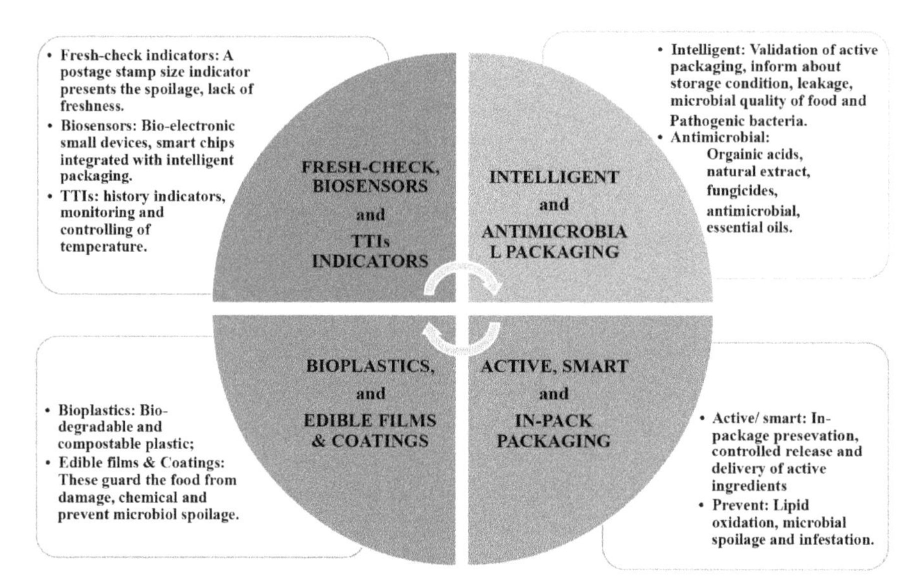

FIGURE 12.1 Novel food packaging systems for preservation and quality.

Generally, carbon dioxide emitter, ethanol emitter, antimicrobial organic acids (inhibition of pathogenic bacteria causing spoilage), sulfur dioxide (to inhibit mold), and butylated hydroxytoluene (to inhibit fat/oil oxidation) are used in active packaging systems. Nowadays, organophosphate, butylated hydroxytoluene, and thioester compounds are also utilized in a limited quantity due to more migration of compounds to pack processed foods and the production of toxins in pack [24]. Hence, the usage of these compounds is replaced with essential-oils, spices, tocopherol, and natural extracts of plants.

Antimicrobial concentration is optimized above minimum inhibition concentration (MIC) to have the influence of antimicrobial agents on the food system [2]. The solubility of the antimicrobial in food matrix is also a critical factor during functional component migration and storage period of the food to be preserved. More solubility facilitates more diffusion and very low solubility generates moderate effect. New development in active packaging is based on stimuli-responsive mechanism of polymers and biopolymers [5, 80]. During this, packaging material interacts with existing internal environment conditions and regulates itself the active release of functional molecules in reaction to external package conditions.

12.2.2 INTELLIGENT PACKAGING

Intelligent packaging contains an indicator to check and evaluate the food freshness being packed. Intelligent packaging provides data of product packaged, transported, and stored and till it reaches to end-consumer. This helps all stockholders about the status of the food product [5, 27, 57]. Intelligent packaging gives information on storage conditions, microbial quality of food, package leak, and information about specific pathogenic bacteria such as *E. coli* in those processed foods which are perishable, chilled/frozen, and have low or high oxygen concentration than the required standard in a pack [6, 82]. There are two indicators (internal and external), which are placed either internal or external on the food pack to regulate and monitor the quality of food. External indicators like time temperature-based indicators and internal indicators like oxygen indicators, microbial growth indicators, etc. are part of the intelligent packaging system [82].

Quality of the food product is measured after direct contact between food material and the package containing the markers. Indicators provide complete information and tack the condition of the package during transportation and storage and handling [60]. Therefore, an intelligent packaging system can monitor, record, and represent the spoilage or being spoiled, and report information to the consumer.

Ripeness indicators, gas leakages, time-temperature based indicators (TTIs), bio-probes, radiofrequency indicators (RIF)/radio frequency identification (RFID), and toxin identification indicators (TII) are some examples of complex indicators [70]. RFID is coupled with intelligent packaging system and the radiofrequency based system is connected to radar to give real-time information of the food product that includes a reader (receiver), label, and finally, radar is used to carry the information as a data carrier to the intelligent package to change the fusion of ingredients [38, 39]. RFID technology helps in finding the location and in addition to the condition of the pack by barcode identification of packaging.

The purpose of intelligent packaging is to monitor the conditions of the food package and do necessary alterations by coordinating with the active packaging system. Actually, intelligent packaging evaluates the effectiveness and active packaging system strength [15]. This concept led to the development of a smart packaging system to the food industry. An intelligent packaging system helps in making decision to consumer to buy the food product and warn of possible spoilage. Intelligent packaging provides the information and knowledge about the product throughout to maintain the quality of food products. The information is presented by adhesive indicators, which are attached to pack or printed on the package [17]. Therefore, the quality of food products can be improved. Based on the ability to sense, intelligent packaging communicates real-time information to the consumer. It helps in the detection and recording of external and internal changes that take place in food products. During the supply of the commodity, RFID improves the tractability.

Also, TTIs communicate about the temperature abuse of food products and indicate if the product at higher temperatures can deteriorate quicker. TTIs are mostly placed in containers during shipping to trace the history of the temperature fluctuations. Individual packages may also be tagged with TTIs within the shipping containers. The color change is the most important indicator of food spoilage to evaluate the shelf-life of food products. Nestle had employed TimeStrip Technology, a TTI based technology, to their food products in the United Kingdom [30]. In addition to this, thermochromic inks are available, which change the color with a change in temperature. This ink can be printed on the surface of the package and a reference color coding would help in checking the condition of the food product. Thermochromic ink will tell us the status of material inside the package if it is cold or hot enough without opening the package. Thermochromic inks have been popularized in the food and beverage industries [59]. Therefore, intelligent packaging can deliver specific information about the 'actual' shelf-life and point toward the existing food quality to safeguard product quality and safety and also to improve supply chain logistics and to lower the food losses.

12.2.3 ANTIMICROBIAL PACKAGING

Antimicrobial packaging is achieved by adding functional ingredients inside package so that packaging will restrict the microbial growth to prevent the food spoilage [26]. In this technique, active functional ingredients to preserve the food spoilage are released at controlled rates. This technology impregnates specific antimicrobials in a specific food packaging film/package and delivers the active ingredients in a stipulated period to kill pathogens, thereby enhance food quality, and help in extension of shelf-life [43]. Formulation preservative ingredients are added in the food directly, however, antimicrobial films are wrapped over the food. Instant addition of antimicrobials in packaging films results in prompt inhibition of undesirable microorganisms. Due to interactions of food with packaging matrix, effect of degradation in antimicrobial effect was observed as the survival population of pathogenic microorganisms continued to accelerate, as soon as the concentration of antimicrobial got declined [14, 20].

Different antimicrobial ingredients/agents possess different activities against each type of pathogenic microorganism, and these agents are either integrated into the food component or attached to polymeric material (pack) to release active functional ingredients. The antimicrobial agent performs microbial static effects, which retard the target growth of pathogens. Commonly used antimicrobial agents are: antioxidants, bacteriocins, enzymes, organic acids, natural extracts, fungicides, antimicrobial, essential oils (EOs), and antimicrobial polymers [43].

Control packaging focuses on the controlled release mechanism. Here, the packaging material is used as a controlled delivery medium to bring the active functional ingredients in the vicinity of the package to preserve processed food. This technique regulates the concentration by the controlled release of functional ingredients in the food matrix at a particular targeted level. Slow-release, time-release, and controlled release are three terminologies that are generally used by medication, antioxidants, and vitamin-controlled release [38]. Many parameters (like polymeric material matrix, surface molecular size, intermolecular binding between particles of polymer and within the food and the properties of functional ingredients) play an imperative part in the controlled discharge of an antimicrobial from a package (polymer/biopolymer) matrix [45].

Lysozyme (LYZ) prevents the evolution of microorganisms more specifically. For example, lactic acid bacteria (LAB) cause of fermentation (malolactic) in PVOH (polyvinyl alcohol) films, and in many cases level of networking of PVOH films affects the rate of release of antimicrobials [10]. Cooksey developed low-density polyethylene (LDPE) films coated with nisin on the natural polymeric matrix for packaging. Coating suppressed the *Staphylococcus Aureus* and *Listeria Monocytogenes* growth to a considerable level [16]. Nisin coated LDPE films are utilized to store the microbiota raw milk showing *M. Luteus ATCC 10240* inhibition. Structural modification of polymers/biopolymers increased the capacity to modulate the controlled release of immobilized functional active compounds. Polyanhydrides are biodegradable polymers gaining popularity in drug delivery and implant coatings [74].

Milk protein films blended with EOs have presented antioxidant and antimicrobial effects by controlling pathogenic bacteria [67]. Films prepared from ethylene-vinyl alcohol (EVA) linear incorporated with potassium sorbate inhibited the growth of microorganisms in cheese [77]. Starch-based chitosan incorporated films have shown increment in mechanical properties and exhibited controlled release of antimicrobial agents [8]. Garlic oil added to edible films made up of chitosan showed antimicrobial activity for most common bacterial strains, i.e., *E. coli, Salmonella, Bacillus cereus, Staphylococcus aureus,* and *L. Monocytogenes* [55].

Antimicrobial agents are blended, immobilized, or coated differently according to the design of the packaging system and requirements of food products. Coating and edible coating on the processed food create barrier layer, which contains antimicrobial agents. Coating material penetrates through the matrix on the surface of food to exhibit the preservation effect. Silver nanoparticles as antimicrobials agents are under trial for new antimicrobial food packaging systems [8]. The optimization of the antimicrobial agent to get released is a critical factor during preservation.

It is concluded that after the application of antimicrobial agents either the agent is released very late or is released faster than the expected cause of food spoilage. In the first instance, if it is released late, then microbes can grow instantly before the activity of the antimicrobial agent or the release of antimicrobial agents. In the second case, if the migration rate of antimicrobial agents through permeable packaging material is faster than the evolution rate of microorganism, then the antimicrobial agent will be consumed and depleted by growing microorganisms, earlier than the required storage of food product [45]. Therefore, the antimicrobial packaging system can retard the growth of microorganisms responsible for food spoilage and thus assure food safety with extended shelf-life.

12.2.4 TIME TEMPERATURE INDICATORS (TTIS)

Monitoring and controlling of temperature in food processing are of immense importance to all food processing industries. During the transportation and storage of commodities as consignment, monitoring of temperature is done as a batch in a container, but the complexity arises when the product is exposed to undesirable temperatures. Therefore, assuring the quality of individual unit/product becomes quite complex because the temperature abuse occurs within the batch and affects the temperature exposed by every single pack throughout the transportation and distribution line [77]. The technical solution to this problem is to monitor the individual product to assure complete safety and quality of food. Therefore, this requirement can be satisfied by time temperature-based indicators. Applications of active, intelligent, and antimicrobial packaging systems in food preservation are indicated in Table 12.1.

TABLE 12.1 Applications of Active, Intelligent, and Antimicrobial Packaging Systems in Food Preservation

Packaging System	Type of Packaging	Purpose of Packaging	Examples of Applications
Active packaging	Sachets, films, corks, and labels: Oxygen; carbon dioxide; ethylene; and humidity absorber.	**Prevent**: mold, yeast, and bacterial growth; bursting of the package, ripening control; excess moisture control.	Ready to eat products, milk powder, cheese; roasted coffee; fruits; bakery products, etc.
Intelligent packaging	**External**: TTIs. **Internal**: Oxygen, carbon dioxide, pathogen indicators. **Internal/External**: Microbial growth indicators.	**External**: Storage conditions. **Internal**: Storage conditions, package leak, and specific pathogenic bacteria. **Internal/External**: microbial quality of food during the storage of specified pathogenic bacteria like *E. Coli* O157.	**External**: Food stored under frozen conditions. **Internal**: food stored with reduced oxygen concentration, CO_2 for MAP, etc. **Internal/External**: perishable foods such as fruits and vegetable products, mil products and meat, fish, and poultry, etc.
Anti–microbial packaging	**Organic Acids:** benzoic, sorbic, propionates, acetic, and propionic acids **Enzymes:** Lysozyme, nisin, EDTA, propyle paraben, immobilized lysozyme, glucose oxidase; Bacteriocines. **Fungicides:** natural extracts and oxygen absorber, etc.	**Organic Acids:** Total bacteria, yeast, and mold reduction **Enzymes:** Bacterial reduction; *E. coli*, L. monocytogenes, S. aureus, Sal. typhimurium and lysozyme activity test. **Bacteriocines:** Reduction in total aerobes, *L. monocytogenes*, *Sal. typhimurium*, *M. flavus*, *M. favus*, etc. **Fungicides (Molds):** natural extracts and oxygen absorber (aerobes).	**Organic Acids:** Culture media, cheese, apple, and water, etc. Enzymes: Culture media, fish processing. **Bacteriocines:** Shredded cheese, processed fruits and vegetable, RTE foods. Fungicides (Bell pepper and cheese); natural extracts (lettuce, cucumber, strawberry,); oxygen absorber (Bread, breakfast cereals).

TABLE 12.1 *(Continued)*

Packaging System	Type of Packaging	Purpose of Packaging	Examples of Applications
Time-temperature indicators (TTIs)	Critical temperature indicators; and time-temperature integrator or indicators (TTIs)	Expose above and below reference temperature: show collective time and temperature expose beyond a reference precarious temperature; TTIs present continuous temperature response of the product's history.	CTI (temperature abuse and defrosting application); CTTI (Indicate breakdowns maintaining temperature in the distribution food chain); TTI provide continuous tile temperature change detail and prevent quality loss

Broadly, these indicators are classified as: critical temperature-based indicators (CTI), critical temperature time-based indicators (CTTI), and time-temperature integrator or sign indicators (TTIs). TTIs are classified as history indicators, i.e., the handling practices with respect to time and temperature being followed during transportation and storage of the product [76]. Temperature is one of the crucial parameters that has strong effect on spoilage of food products. Temperature abuse sometime becomes cause of microbial spoilage. Conventionally, an expiry or use by date is mentioned on the food product to aware the retailers and consumers on the storage life of foods. However, it does not ensure the handling practices being followed during the distribution and storage of the products. Keeping the products above their recommended storage temperature can seriously impair the quality and shelf-life. Therefore, to ensure food safety and authenticity of food goods, TTIs have an important role in the modern food packaging systems.

The first patent of TTI was in 1930s; and since then number of patents have been granted for TTIs. Despite the number of patents, the commercial applications of TTIs have been negligible till 1980s [75].

The working principle of TTIs is based on the change in physical properties, viz.: color, light reflection, or mechanical deformation of the indicator due to change in temperature for that time [78]. TTIs are further categorized as: partial history TTIs or full history TTIs depends on the responding temperature at which the properties of the indicator changes. Partial history TTIs operate when the product is stored above the predefined critical temperature while full history TTIs works as the product is stored above recommended temperature. Currently, commercial TTIs are either based on enzymatic change in color, diffusion of colorant or a polymerization reaction in the solid matrix to generate highly colored polymer [76, 78].

In enzymatic-based indicators, a change in color is achieved by a change in pH when the enzyme acts on the substrate. Enzyme and substrate are separated by a thin seal and when the product is exposed to higher temperatures, the seal separating the substances is broken resulting in the mixing of contents. VITSAB is the commercial example of an enzyme-based indicator in which hydrolysis of lipids causes the change in pH and thus the color (US Patent 4043871). Diffusion type TTIs works on the principle of flow of a colorant along the wick due to change in temperature at that specified time. The commercial example of a diffusion-based indicator is 3M Monitor Mark (US patent 3954011). A mechanism of the polymerization-based indicator changes in reflectance due to change in the color of the polymer. Fresh-check indicators are a commercial example of polymerization-based indicators (US patent 3999946). To select the TTIs for the specific food product, information of reaction kinetics of deterioration is essential to guess the storage life cycle of packaged food.

The knowledge regarding product deterioration kinetics is fundamental to optimally design the TTIs for the food product. A TTI should precisely simulate the change in quality attributes of the food product to be observed. Therefore, an initial step in the design of a TTI is to specifically define the quality characteristics of a product that are mainly responsible for deterioration. After defining the quality attributes, the study of a change in the kinetics of that particular characteristic of food product and TTI when exposed to the same change in temperature for that specific time is unconditionally essential to appropriately predict the storage life [79].

TTIs are intelligent packaging systems to manage the FIFO (first in first out) and in special cases, where product quality is deteriorating at accelerated rates in LIFO (last in first out) policy in the food industry. This helps to reduce the food wastage and ensure the quality of food to the end consumer.

12.2.5 FRESH-CHECK INDICATORS

Fresh-check indicators are mostly of size of a postage stamp and present spoilage, lack of freshness, temperature abuse and package leak causing food spoilage inside the package. A fresh check indicator shows the superiority of the food product straight away as a color sign than the leak indicator and time temperature indicator. Microbial quality is usually indicated based on the reaction between the metabolites and indicator as the microorganism starts growing inside the food product. Therefore, a fresh-check indicator provides to the consumer knowledge about right from very beginning of packaging up to time of consumption of food as a whole or processed.

The fresh-check indicator works on the principle that biochemical reaction occurs due to the temperature increase, causing a change in the color of the adhesive agent attached to packaging material [67]. As the active adhesive agent is exposed to the temperature, over time it changes the color of the active adhesive agent (tag) to show the degree of freshness or food spoilage. The active agent of the fresh-check indicator darkens quickly when exposed to higher temperatures [7, 29]. This helps the consumer to decide about the product to use or to discard within the prescribed product date codes. Fresh-check indicators are classified as indicators sensitive to: pH, volatile nitrogen compounds, hydrogen sulfide, microbial metabolites, and pathogen, etc. [26].

World Health Organization (WHO) applied this technology for real-time quality assessment to vaccines in Africa, which helped the workers to prevent the intake of those medicines with lost effectiveness due to exposure to higher temperature. This concept and technology was well-taken and revolutionized by pharmaceutical and food processing sectors to check the real time condition of their products after packaging [71].

Nowadays, fresh-check technology has expanded to applications in the fresh food industry. The reason "is to use by date or expiration date" is not food safety date as the product may spoil within this period during the whole food supply chain so that fresh-check indicators are considered good-faith promise of freshness [23]. Freshness plays a significant role to assess quality in totality.

12.2.6 MODIFIED/CONTROLLED ATMOSPHERIC PACKAGING

Modified atmosphere packaging (MAP) and controlled atmospheric storage/packaging (CAS/P) is based on the principle of modification of the environment in food contact to enhance shelf-life of fresh goods. The change in packaging environment generally consists of modification in concentration of gases (viz. oxygen, carbon dioxide, and nitrogen) to retard the physiological, microbiological, and deteriorative processes occurring in the product [1]. However, occasionally gases like CO, SO_2, and Arare also used to prolong the storage stability of fresh produce. In conventional preservation methods, the deterioration of spoilage causing bacteria and inactivation of enzymes is attained by high temperature [22]. Though in MAP, an increase in shelf-life is accomplished by retarding the growth of spoilage causing microorganisms and enzymes through alteration in their environmental gases. The extended shelf-life by MAP might be due to several possible factors, viz.: respiration rate retardation, less ethylene production, and bacteriostatic and bactericidal effects of carbon dioxide.

The basic difference between MAP and CAS is the way both conserve and generate the conditions around the product. In controlled atmospheric storage (CAS), continuous flushing of air (inert gas) can create and maintain a desired concentration of gases. MAP is maintained by respiration of agricultural-produce and permeation of unwanted gases within permeates packaging material [81]. Moreover, the MAP is more economical than CAS as continuous flushing of gas is not required. MAP is further categorized into [48]:

- In active modification, the concentration of desired gases is achieved by flushing the mixture of gases in the headspace of packaging material.
- In passive packaging, it is achieved by natural respiration of the products inside the packaging material.

The selection of concentration of gases for the extension of shelf-life is very crucial and varies from product to product. Generally, decrease in oxygen and increase in concentration of other gases is employed to reduce the rate of deterioration. Oxygen is not only essential for the growth of spoilage causing microorganisms but it also causes undesirable changes like oxidation of fat and change in color of food products. Carbon dioxide decreases the spoilage of food products by reducing the rate of respiration of some fresh produce and also by reducing the ripening process via suppressing the activation of ethylene [48, 81].

Moreover, the concentration >10–15% of CO_2 can reduce the growth of several spoilage causing microflora. Nitrogen is a non-reactive gas and is mainly flushed to eliminate other more reactive gases from the packaging material.

12.2.7 BIOPLASTIC (BIODEGRADABLE) PACKAGING

Many goods made of polymeric plastic materials are increasing plastic usage and increase the ecological problems, however bulk of plastic is landfilled and burnt [35]. Starch, protein, cellulose, gums, lipids, pectinates, and fiber-based biodegradable and edible films are gaining popularity to wrap up the food components and to prevent moisture loss and to create the glossy surface of the product [34, 65]. Growing environmental awareness motivates the scientists and researchers to look for eco-friendly biodegradable and edible films/coating.

Starch-based and polylactic acid (PLA)-based biodegradable polymeric materials are being commercialized in the market and research is going on. Native starch modified to thermoplastic starch (TPS) is known as destructurized starch (DS). TPS is made from native starch by swelling the starch in water or relevant solvent and plasticizer is added and mixed solution is kneaded and heated. This destrurization of starch results in thermo-mechanical transformation into amorphous polymeric material [64]. Thermoplastics are utilized as secondary packaging material to hold the primary pack or a whole fruit and vegetable/food.

12.2.8 CLASSIFICATION OF BIOPLASTICS

Plastic-based renewable plastic is biodegradable and compostable. Many renewable sources like corn/potato starch, proteins, and fibers are utilized in the development of bioplastics [65]. Corn and potato crops are major starch sources for bioplastics, and different starches from different sources are utilized in various food and non-food applications [25, 63, 68]. Bacterial microorganisms and nanoparticles especially carbohydrate chains origin can also be used to develop bioplastics [31, 69]. Bioplastics are classified as starch-based chemically synthesized [4, 18, 46, 49], polylactic acid (PLA) [32, 56, 58] and genetically modified bioplastics [28, 47, 83].

12.3 NANO PACKAGING SYSTEM

Nano particles-based biopolymers affect the water solubility, water permeation rates, tensile strength, and related stability. Nanocomposite films presented a strong degree of integration, improved strength, stiffness, toughness, and can decrease the water permeation and water solubility index to a significant level. Nanocomposites for the food have the potential to future plastics for packaging [66]. The concept of nano packaging includes: nanostructured multilayer films, nano-composites, nano-modulated lattice films, nano-crystalline, and nano-graded films and these options are new trends in preparation of nano-packaging material for food/non-food applications.

12.4 NANOBIOACTIVE PACKAGING

The bioactive packaging system is a novel technology, which changes the package condition to have a positive effect on our health. Bioactive packaging involves encapsulation, nano-encapsulations, enzyme, and micro-encapsulation immobilization. The desired concentration of bioactive components is maintained during the active component-controlled release. Fast dispersion rates of the active functional agent in packaged food are regulated during a complete storage period till the product packs are not opened [42]. During the bioactive packaging, biodegradable material is used to entrap the bioactive ingredient by encapsulation. This technique is utilized to develop probiotics and prebiotics, bioavailable flavonoids, and phytochemicals. Growing food habits to have functional food will boost the delivery of nano-bioactive ingredients through packaging.

12.5 EDIBLE PACKAGING SYSTEM

Edible films or coatings are a thin layer of coatings of eatable ingredients applied onto food products, which help in preservation, delivery, promotion, and marketing. Edible coatings functions may guard the food from microbiological, chemical, and physical spoilage. Mostly, polysaccharides are casted to form edible coatings of starch, protein, and food-grade additives like pectinates, gums, and cellulose, etc. Edible coatings/films can chiefly mend the functional requirements of processed foods, such as lower the respiration of fruits/vegetables, minimize the moisture loss and air movement, and improve the sensory qualities of foods. Nowadays, the concept to develop clear edible films is gaining momentum due to enhanced transparency, which can benefit in showing the actual image and the physical condition of food products during transportation, distribution, and storage till the product reaches to end-consumer [12].

Potential of edible films/coatings to fulfill the need of customers in terms of eco-friendly and being natural is the driving thrust in the development of such films. In addition to packaging, edible films fulfill extra quality to the food packaged. Biodegradable packaging is made of starch, protein, and other agricultural residue and/or commodities of industrial waste production; and this could help in elimination of waste. Edible films/coatings help in prevention of spoilage of food in addition to packaging. Edible coatings/films are prepared from edible biopolymer as base material and then additives (hydrocolloids) are added (cellulose, alginates, gums, etc.). More functional composite films (two or more edible base material) are being developed to better entrap and control the release of food active ingredients and additives [11, 50]. Proteins-based edible films of improved strength could be produced by using the extrusion process [46].

Thus, thermo-plasticization of proteins is still under research and requires large scale production, while optimizing the formulation and processing parameters [49]. Edible films are classified as (Table 12.2): polysaccharides based films (cellulose and cellulose derivatives, chitosan, and chitin, starch, and starch derivatives), lipids based films (waxes and parafines, acetoglycerides, shellac resins), protein films (gelatin, corn protein, soy protein and most frequently used wheat gluten) and composite edible films [82]. Therefore, the development of edible films/coatings will present a stimulating route-to create new packaging material.

TABLE 12.2 Biodegradable (Bio-Plastic), Nano Packaging and Edible Packaging System

Packaging System	Type of Packaging	Purpose of Packaging	Examples of Applications
Bioplastic	**Biopolymer**: PLA-based (cup, bowl, bags, jars, paper bags); starch-based (starch-based tray and packaging); cellulose-based packaging and metalized cellulose films	Biodegradable, biocompatible, environment-friendly, low cost of production.	PLA-based (beverage, potato, fresh salad, bread, etc.); Starch-based (milk chocolates and organic tomatoes); cellulose-based (kiwi, potato chips, sweets, etc.).
Edible packaging/ coating	**Primary packaging films and edible coatings:** starch, protein, cellulose, and lipid-based	Prevent physical-chemical and microbiological spoilage, moisture retention, etc.	Enrobed nuts, fruits, vegetables, dairy (cheese, paneer, etc.)
Nano packaging	**Nano particles-based packaging:** nanocomposites, nano-emulsions, nano-coatings, nano-structured materials, nanomaterial-based active/intelligent packaging films, and bionano composites	Higher stability, reinforcement to package strength by strong degree of integration	Nano-sized solubilisates and nanoparticle size-based pectin films application in dairy and food industry.

Hot pressed, extruded films showed advantages of transparency and homogeneous structure compared to biodegradable films made from thermoplastics starch [84]. However, the problem of solubility, high water permeation rates, tensile strength, and production of biopolymer exists but market demand of biodegradable plastic is increasing day by day. On the other hand, starch-based paper foam in food packaging is gaining momentum. As starch has sufficient moisture, stable foam is utilized by McDonalds for developing tray. Particles of small size have resulted in the development of starch viscous suspensions, cellulose fibers with water as solvent. This slurry is heated to 200°C, and the starch gets gelatinized and form tray when suspended and dried in mold. This is biodegradable and is easily recyclable.

Further proteins are utilized in the development of biodegradable packaging. Vegetable proteins and cereal proteins, milk proteins are natural proteins that are generally used for the development of biodegradable material [28]. In addition to starch-based biodegradable films, protein-based films are transparent and more water-resistant.

12.6 OPTIMIZATION OF FOOD PACKAGING

Optimized packaging system should offer convenience to consumers like easy to seal, handle, carry, easy-opening, recyclability all the basic attributes of the packaging system in addition to basic packaging properties. Packaging system optimization is done to create favorable intrinsic and extrinsic conditions so that food can be protected and spoilage of food products can be minimized. In addition to this, an *optimized packaging system* may also reduce food supply chain losses, and wholesale/retailer costs to a greater extent. Optimization of the food packaging system ensures the safety of the food and helps in improving the quality. While choosing a package for food material, its effect on environmental conditions is also considered. Therefore, optimization of package design is of immense importance for better containment, preservation, and distribution as an integral part of processing and distribution.

Package optimization in terms of the type of material (polymeric and biopolymeric) is a function of food character, processing, transportation, distribution, and retailing conditions, etc. In addition to these parameters, the critical factor in optimization is how long the food must be preserved compared to its real shelf-life [61]. Basically, all these factors in the decision matrix are taken into consideration when a new product must be launched in the market. Based on the techno-feasibility of the packaging system, a balance between the quality and cost of the packaging system will help to deliver cost-effective packaging solutions to consumers (Figure 12.2).

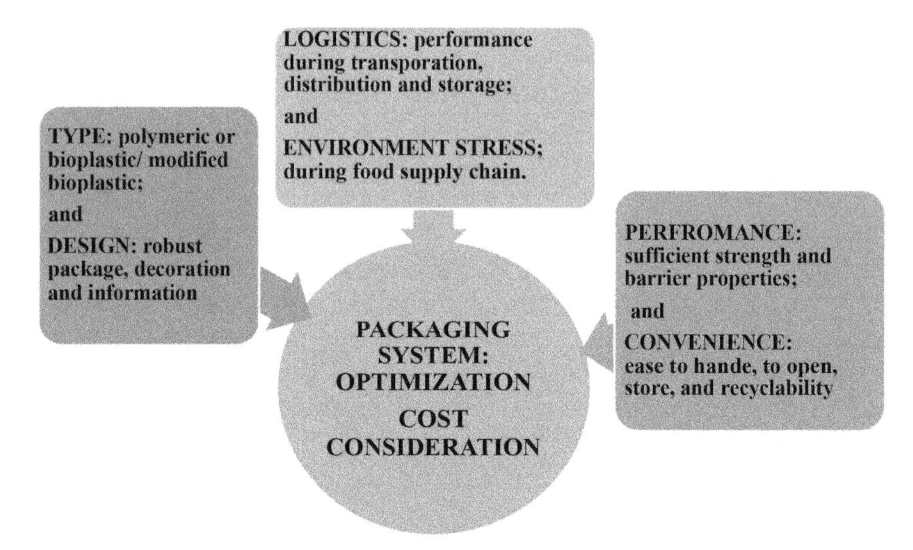

FIGURE 12.2 Optimization criteria for the food packaging system.

12.7 INTEGRATION OF BIOSENSORS IN THE PACKAGING SYSTEMS

Biosensors are analytical bioelectronic small devices to sense the signal by a bioreceptor and these signals are amplified to combine with a transducer to display measurable signals. Nanomaterial and biomaterial-based biosensors are being used as a novel sensing technology to see the food quality and safety [37]. This technology is non-destructive and non-contact to food and helps in assuring the rapid quality assessment of food without disturbing the contents in the package [4]. Miniaturized, nano, and microsensors are being developed to sense the biochemical and microbial changes occurring in the food product during its transportation and storage. Foodborne diseases have alarmed the need of biosensors for proper traceability and authentication of food products [73].

A biological sensor produces a voltage, exothermic reaction (heat), current of a reasonable magnitude in response to detect the target analyte. In the case of a biological sensor, a receptor (e.g., enzyme, antibody, and ligand) binds to a specific analyte/target molecule in a food sample and represents measurable signals. Biosensors are classified as: mechanical (resonant), optical detection biosensor, surface plasma resonance (SPR), electrochemical biosensor, conductometric biosensor, amperometric biosensors, potentiometric biosensors, cell-based biosensors, lab on chip type biosensors, and DNA based biosensors.

The sensor integrated packaging system monitors the product's internal condition and performs data exchange between internal and external environmental conditions [13]. This in turn provides knowledge on the real state of the food product. The integration of biobased sensitive elements material with the package and/or labels will flow information to the end-consumer. Such integration helps in the execution of online control of package environmental conditions. Biosensors can be created by embedding microdevices that are generally called "smart chips," either by depositing micro/nano-sensitive component onto the label or the package itself [4, 6].

12.8 SUMMARY

The novel packaging system has a vital role in preservation, food safety assessment, and containment. The chapter focuses on the novel food packaging systems, requirements of innovation in packaging systems, the food quality during the whole food supply chain. Packaging engineers, technologists, and food scientists together have developed novel packaging systems to fulfill the demand of consumers. Research in food packaging systems has enhanced the functionality of package manifolds by integrating the already existing packaging techniques with new packaging materials. Therefore, the topics discussed in this chapter will help in better understanding the food packaging systems. The applications of novel packaging systems will surely be of great importance in the entire food sector.

KEYWORDS

- **antimicrobial packaging**
- **critical temperature-based indicators**
- **edible packaging**
- **intelligent packaging**
- **modified atmosphere packaging**
- **time temperature indicators**

REFERENCES

1. Ahvenainen, R., (2003). *Novel Food Packaging Techniques* (1st edn., p. 400). Cambridge: Woodhead Publishing.
2. Appendini, P., & Hotchkiss, J. H., (2002). Review of antimicrobial food packaging. *Innovative Food Science and Emerging Technologies*, *3*(2), 113–126.
3. Aung, M. M., & Chang, Y. S., (2014). Traceability in a food supply chain: Safety and quality perspectives. *Food Control*, *39*, 172–184.
4. Awad, T. S., Moharram, H. A., Shaltout, O. E., Asker, D., & Youssef, M. M., (2012). Applications of ultrasound in analysis, processing, and quality control of food: A review. *Food Research International*, *48*(2), 410–427.
5. Bawa, P., Pillay, V., Choonara, Y. E., & Du Toit, L. C., (2009). Stimuli-responsive polymers and their applications in drug delivery. *Biomedical Materials*, *4*(2), 022001.
6. Biji, K. B., Ravishankar, C. N., Mohan, C. O., & Gopal, T. S., (2015). Smart packaging systems for food applications: A review. *Journal of Food Science and Technology*, *52*(10), 6125–6135.
7. Bremner, A., (2007). *Review of Traceability and Product Sensor Technologies Relevant to the Seafood Industry* (p. 48). Altona North, Vic-Au: Australian Sea Food Research Center. http://www.frdc.com.au/Archived-Reports/FRDC%20Projects/2007-710-DLD.PDF (accessed on 25 May 2020).
8. Brody, A. L., Bugusu, B., Han, J. H., Sand, C. K., & McHugh, T. H., (2008). Scientific status summary: Innovative food packaging solutions. *Journal of Food Science*, *73*(8), R107–R116.
9. Brody, A. L., Zhuang, H., & Han, J. H., (2010). *Modified Atmosphere Packaging for Fresh-Cut Fruits and Vegetables* (p. 320). New York: Wiley-Blackwell.
10. Buonocore, G. G., Del Nobile, M. A., Panizza, A., Corbo, M. R., & Nicolais, L., (2003). General approach to describe the antimicrobial agent release from highly swellable films intended for food packaging applications. *Journal of Controlled Release*, *90*(1), 97–107.
11. Campos, S., Doxey, J., & Hammond, D., (2011). Nutrition labels on pre-packaged foods: A systematic review. *Public Health Nutrition*, *14*(8), 1496–1506.
12. Chandla, N. K., Saxena, D. C., & Singh, S., (2017). Amaranth (*Amaranthus spp.*) starch isolation, characterization, and utilization in development of clear edible films. *Journal of Food Processing and Preservation*, *41*(6), e-article ID 13217.
13. Cheng, S., Azarian, M. H., & Pecht, M. G., (2010). Sensor systems for prognostics and health management. *Sensors*, *10*(6), 5774–5797.
14. Chi-Zhang, Y., Yam, K. L., & Chikindas, M. L., (2004). Effective control of *Listeria monocytogenes* by combination of nisin formulated and slowly released into a broth system. *International Journal of Food Microbiology*, *90*(1), 15–22.
15. Coles, R., McDowell, D., & Kirwan, M. J., (2004). *Food Packaging Technology* (Vol. 5, p. 346). Oxford, UK: Blackwell. Free online access.
16. Cooksey, K., (2000). Utilization of antimicrobial packaging films for inhibition of selected microorganisms. *Food Packaging: Testing Methods and Applications*, *753*, 17–25.
17. Dainelli, D., Gontard, N., Spyropoulos, D., Zondervan-Van, D. B. E., & Tobback, P., (2008). Active and intelligent food packaging: Legal aspects and safety concerns. *Trends in Food Science and Technology*, *19*, S103–S112.

18. Demirgöz, D., Elvira, C., Mano, J. F., Cunha, A. M., Piskin, E., & Reis, R. L., (2000). Chemical modification of starch based biodegradable polymeric blends: Effects on water uptake, degradation behavior, and mechanical properties. *Polymer Degradation and Stability*, *70*(2), 161–170.
19. Dobrucka, R., & Cierpiszewski, R., (2014). Active and intelligent packaging food-research and development: A review. *Polish Journal of Food and Nutrition Sciences*, *64*(1), 7–15.
20. Dutta, P. K., Tripathi, S., Mehrotra, G. K., & Dutta, J., (2009). Perspectives for chitosan based antimicrobial films in food applications. *Food Chemistry*, *114*(4), 1173–1182.
21. Flores, S., Famá, L., Rojas, A. M., Goyanes, S., & Gerschenson, L., (2007). Physical properties of tapioca-starch edible films: Influence of filmmaking and potassium sorbate. *Food Research International*, *40*(2), 257–265.
22. Floros, J. D., & Matsos, K. I., (2005). Introduction to modified atmosphere packaging, Chapter 10. In: Jung, H. H., (ed.), *Innovations in Food Packaging* (pp. 159–172). Elsevier.
23. Galagan, Y., & Su, W. F., (2008). Fadable ink for time-temperature control of food freshness: Novel new time-temperature indicator. *Food Research International*, *41*(6), 653–657.
24. Gomez-Estaca, J., Lopez-De-Dicastillo, C., Hernández-Muñoz, P., Catalá, R., & Gavara, R., (2014). Advances in antioxidant active food packaging. *Trends in Food Science and Technology*, *35*(1), 42–51.
25. Guilbert, S., Cuq, B., & Gontard, N., (1997). Recent innovations in edible and/or biodegradable packaging materials. *Food Additives and Contaminants*, *14*(6–7), 741–751.
26. Han, J. H., (2002). *Active Food Packaging* (pp. 11–24). Winnipeg, Canada: SCI Publication & Communication Services.
27. Han, J. H., & Flores, J. D., (1997). Casting antimicrobial packaging films. *J. Plastic Film and Sheeting*, *13*, 287–298.
28. Haugard, V. K., Danielsen, B., & Bertelsen, G., (2003). Impact of polylactate and poly on food quality. *European Food Research and Technology*, *216*(3), 233–240.
29. Hempel, A. W., (2014). *Use of Oxygen Sensors for the Non-Destructive Measurement of Oxygen in Packaged Food and Beverage Products and its Impact on Product Quality and Shelf-Life* (p. 270). PhD Dissertation, Department of Agriculture, Food and the Marine, University of College Cork, Ireland.
30. Huff, K., (2008). *Active and Intelligent Packaging: Innovations for the Future* (pp. 1–13). Department of Food Science and Technology. Virginia Polytechnic Institute and State University, Blacksburg, VA.
31. Jabeen, N., Majid, I., & Nayik, G. A., (2015). Bioplastics and food packaging: A review. *Cogent Food and Agriculture*, *1*(1), 8, e-article ID 1117749.
32. Jamshidian, M., Tehrany, E. A., Imran, M., Jacquot, M., & Desobry, S., (2010). Poly-lactic acid: Production, applications, nano composites, and release studies. *Comprehensive Reviews in Food Science and Food Safety*, *9*(5), 552–571.
33. Jin, T., & Zhang, H., (2008). Biodegradable polylactic acid polymer with nisin for use in antimicrobial food packaging. *Journal of Food Science*, *73*(3), M127–M134.
34. Kale, G., Auras, R., & Singh, S. P., (2006). Degradation of commercial biodegradable packages under real composting and ambient exposure conditions. *Journal of Polymers and the Environment*, *14*(3), 317–334.

35. Kirwan, M. J., & Strawbridge, J. W., (2003). Plastics in food packaging: Chapter 7. In: Coles, R., McDowell, D., & Kirwan, M. J., (eds.), *Food Packaging Technology* (pp. 174–240). Blackwell Publication, CRC press.
36. Kruijf, N. D., Beest, M. V., Rijk, R., Sipiläinen-Malm, T., Losada, P. P., & Meulenaer, B. D., (2002). Active and intelligent packaging: Applications and regulatory aspects. *Food Additives and Contaminants*, *19*(S1), 144–162.
37. Kuswandi, B., Wicaksono, Y., Abdullah, A., Heng, L. Y., & Ahmad, M., (2011). Smart packaging: Sensors for monitoring of food quality and safety. *Sensing and Instrumentation for Food Quality and Safety*, *5*(3–4), 137–146.
38. LaCoste, A., Schaich, K. M., Zumbrunnen, D., & Yam, K. L., (2005). Advancing controlled release packaging through smart blending. *Packaging Technology and Science: An International Journal*, *18*(2), 77–87.
39. Lee, S. J., Choi, D. S., & Hur, S. J., (2015). Current topics in active and intelligent food packaging for preservation of fresh foods. *J. Sci. Food Agric.*, *95*(14), 2799–2810.
40. López-de-Dicastillo, C., Catalá, R., Gavara, R., & Hernández-Muñoz, P., (2011). Food applications of active packaging EVOH films containing cyclodextrins for the preferential scavenging of undesirable compounds. *Journal of Food Engineering*, *104*(3), 380–386.
41. Lopez-Rubio, A., Almenar, E., Hernandez-Muñoz, P., Lagarón, J. M., Catalá, R., & Gavara, R., (2004). Overview of active polymer-based packaging technologies for food applications. *Food Reviews International*, *20*(4), 357–387.
42. Lopez-Rubio, A., Gavara, R., & Lagaron, J. M., (2006). Bioactive packaging: Turning foods into healthier foods through biomaterials. *Trends in Food Science and Technology*, *17*(10), 567–575.
43. Malhotra, B., Keshwani, A., & Kharkwal, H., (2015). Antimicrobial food packaging: Potential and pitfalls. *Frontiers in Microbiology*, *6*, 611.
44. Marsh, K., & Bugusu, B., (2007). Food packaging: Role, materials, and environmental issues. *Journal of Food Science*, *72*(3), R39–R55.
45. Mastromatteo, M., Mastromatteo, M., Conte, A., & Del Nobile, M. A., (2010). Advances in controlled release devices for food packaging applications. *Trends in Food Science and Technology*, *21*(12), 591–598.
46. Mensitieri, G., Di Maio, E., Buonocore, G. G., Nedi, I., Oliviero, M., Sansone, L., & Iannace, S., (2011). Processing and shelf-life issues of selected food packaging materials and structures from renewable resources. *Trends in Food Science and Technology*, *22*(2–3), 72–80.
47. Modi, S. J., (2010). Assessing the feasibility of poly-(3-hydroxybutyrate-co-3-hydroxyvalerate) (phbv) and poly-(lactic acid) for potential food packaging applications. *Master's Thesis* (p. 123). The Ohio State University, Columbus-OH.
48. Mullan, M., & McDowell, D., (2003). Modified atmosphere packaging. In: Coles, R., McDowell, R., & Kirwan, M. J., (eds.), *Food Packaging Technology* (pp. 303–339). Blackwell Publishing, Oxford.
49. Müller, C. M., Laurindo, J. B., & Yamashita, F., (2009). Effect of cellulose fibers addition on the mechanical properties and water vapor barrier of starch-based films. *Food Hydrocolloids*, *23*(5), 1328–1333.
50. Oliviero, M., Verdolotti, L., Di Maio, E., Aurilia, M., & Iannace, S., (2011). Effect of supramolecular structures on thermoplastic zein-lignin bio-nanocomposites. *Journal of Agricultural and Food Chemistry*, *59*(18), 10062–10070.

51. Oussalah, M., Caillet, S., Salmiéri, S., Saucier, L., & Lacroix, M., (2004). Antimicrobial and antioxidant effects of milk protein-based film containing essential oils for the preservation of whole beef muscle. *Journal of Agricultural and Food Chemistry*, *52*(18), 5598–5605.
52. Park, S. K., Rhee, C. O., Bae, D. H., & Hettiarachchy, N. S., (2001). Mechanical properties and water-vapor permeability of soy-protein films affected by calcium salts and glucono-δ-lactone. *Journal of Agricultural and Food Chemistry*, *49*(5), 2308–2312.
53. Pavelkova, A., (2016). Time temperature indicators as devices intelligent packaging. *Acta Universitatis Agriculturae et Silviculturae Mendelianae Brunensis*, *1*(1), 245–251.
54. Peltzer, M., Wagner, J., & Jiménez, A., (2009). Migration study of carvacrol as a natural antioxidant in high-density polyethylene for active packaging. *Food Additives and Contaminants*, *26*(6), 938–946.
55. Pranoto, Y., Salokhe, V. M., & Rakshit, S. K., (2005). Physical and antibacterial properties of alginate-based edible film incorporated with garlic oil. *Food Research International*, *38*(3), 267–272.
56. Rasal, R. M., Janorkar, A. V., & Hirt, D. E., (2010). Poly (lactic acid) modifications. *Progress in Polymer Science*, *35*(3), 338–356.
57. Restuccia, D., Spizzirri, U. G., Parisi, O. I., Cirillo, G., Curcio, M., Iemma, F., & Picci, N., (2010). New EU regulation aspects and global market of active and intelligent packaging for food industry applications. *Food Control*, *21*(11), 1425–1435.
58. Rhim, J. W., Hong, S. I., & Ha, C. S., (2009). Tensile, water vapor barrier and antimicrobial properties of PLA/nanoclay composite films. *LWT-Food Science and Technology*, *42*(2), 612–617.
59. Robertson, G. L., (2012). Active and intelligent packaging: Chapter 15. In: Robertson, G. L., (ed.), *Food Packaging: Principles and Practice* (3rd edn., pp. 333–360). Boca Raton, FL: CRC Press.
60. Robertson, G. L., (2009). *Food Packaging and Shelf-Life: A Practical Guide* (1st edn., p. 404). Boca Raton, FL, CRC Press.
61. Robertson, G. L., (2012). *Food Packaging: Principles and Practice* (3rd edn., p. 733). Boca Raton, FL: CRC Press.
62. Rooney, M. L., (1995). Active packaging in polymer films. In: *Active Food Packaging* (pp. 74–110) Australia: Springer Science Business Media, B. V.
63. Scott, G., & Wiles, D. M., (2001). Programmed-life plastics from polyolefin's: A new look at sustainability. *Biomacromolecules*, *2*(3), 615–622.
64. Shanks, R. A., Li, J., & Yu, L., (2000). Polypropylene-polyethylene blend morphology controlled by time-temperature-miscibility. *Polymer*, *41*, 2133–2139.
65. Siracusa, V., Rocculi, P., Romani, S., & Dalla, R. M., (2008). Biodegradable polymers for food packaging: A review. *Trends in Food Science and Technology*, *19*(12), 634–643.
66. Sisti, L., & Totaro, G., (2016). *Biodegradable and Biobased Polymers for Environmental and Biomedical Applications* (p. 515). New York: John Wiley & Sons.
67. Smolander, M., Alakomi, H. L., Ritvanen, T., Vainionpaa, J., & Ahvenainen, R., (2004). Monitoring of the quality of modified atmosphere packaged broiler chicken cuts stored in different temperature conditions. A time-temperature indicators as quality-indicating tools. *Food Control*, *15*(3), 217–229.
68. Song, J. H., Murphy, R. J., Narayan, R., & Davies, G. B. H., (2009). Biodegradable and compostable alternatives to conventional plastics. *Philosophical Transactions of the Royal Society of London B: Biological Sciences*, *364*(1526), 2127–2139.

69. Sorrentino, A., Gorrasi, G., & Vittoria, V., (2007). Potential perspectives of bio-nanocomposites for food packaging applications. *Trends in Food Science and Technology*, *18*(2), 84–95.
70. Stauffer, L., Wursch, A., Gächter, B., Siercks, K., Verettas, I., Rossopoulos, S., & Clavel, R., (2005). A surface-mounted device assembly technique for small optics based on laser reflow soldering. *Optics and Lasers in Engineering*, *43*(3–5), 365–372.
71. Stuppa, G., (2007). *The Fresh-Check Indicator a Unique tool for Brand Differentiation and Customer Loyalty: Review* (p. 18). Morris Plains, NJ: TempTime Corporation.
72. Suppakul, P., Miltz, J., Sonneveld, K., & Bigger, S. W., (2003). Antimicrobial properties of basil and its possible application in food packaging. *Journal of Agricultural and Food Chemistry*, *51*(11), 3197–3207.
73. Takhistov, P., George, B., & Chikindas, M. L., (2009). *Listeria monocytogenes'* step-like response to sub-lethal concentrations of nisin. *Probiotics and Antimicrobial Proteins*, *1*(2), 159.
74. Tamada, J., & Langer, R., (1992). The development of polyanhydrides for drug delivery applications. *Journal of Biomaterials Science, Polymer Edition*, *3*(4), 315–353.
75. Taoukis, P. S., & Labuza, T. P., (1989). Applicability of time-temperature indicators as shelf-life monitors of food products. *Journal of Food Science*, *54*(4), 783–788.
76. Taoukis, P. S., & Labuza, T. P., (2003). Time-temperature indicators (TTIs): Chapter 6. In: Ahvenainen, R., (ed.), *Novel Food Packaging Techniques* (pp. 103–126). Woodland publishing, Elsevier.
77. Ture, H., Gällstedt, M., & Hedenqvist, M. S., (2012). Antimicrobial compression-molded wheat gluten films containing potassium sorbate. *Food Research International*, *45*(1), 109–115.
78. Wells, J. H., & Singh, R. P., (1988). A kinetic approach to food quality prediction using full-history time-temperature indicators. *Journal of Food Science*, *53*(6), 1866–1871.
79. Welt, B. A., Sage, D. S., & Berger, K. L., (2003). Performance specification of time-temperature integrators designed to protect against botulism in refrigerated fresh foods. *Journal of Food Science*, *68*(1), 2–9.
80. White, E. M., Yatvin, J., Grubbs, III. J. B., Bilbrey, J. A., & Locklin, J., (2013). Advances in smart materials: Stimuli-responsive hydrogel thin films. *Journal of Polymer Science Part B: Polymer Physics*, *51*(14), 1084–1099.
81. Yam, K. L., & Lee, D. S., (2018). Design of modified atmosphere packaging for fresh produce: Chapter 3. In: Rooney, M. L., (ed.), *Active Food Packaging* (pp. 55–73). Switzerland AG: Springer Nature.
82. Yam, K. L., Takhistov, P. T., & Miltz, J., (2005). Intelligent packaging: Concepts and applications. *Journal of Food Science*, *70*(1), R1–R10.
83. Yoon, Y. J., Lu, Y., Frye, R. C., & Smith, P. R., (1999). Spiral transmission-line baluns for RF multichip module packages. *IEEE Transactions on Advanced Packaging*, *22*(3), 332–336.
84. Zhang, Q. X., Yu, Z. Z., Xie, X. L., Naito, K., & Kagawa, Y., (2007). Preparation and crystalline morphology of biodegradable starch/clay nano-composites. *Polymer*, *48*(24), 7193–7200.

CHAPTER 13

ROLE OF BIOSENSORS IN QUALITY AND SAFETY OF DAIRY FOODS

H. V. RAGHU, AJEET SINGH, and NARESH KUMAR

ABSTRACT

The dairy industry is one of the many industries that are concerned with the occurrence of pathogenic bacteria, where a pathogenic microbe further leads to a destructive influence on human health. The food-borne outbreaks of diseases from bacterial pathogen contamination and chemicals are very common around the world. Improper pasteurization and even incidents of the presence of contaminating microbes after pasteurization occurs most often. Dairy products are consumed in large volumes so infrequent contamination of commercially distributed products may end up in several sicknesses. Biosensors can help in the analysis of food quality. Biosensors facilitate to hold out the procedures that are selective, sensitive, rapid, cost-efficient, and moveable. These devices are excellent substitutes for the current conservative techniques. The current trend is a biosensor application within the food processing discipline. The fundamental principle, biosensor development, history, and their classification are discussed in this chapter. Primary contaminants in food processing can be identified with the use of biosensors.

13.1 INTRODUCTION

One of the primary concerns in human health is food safety and food-borne infections. Although the level of food safety has improved the food-borne outbreaks from microbial spoilage, chemicals, yet these are still frequent [34]. In food spoilage, poultry (18%), dairy (18%), and beef (13%) commodities are primary culprits. The dairy industry has been a common source of health complaints (14%) and deaths (10%) from spoiled dairy products. The process of pasteurization removes most of the pathogens in milk and milk products, however, sometimes inappropriate pasteurization and chances of contamination after pasteurization usually occur [24].

Dairy products are consumed on a large scale around the globe, which may lead to infrequent spoilage of commercially distributed dairy products resulting in various illnesses [25]. In the dairy industry, a large number of outbreaks of diseases and illnesses are associated with *Campylobacter* spp. in raw milk [17]. In many of the industrialized countries, 30% of the total population every year suffer from food-borne diseases (FBD) mainly associated with *L. monocytogenes, Campylobacter*, *E. coli* O157:H7, *Salmonella, S. aureus, B. cereus, E. faecalis,* etc., compared with other uncommon food-borne bacterial pathogens [2, 6].

Milk is one of the important constituents in our daily diet; therefore its assurance for quality and safety is an essential component for the welfare of humanity. Dairy producers will be benefited with products that will have consistency in organoleptic and compositional features, longer shelf-life, and most importantly safe for human consumption. These unique qualities drive to increase the use of superior grade products.

In India, the quality of milk and milk products are governed by FSSAI (Food Safety and Standards Authority of India) regulations for both microbial and non-microbial contaminants. Therefore, it is mandatory to test the dairy products for the presence of pathogenic agents before transporting into the market. Presently, conventional testing methods have been used for the evaluation of microbial and non-microbial contaminants in dairy products. There are shortcomings in these conventional methods, which take several days to complete the testing of products and are costly and laborious. These time-consuming processes result in the delivery of partially processed products in the market as a result of which the industry has the only option of banning and recalling the product.

The industries are under tremendous pressure to meet the consumer requirement and there is an urgent need for substitute methods to replace conventional techniques. The alternate methods must be cheap, rapid, easy, and validated. Primary contaminants (like antibiotics, pesticides, heavy metals, bacterial pathogens) can be identified using the biosensor method in the dairy supply chain.

This chapter discusses the basic working principle of a biosensor along with its developmental history and classification.

13.2 BIOSENSORS

In the biosensor, the biochemical moiety is integrated with the transducer or electrode to produce a physical signal that is then translated to indicate the quality and safety of dairy products. Procedures involving biosensors are very selective, rapid, sensitive, cost-effective, and portable. In the dairy industry, biosensor devices are excellent substitutes for the existing conventional techniques [29].

13.2.1 TYPES OF BIOSENSORS

The types of biosensors are listed in Table 131.1.

TABLE 13.1 Classification of Biosensors

Signal Transduction	
Electrochemical	Amperometric, potentiometric, Conductometric, and Impidometric
Mass Sensitive	Piezoelectric and Cantilever biosensors
Optical	Optical density, Bioluminescence, Chemiluminescence, Fluorescence, Phosphorescence, Refractive index, Surface Plasmon/Total Internal Reflection, Diffraction, and Polarization
Thermometric	Energy, exo-thermic, and endo-thermic reactions
Biorecognition Molecule	
Antibodies	Monoclonal, and Polyclonal
Enzyme	Catalytic types and inhibition types
Lectin	Carbohydrate-binding proteins
Nucleic acid	Hybridization, and Low weight compound interaction
Protein	Metatropic receptors, and Ionotropic receptors
Spores	Bacterial endospores-inhibition type and germination types
Whole-cell	Micro-organisms, immune cells, and tissues

13.2.2 REQUIREMENTS OF A BIOSENSOR

The development of a biosensor requires the following features for its potential use in dairy plants [28].

- **Limit of Detection (LOD):** Minimum concentration or analyte detection ability with least number of steps for analysis.

- **Quick Response Time and Recovery Time:** Real-time monitoring and small recovery time.
- **Reproducibility of Signal Response:** It should show the same reaction/signal.
- **Selectivity or Specificity:** Biosensors should have no interference with analyte having an analogous structure.
- **Stability and Operating Life:** Biochemical and bioaffinity activity is retained for a longer period.
- The linearity of the response of biosensors reaction must cover the concentration range.

13.2.3 PRINCIPLE OF OPERATION OF A BIOSENSOR

As shown in Figure 13.1, the transducer, which creates a physical variation associated with the biochemical or bioaffinity intercation of a ligand with an analyte has been explained in detail by Thakur et al. [29] as follows:

- Exo-thermic or endo-thermic reaction, i.e., enzyme thermister.
- Oxidation-reduction reaction, i.e., Amperometric biosensors.
- Change in the transducer physico-optical characteristics during biochemical or biomolecular interaction, i.e., Optical-based biosensors.
- Change in the oscillation of vibrating material immobilized with biomolecules, i.e., Piezo-electric-based biosensors.

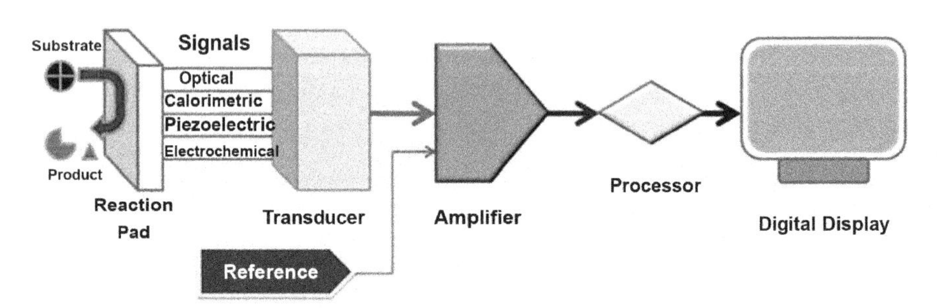

FIGURE 13.1 Graphical representation of biosensor.

13.2.4 HISTORY OF BIOSENSOR TECHNOLOGY

In 1956, Leland C. Clark Jr. (the forerunner in an arena of biosensor research) published a paper on electrode to measure the oxygen concentration in the blood [7]. Later during 1967, Updike, and Hicks presented their first oxygen sensor, wherein the glucose oxidase (GOX) was immobilized on functional electrode. Guilbault and Montalvo have described an earliest potentiometric electrode for urea estimation by urease enzyme in 1970 [12]. In 1973, Guilbault, and Lubrano [11] described a platinum electrode using glucose and a lactate enzyme for the detection of hydrogen peroxide. In 1974, a heat-sensitive enzyme sensor termed 'thermistor' was developed by Klaus Mosbach [21].

In 1983, Liedberg used SPR technique to monitor affinity reactions in real time [18]. In 1987, MediSense at Cambridge University developed a pen-sized meter based on screen-printed enzyme electrodes for the monitoring of glucose level at home. Further, the concept was improved into card and computer mouse-style formats, which caused a profit of US$175 million to the company by 1996.

13.2.5 CLASSIFICATION OF BIOSENSORS

The biosensor may be categorized based on the use of biological recognition mechanism or the mode of signal transduction [4].

- **By Biorecognition/Biomolecule:** It is a biological species or a living biological system or ligand that employs biochemical reactins for recognition. It can include: nucleic acid, enzyme, antibody, proteins, lectins, bacterial spores, a cellular organelle, microorganism, whole cell, and tissue, etc.
- **By the Transduction System:** It is used to translate the biological recognition into the signal, which is measurable and can be detected and displayed.

13.2.5.1 BIORECOGNITION MOLECULE

The construction of the biorecognition element forms the fundamental aspect of designing of a biosensor. Functional biological moieties having affinity and specificity include enzymes, antibodies, whole cells, and functional oligo-nucleotides [19, 26]:

- **Enzymes or Biocatalysts:** These are protein molecules, which are specific to substrates, catalyze a specific biochemical reaction. In the last decade, this type of biosensor has been extensively explored. In the dairy and pharmaceutical industry, an environmental pollution control and monitoring are primary concerns [19]. In the enzyme-based optical biosensor, enzymes immobilized on solid surfaces are used to improve half-life and sensitivity. Furthermore, the use of optical transducers improves self-containment and compactness. These enzyme-based optical biosensors can control environmental pollution. Although remarkable improvements have been made to expand the quality of enzyme-optical based biosensors to extend their competences for sky-scraping selectivity, sensitivity, and response time, yet there are limitations in the detection of environmental pollution for early warning [19].

- **Antibodies:** Immuno-sensors are potential tools in the area of quality and safety monitoring and diagnosis in the dairy industry exploring the specific biological interactions between specific antigen and antibody [19]. Optical or electronic transducers are used to monitor the interaction of antibody (contains two binding sites) with a particular target [8]. Hence, the antigen-antibody based immunosensor provides a highly specific interaction set-up to identify particular analyte in the food matrix. In an immunoassay, an antibody is defined by its selectivity and superiority. Therefore, it is widely known by the outcome of affinity between hapten binding to the carrier protein referred as antigen. Immunosensors are superior than enzyme-linked immune sorbent assay (ELISA), which differs with respect to its ability to regenerate and specificity. Also, there are limitations like: configuration and placing of immune cells on the exterior sensor chip are challenging [1]. The use of acids for regeneration in the antibody-based biosensor reduces the capability to recognize immobilized ligand after sensor surface reuse. This process provides firmness and consistency of an immunosensor. There is also a limit of regeneration up to several attempts, and in a repeated use, there is decrease in binding activity of ligand, which may yield inaccurate results.
- **Aptamers:** -based biosensors use sequences of single-stranded DNA or RNA, respectively, which are selected by systematic evolution of ligands by exponential (SELEX) enrichment [2]. The mode of aptamer binding exploits its target selectively by folding into a tertiary structure [8]. The structural compatibility, piling of aromatic rings, electrostatic, and Vander Waals interfaces, hydrogen bonding, or a blend of all these effects are most common interactions of an aptamer

to its target [1]. Because of their unique characteristics, aptamers serve as a useful alternative to antibodies used in immunesensors as sensing molecules [5]. Aptamers can be chemically synthesized, which minimize the batch-to-batch variations. Furthermore, chemical synthesis can also be used to modify these molecules to enhance the affinity, specificity, and stability. These molecules are also more stable, resistant to degradation and denaturation than antibodies [20].

- **DNAzymes or Deoxyribozymes or Catalytic DNAs:** These are nucleic acids (NAs) folding into a distinct 3-D construction with specific binding pocket [33]. These catalytic DNAs are selected by *in vitro* screening, made to work in the existence of a specific object. DNAzymes along with aptamers bind to a broad series of molecules, which in turn create a new class of functional NAs. [14].
- **Whole Cells:** These are excellent sources to detect toxic compounds. Large number of microbe-based whole-cell biosensors have been developed, which work on the fluorescing ability of the molecule to identify toxicity and pollutants [23].

13.3 TYPES OF TRANSDUCERS

1. **Electrochemical-Based Transducers:** This measure changes in electron transfer caused by a redox reaction involving the analyte at the surface of a suitable electrode [30]. These are further classified into:

    - Amperometric Transducer: It can detect the variations in current as a means of electroactive species concentration, e.g., solid electrolyte gas sensors electronic noses. These transducers work on the basis of an existing linear relationship between analyte concentration and current. Amperometric biosensor for Hemorrhagic *Escherichia coli (E. coli* O157:H7) was established by Varshney et al. [32] having a LOD of 8×10^1 cfu/ml in 6 h after pre-enrichment.
    - Potentiometric Transducer: It depends on changes in the potential of a system at constant current (I=0) or detects the change in the distribution of charge, e.g., ion-selective electrode (such as pH meter), ion-selective field-effect transistor, LAPS. A biosensor for *E. coli* based on this system was developed by Ercole et al. having a LOD of 7.1×10^2 cfu/ml in 30 min [9].

- Conductometric Transducer: It measures the change in conductance (the migration of ions). These are used for the study of enzyme-linked reactions, which are the result of a change in ion concentration.

2. **Piezoelectric Biosensors:** It changes in oscillation rate directly relative to the weight immersed by the material or sensitive to changes in density, mass, viscosity, and acoustic coupling phenomena, e.g., surface acoustic wave sensors. The general principle is based on surface immobilizing with a ligand on piezoelectric biosensors/ quartz crystal. Quartz crystal microbalance immunosensor can detect Entero-hemorrhagic of *E. coli* based on beacon immunomagnetic nanoparticles with a LOD of 23 CFU/ml and 53 CFU/ml in phosphate buffer saline (PBS)/milk with an overall sensing period of about 4 h with an additional time of 24 h for pre-enrichment at 37°C.
3. **Optical Biosensor:** It converts a biorecognition or biomolecular event into a measurable signal by an optical transducer. A layer is formed by placing a biomolecule or system in connection with the exterior of a transducer [10]. The interaction between the two produces an optical change in the reflected light. The transducer detects these changes and quantifies the interactions and the induced changes [31].
4. **Surface Plasmon Resonance:** It works on the basis of the biomolecular interaction of ligand molecule with an analyte on the surface of the gold chip. Hence, this interaction forms a critical criterion for designing a useful biosensor. Liedberg et al. first demonstrated the use of SPR for biosensing in 1983 [18]. It is a potent non-label technique for the study of the biomolecular interactions between the analyte and bio-recognition molecules. In this technique, when a photon light is an incident on a gold-coated surface, which can combine with the metal electrons and excites the density waves, instigating those waves into a single beam of electrical waves known as plasmon proliferate equivalent to the metal surface [8]. An oscillation of electron density waves creates an electrical meadow extending nearly 300 nm around the metal surface [13]. The waves formed at the interface of a metal and dielectrics are surface plasmons (SPs), which exploit the changes in the refractive index [15].

13.4 SUMMARY

The current trend of the food processing field is the application of biosensors. The basic working principle of a biosensor along with its developmental history and classification of its various generations are discussed in this chapter. Primary contaminants (like antibiotics, pesticides, heavy metals, bacterial pathogens) can be identified using this technique in the dairy supply chain.

KEYWORDS

- **biosensors**
- **enzyme-linked immune sorbent assay**
- **food pathogens**
- **food safety**
- **nucleic acids**
- **surface plasmon resonance**

REFERENCES

1. Algar, W. R., Tavares, A. J., & Krull, U. J., (2010). Beyond labels: A review of the application of quantum dots as integrated components of assays, bioprobes, and biosensors utilizing optical transduction. *Anal. Chim. Acta*, *673*(1), 1–25.
2. Alocilja, E. C., & Radke, S. M., (2003). Market analysis of biosensors for food safety. *Biosons. Bioelectrons.*, *18*(5), 841–846.
3. Anonymous. *Global Industry Analysts, Inc. Global Biosensors Market to Reach $12 Billion by 2015*. PR Web2012. http://www.prweb.com/ (accessed on 25 May 2020).
4. Belluzo, M. S., Ribone, M. É., & Lagier, C. M., (2008). Assembling amperometric biosensors for clinical diagnostics. *Sensors*, *8*(3), 1366–1399.
5. Borisov, S. M., & Wolfbeis, O. S., (2008). Optical biosensors. *Chem. Rev.*, *108*(2), 423–461.
6. Chemburu, S., Wilkin, E., & Abdel-Hamis, I., (2005). Detection of pathogenic bacteria in food samples using highly dispersed carbon particles. *Biosens. Bioelectron.*, *21*(3), 491–499.
7. Clark, L. C., (1956). Monitor and control of blood and tissue oxygen tensions. *Trans Am. Soc. Art. Int. Org.*, *2*(1), 41–48.
8. Dorst, B. V., Mehta, J., Bekaertb, K., Rouah-Martin, E., Coen, W. D., Dubruelc, P., Blusta, R., & Robbens, J., (2010). Recent advances in recognition elements of food and environmental biosensors: A review. *Biosen. Bioelectron.*, *26*(4), 1178–1194.
9. Ercole, C., Del Gallo, M., Mosiello, L., Baccella, S., & Lepidi, A., (2003). *Escherichia coli* detection in vegetable food by a potentiometric biosensor. *Sensors and Actuators B: Chemical*, *91*(1–3), 163–168.
10. Fang, Y., (2006). Label-free cell-based assays with optical biosensors in drug discovery. *Assay Drug Dev. Technol.*, *4*(5), 583–595.

11. Guilbault, G. G., & Lubrano, G. J., (1973). An enzyme electrode for the amperometric detection of glucose. *Ana. Chim. Acta*, *64*(3), 439–455.
12. Guilbault, G. G., & Montalvo, J. G., (1970). An enzyme electrode for the substrate urea. *J. Am. Chem. Soc.*, *92*(8), 2533–2538.
13. Hoa, X. D., Kirk, A. G., & Tabrizian, M., (2007). Towards integrated and sensitive surface plasma resonance biosensors: A review of recent progress. *Biosens. Bioelectron*, *23*(2), 151–160.
14. Hollenstein, M., Christopher, H., Curtis, L., David, D., & David, M. P., (2008). A highly selective DNAzyme sensor for mercuric ions. *Angewandte Chemie.*, *120*(23), 4418–4422.
15. Homola, J., (2003). Present and future of surface plasmon resonance biosensors. *Anal. Bioanal. Chem.*, *377*(3), 528–539.
16. Homola, J., (2008). Surface plasma resonance for detection of chemical and biological species. *Chem. Rev.*, *108*(2), 462–464.
17. Langer, A. J., Ayers, T., Grass, J., Lynch, M., Angulo, F. J., & Mahon, B. E., (2012). Non-pasteurized dairy products, disease outbreaks, and state laws-United States (1993–2006). *Emerg. Infect. Dis.*, *18*, 385–391.
18. Liedberg, B., Nylander, C., & Lundstrom, L., (1983). Surface plasmon resonance for gas detection and biosensing. *Sens. Actuat.*, *4*, 299–304.
19. Long, F., Zhu, A., Gu, C., & Shi, H., (2013). Recent progress in optical biosensors for environmental applications. In: Rinken, T., (ed.), *State of the Art in Biosensors: Environmental and Medical Applications* (pp. 4–28). In Tech Open: Rijeka, Croatia, Chapter 1.
20. Mehta, J., Rouah-Martin, E., Van, D. B., Maes, B., Herrebout, W., Scippo, M. L., Dardenne, F., et al., (2012). Selection and characterization of PCB-binding DNA aptamers. *Anal. Chem.*, *84*(3), 1669–1676.
21. Mosbach, K., & Danielsson, B., (1974). An enzyme thermistor. *Biochim. Biophys. Acta*, *364*(11), 140–145.
22. Mullet, W., Lai, E. P. C., & Yeung, J. M., (1998). Immunoassay of fumonisins by a surface plasmon resonance biosensor. *Anal. Biochem.*, *258*(2), 161–167.
23. Olaniran, A. O., Lettisha, H., & Balakrishna, P., (2011). Whole-cell bacterial biosensors for rapid and effective monitoring of heavy metals and inorganic pollutants in wastewater. *J Environ. Monit.*, *13*(10), 2914–2920.
24. Olsen, S. J., Ying, M., Davis, M. F., Deasy, M., Holland, B., & Iampietro, L., (2003). Multidrug-resistant *S. Typhimurium* infection from milk contaminated after pasteurization. *Emerg. Infect. Dis.*, *10*(5), 932–935.
25. Ryan, C. A., Nickels, M. K., Hargrett-Bean, N. T., Potter, M. E., Endo, T., & Mayer, L., (1987). Massive outbreak of antimicrobial-resistant *Salmonellosis* traced to pasteurized milk. *JAMA*, *258*(22), 3269–3274.
26. Shankaran, D. R., Gobi, K. V., & Miura, N., (2007). Recent advancements in surface plasmon resonance immune sensors for detection of small molecules of biomedical, food and environmental interest. *Sens. Actuators B.*, *121*(1), 158–177.
27. Stenberg, E., Persson, B., Ross, H., & Urbaniczky, C., (1991). Quantitative determination of surface concentration of protein with surface plasmon resonance by using radio labeled proteins. *J. Colloid Interface Sci.*, *143*(2), 513–526.
28. Thakur, M. S., & Raghavan, K. V., (2013). Biosensors in food processing. *J. Food Sci. Technol.*, *50*(4), 625–641.

29. Thapar, P., Salooja, M. K., & Malik, R. K., (2018). Application of biosensors for detection of contaminants in milk and milk products. *Acta Scientific Microbiology, 1*(12), 17–24.
30. Thévenot, D. R., Klara, T., Richard, A., Durst, S., & George, S. W., (2001). Electrochemical biosensors: Recommended definitions and classification. *Analytical Letters*, *34*(5) 635–659.
31. Van, D. G. J., & McBurney, R. N., (2005). Rescuing drug discovery: *In vivo* systems pathology and systems pharmacology. *Nat. Rev. Drug Discovery*, *4*, 961–967.
32. Varshney, M., Liju, Y., Xiao-Li, S., & Yanbin, L., (2005). Magnetic nanoparticle-antibody conjugates for the separation of *Escherichia coli* O157: H7 in ground beef. *J. Food Prot.*, *68*(9), 1804–1811.
33. Wang, X. D., & Wolfbeis, O. S., (2012). Fiber-optic chemical sensors and biosensors. *Analytical Chemistry*, *85*(2), 487–508.
34. WHO, (2007). *Food Safety and Food Borne Illness* (p. 10). Fact Sheet No. 237, World Health Organization (WHO), Geneva.

CHAPTER 14

POTENTIAL USE OF LACTIC ACID BACTERIA (LAB): PROTECTIVE CULTURES IN FOOD BIOPRESERVATION

MANJU GAARE and SANTOSH K. MISHRA

ABSTRACT

Lactic acid bacteria (LAB) as protective cultures (PC) have emerged as a promising alternative to chemical preservatives due to their natural and food-grade benefits. Application in food products demonstrated the potential to control pathogenic and spoilage microorganisms. There is significant scope in biopreservation of fermented and minimally processed food products. The careful selection of culture and further optimization in food products enables the protection of food from spoilage, when storage temperature-abuse takes place.

14.1 INTRODUCTION

The risk of food safety and loss of quality due to post-processing contamination with pathogens and spoilage microorganisms has increased besides the emergence of new food-borne pathogens. The storage of foods at low temperatures has been followed to preserve the foods without spoilage. However, the psychographic bacteria including pathogenic (*L. monocytogenes*) and spoilage (*Pseudomonas*) continue to grow at these temperatures. These days, the world food market is driven by consumers, who desire foods that are more convenient to store, ready-to-consume, least processed, and natural like a wide variety, nutritionally sound, and health-promoting [39]. Thus, the chemical preservatives or artificial antimicrobials are frowned on as it is not compatible with natural image, which poses challenges to food processors. Thus, the approach to meet these demands is to use alternatives to chemical preservatives. Bacteriocins in pure or semi-pure forms are incorporated into foods as a natural preservative [40].

The application of bacteriocins in foods has popularized due to specific antimicrobial activity against pathogens. These are active at a wide pH range, tolerate high temperature, and believed to be degraded by enzymes in the human gastrointestinal tract (GIT). In some cases, incorporation of bacteriocins or other antimicrobial metabolites into foods adsorbs to the food constituents leading to loss of antimicrobial activity. Therefore, an alternative approach is the inoculation of food with food-grade bacteria capable to produce bacteriocins or other antimicrobial metabolites [30].

The importance and applications of lactic acid bacteria (LAB) in biopreservation of food through *in situ* production of antimicrobials are described in this chapter.

14.2 PROTECTIVE CULTURES (PC)

The benefit of using protective cultures (PC) as additives is chemical-free preservation, whenever the abuse of storage temperature takes place through the production of antimicrobial compounds within the food thus protect from microbial spoilages. The *in-situ* production of bacteriocins by viable bacteria enables to maintain desired concentration of bacteriocins with foods. It can overcome the problem of degradation of bacteriocins and unavailability due to binding to food constituents, when purified form is used as an additive [9, 28, 47]. Besides PC, being natural can tackle the emerging problem of antibiotic and preservative resistance in food chain. Moreover, the utilization of whole live PC could replace purified form of bacteriocin as the high expenses for isolation and purification restrict the wider use of novel bacteriocins on commercial scale and no specialized equipment are required [16]. Incorporation of PC in minimally processed foods helps to ensure food safety and preserve the nutritional quality.

The PC are "live microorganisms that are deliberately inoculated into a product to control the microbiological status without significantly altering the technological and sensory quality." PC may differ from starters by their lack of, or limited ability to transform the product. They produce antagonistic *in situ* that increases the competitiveness of producer, thereby contributes to control pathogen and food spoilage microorganisms. The principles for PC application in food products are [25]:

- **Competitive Exclusion:** This competes with indigenous microflora for readily fermentable nutrients and or binding sites on substrates, better adaptability to oxygen levels.
- **Production of Antimicrobial Substances**: Bacteriocins (bacteriocinogenic strains) and other fermentation products (organic acids, low molecular weight (MW) bioactive compounds, etc.).
- **Combination of Above Principles:** Such bacteria are considered important to provide protection to food from spoilages by microorganisms and control pathogen.

To distinguish between bacterial cultures for usage as PC, the desirable properties are:

- Cost factors;
- Predictable metabolic activity under the given conditions;
- Reliable production of antagonistics with broad spectrum of antimicrobial activity;
- Serve as indicator for abused situation;
- Should be able to grow at high levels by withstanding the intrinsic factors of food;
- Should not affect product sensory;
- Should not be harmful to humans: Non-pathogenic; Should not produce toxins, biogenic amines or any other metabolites;
- Should show good growth at varying temperatures;
- Should tolerate adverse conditions of processing and storage.

14.3 LACTIC ACID BACTERIA (LAB)

LAB constitute ecologically group of various bacteria that produce lactic acid as major metabolite of lactose metabolism. These are derived from different plant habitats, raw, and fermented products of milk, meat, and vegetables. Gram-positive, non-spore forming, facultative anaerobic, rod to spherical-shaped, possess genome size ranging from 1.8 to 3.2 Mb and DNA has low G+C content of (below 55 mol%). LAB is incapable of producing iron-containing porphyrin, such as catalase (CAT) and cytochrome oxidase and members are nutritionally fastidious; often require specific growth factors (viz., amino acids, nucleic acid derivatives, and vitamins).

Basically, the biochemical pathways of carbohydrate metabolism by LAB that influence fermentation can be classified as:

- **Homolactic Fermentation:** It follows the route of Embden-Meyerhof-Parnas (EMP) pathway or glycolysis for glucose hydrolysis yield lactic acid (>85%) either D (-) or L (+) or both.
- **Heterolactic Fermentative:** bacteria lack fructose 1,6-bisphosphate aldolase enzyme follows the route of pentose-phosphate (6-phospho-gluconate/phosphoketolase) pathway for glucose hydrolysis, therefore producing a mixture of lactic acid, acetate, ethanol, and CO_2.

14.4 LAB AS PROTECTIVE CULTURES (PC)

Certain LAB can inhibit bacteria including foodborne pathogens; therefore it opened a new opportunity to exploit their use in food safety and preservation. Such applications were coined as "bioprotection" and the microbial preparations as "PC." Therefore, many PC available for use are from the microorganisms grouped as starter cultures [42]. The benefits of using LAB as PC are: LAB is recognized as food-grade bacteria as many of them have the status of GRAS and QPS. Apart from this, LAB has been known for their vital ability to dominate the niches by the production of antimicrobials, thereby changing the microenvironment unfavorable to unwanted microorganisms. Some LAB can cause spoilage of food and may utilize some amount of food components, which could affect the sensory parameters. Therefore, the physicochemical changes caused by the culture should be considered.

Based on the available scientific reports, the antimicrobial producing LAB for applications in food biopreservation (Figure 14.1) can be grouped as below:

- **Bacteriocin-Producing (Bacteriocinogenic) LAB Cultures:** These cultures mainly rely on bacteriocins to exhibit antagonism. Mode of action depends on the release of bacteriocins, which are peptides of low MW with a narrow mode of antibacterial action [3]. They may have narrow (taxonomically close bacteria) to broad-spectrum (wide variety of bacteria) of activity. For example, nisin, pediocin, entocin AS-48, and lactocin 3147 are often used in food biopreservation either single or in synergies with other preservation methods. Bacteriocinogenic LAB has gained tremendous research attention as a potential PC:
- **Antifungal LAB to Delay the Spoilages:** Mode of action mainly depends on the production of different antifungal compounds, e.g., PLA, reuterin, fatty acids, etc. These cultures are mainly applied to safeguard the foods from decay and enhance the shelf-life.

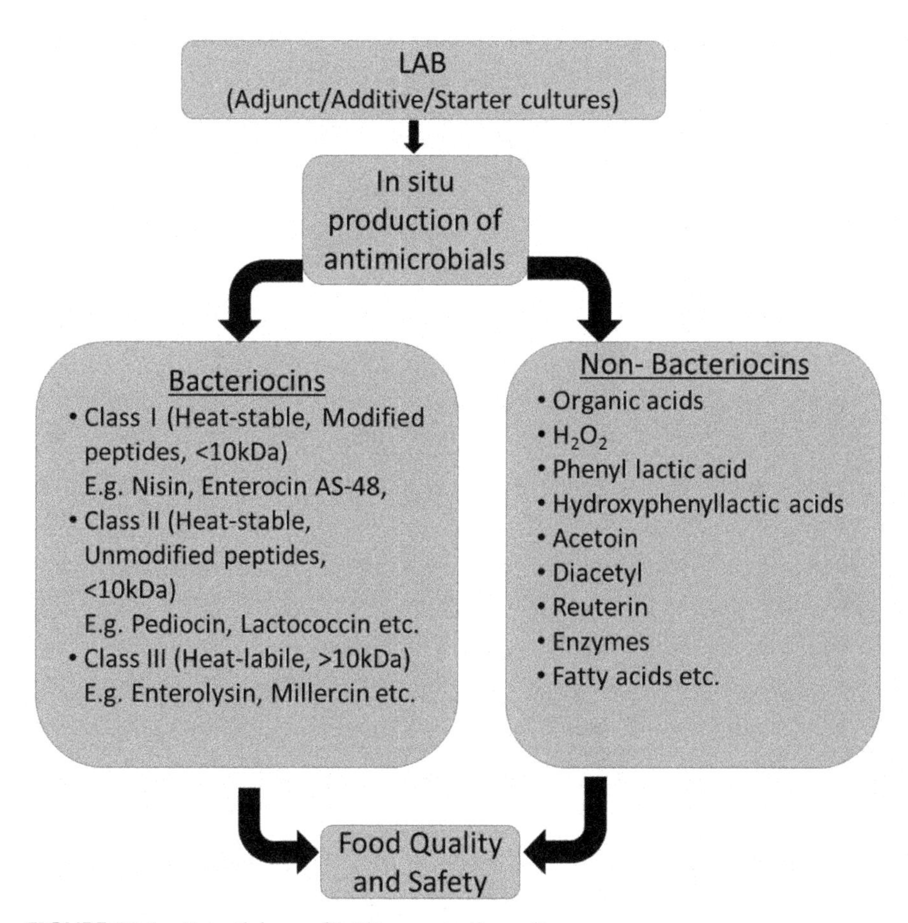

FIGURE 14.1 Potential use of LAB as protective cultures.

- **LAB with Non-Proteinaceous Low Molecular Weight Compounds:** This relies on the production of different metabolites other than bacteriocins, e.g., the protective effect of cultures on the metabolites of lactic acid, acetic acid, H_2O_2, and depletion of oxygen, etc.

14.5 AVAILABILITY OF PROTECTIVE CULTURES (PC) LAB

LAB plays a vital role in food fermentation and is exploited as a starter culture in the manufacturing of number of fermented products from dairy, meat, bakery, fruits, and vegetables. In mixed natural fermentations and other competitive niches, the incidence of fresh isolates with the potential to use as PC is greatest. The antimicrobial production by one bacterium is to strive against other bacteria. Bacteriocin productions appear to compete over non-producing bacteria, which are either closely-related or co-exist in the same ecological niche. The selected strains of antimicrobial producing LAB offer a selective advantage to be exploited as efficient alternatives to chemical preservatives. The bacteriocin producing strains LAB isolated from different sources is listed in Table 14.1.

TABLE 14.1 Species of LAB Producing Bacteriocin Isolated from Different Sources

Bacteriocin Producer Strains	Source	Bacteriocin
Enterococcus faecium CN-25	fermented fish	Enterocin A and B
Enterococcus faecium ST5Ha	Smoked salmon	Bacteriocin ST5Ha
Lactobacillus brevis UN	Dhulliachar	Bacteriocins UN
Lactobacillus gasseri KT7	Infant feces	Gassericin KT7
Lactobacillus helveticus G51	Whey	Helveticin J
Lactobacillus paracasei BGBUK2-16	white-pickled cheese,	Bacteriocin 217
Lactobacillus paraplantarum FT259	Cheese	Plantaricin NC8
Lactobacillus plantarum J23	Grape must	Plantaricin
Lactobacillus plantarum MBSa4	Brazilian salami	Plantaricin W
Lactobacillus plantarum ST16Pa	Papaya	Bacteriocin ST16Pa
Lactobacillus sakei ST154Ch	fermented meat	Curvacin A
Lactococcus lactis KU24	Kimchi	Bacteriocin KU24
Lactococcus lactis ssp. *lactis* BGBM50	semi-hard cheese	Lactococcin G
Pediococcus acidilactici ITV 26	Fermented sausage	Pediocin
Streptococcus thermophilus SBT1277	Raw milk	Thermophilin 1277
Weissellaparames enteroides J1	Chicken gizzard	Bacteriocins BacJ1

14.6 ISOLATION OF PROTECTIVE CULTURES (PC)

LAB with potential PC characteristics can be obtained by first classical step, i.e., pour plating the sample on selective media most popularly known as MRS-deMan, Rogosa, and Sharpe. Other media M17, Lactobacillus selective broth, litmus milk, tryptic soy broth (TSB) are preferred for the production of bacteriocin. After morphological examinations, typical colonies are isolated, and the isolates are identified by dedicated methods. PCR amplification and partial or complete sequencing of 16S ribosomal RNA is performed. The isolates are screened for antagonism based on inhibitory activity against food pathogen and spoilage microorganisms (Figure 14.2). Some of the sensitive test organisms that have been used by different microbiologists are:

- *Bacillus sp.;*
- *Clostridium* sp.;
- *E. coli;*
- *Enterococcus faecalis* MB1;
- *L. casei;*
- *L. plantarum;*
- *Leuconostocmes enteroides;*
- *Listeria innocua;*
- *Listeria monocytogenes;*
- *Micrococcus;*
- *Salmonella typhimurium*;
- *Staphylococcus aureus;*
- Streptococcus.

Several researchers have found differences between observation of *in vitro* (laboratory medium) and actual results *in situ* (food matrix) and sometimes adversely affect the product sensory parameters. The difference is associated with the food to be protected, target microorganisms and storage conditions, etc. [18, 28, 51]. Hence, it is necessary that the effectiveness of isolates selected based on *in-vitro* screening are appraised *in situ* in real foods for bioprotective ability through challenging studies [25].

14.7 POTENTIAL USE OF BACTERIOCIN PRODUCING CULTURE IN FOODS

The application of bacteriocin producing LAB has been demonstrated in different foods (Table 14.2). The unique advantages of protective include natural as they are food grade bacteria, bacteriocins produced when abuse in storage temperature arises. The continuous releases of bacteriocins by live culture compensate the decomposition of bacteriocins and binding to food constituents when added in purified form.

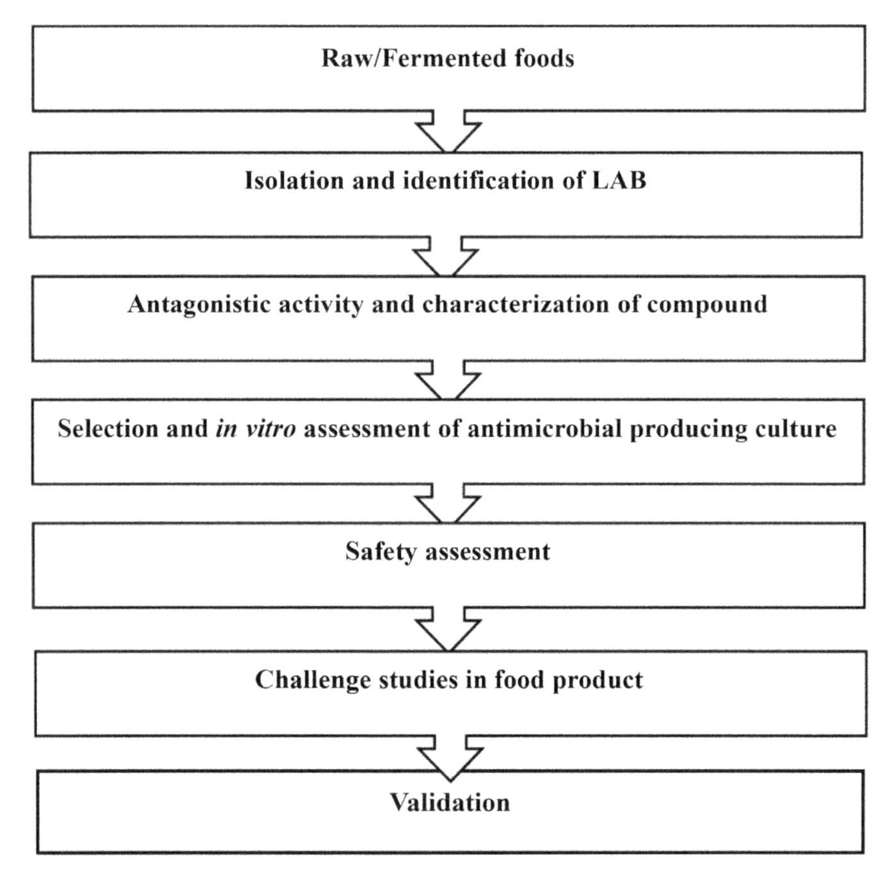

FIGURE 14.2 Strategy of selection of protective cultures. (Source: Modified significantly from Leroi et al. [32].)

TABLE 14.2 Research Studies on the Effectiveness of Protective Culture LAB in Foods

Food	Cultures	Tested Against	References
Bread	*Lb. plantarum* UIG 121	*Aspergillus* sp.	[43]
		Penicillium sp.	
Cereal products	*Lb. plantarum*	*Penicillium* sp., *Aspergillus* sp.	[41]
		Aflatoxin	
Cheese	*Lc. lactis* INIA 415	Acceleration of ripening	[6]
Cheese	*Lc. lactis*	*L. monocytogenes*	[28]
Cheese	*Lb. gasseri* K7	*Cl. tyrobutyricum*	[10]
Cooked bacon	*Lb. sakei* and *Lc. lactis*	*Leu. mesenteroides*	[12]
Cottage cheese	*Lb. amylovorus* DSM 19280	*Penicillium expansum*	[35]

TABLE 14.2 *(Continued)*

Food	Cultures	Tested Against	References
Fermented pork meat	*Lb. plantarum* PCS20	*Clostridium* spp.	[17]
Fermented soy milk	*Lb. helviticus*YML014	*Penicillium* sp.	[9]
Fish sauce	*Stap. carnosus* FS19	Biogenic amines	[51]
Letuce leaf	*Leuconostoc* spp.	*L. monocytogenes*	[47]
Refrigerated pea soup	*Lb. plantarum* ATCC 8014	*Cl. botulinum*	[45]
Salmon	*Lb. plantarum and C. piscicola*	H_2S producer and yeast and mold	[31]
Salmon	*Lb. sakei* CTS494	*L. monocytogenes*	[7]
Sausage	*Lb. sakei*	*L. monocytogenes, Salmonella spp.* and amines	[18]
Tomato puree	*Lb. fermentum* YML014	Yeast and mold	[1]
Yogurt	*Lb. rhamnosus, Lb. zeae, Lb. harbinensis*	Yeast and mold	[16]
Yogurt	*Lb. casei* AST18	*Penicillium* sp.	[34]

Several PC of LAB cannot tolerate heat treatment; hence the addition of LAB as PC to foods should be after heat treatment. In the production of fermented products, PC are inoculated into the food as the main starter or as adjunct to starter culture meant for acidification. In this case, the acidifying bacteria should not be affected significantly by bacteriocin producing one. In postharvest and bakery foods, the PC or their products are sprayed on the surface. The effectiveness of PC often relies on inoculum level, concertation of metabolites, which can adversely affect the sensory acceptability of foods. Therefore, once the isolation of LAB and characterization of antimicrobials has been conducted, the results of *in vitro* studies should be confirmed in actual foods.

14.7.1 APPLICATIONS OF LAB IN DAIRY PRODUCTS

LAB is the most frequently used bacteria as a starter culture for acidification and flavor development in milk fermentation. Hence, scientists have studied using live bacteriocinogenic LAB for *in situ* production of bacteriocins during the production and storage of dairy products. Several research reports have mentioned the use of LAB to control "late blowing" caused by *Clostridium* spp.

in cheese. For example, the K7 bacteriocin producing *Lactobacillus gasseri* K7 [10], lacticin 3147 *Lactococcus lactis* IFPL 3593 [11], reuterin producing *Lactobacillus reuteri* INIA P572 [23] have shown to inhibit the spore germination and prevent "late blowing" and been proposed as an alternative approach to use of potassium nitrate. Likewise, *in situ* production of nisin-Z by *L. diacetylactis* UL 719 in a mixed culture with *L. cremoris* and *L. lactis* in cheddar cheese was effective to inhibit the growth of *Listeria innocua* throughout the ripening period (6 months). Nisin concentration of 300 IU/ml was observed [8].

In Camembert cheese, although a 3log-reduction in *L. monocytogenes* was observed until second week, yet a regrowth was observed for 6 weeks ripening period [36]. Numerous investigations have also studied the development of bacteriocin producing cultures to improve the maturation and quality of cheese. Bacteriocin-producing LAB may also serve as agents to induce lysis of cell-wall of starter and/or non-starter culture to cause initiation of proteolysis and the release of amino acids can further contribute to synthesis of flavor compounds. The lacticin 3147 producing *L. lactis* was used to hasten the ripening process of cheese and to maintain the significant lower levels of nonstarter LAB for 6 months of ripening [37].

14.7.2 APPLICATIONS OF LAB IN MARINE FOODS

Many psychro-trophic spoilage and pathogenic bacteria can survive in fresh fish. Cold smoked salmon undergoes spoilage in 3–4 weeks mainly due to microbial growth, which limits its commercial popularity. Leroi et al. [32] demonstrated a pool of LAB strains for the prevention of off-odor and acidification of cold-smoked salmon. *L. monocytogenes* is a major concern as it is a frequent contaminant and it can tolerate cold smoke (<30°C), salting, and remain active in the product at chilled temperatures. In this regard, the bacteriocin-producing LAB proved promising as PC isolated from seafood products. The co-culture inoculation of sakacin producing *Lb. sakei* with *L. monocytogenes* was able to demonstrate a 4-log reduction of *Listeria innocua* in cold-smoked salmon after 14 days of storage at 4°C [27]. Interestingly, the antilisterial activity of *Lb. sakei* CTS494 was not affected by the type of salmon having different physicochemical characteristics [7].

14.7.3 APPLICATIONS OF LAB IN MEAT PRODUCTS

The bacteriocin-producing LAB has shown to control foodborne pathogens in sausages. Recently, the addition of *Lb. plantarum* PSC 20 as protective culture was demonstrated as a feasible method to lower the *L. monocytogenes* counts in fermented meat [38]. Therefore, bacteriocin-producing LAB is gaining importance in the manufacture of fermented meat products to control spoilage microorganisms. The *Clostridium* sp. in fermented meat is controlled by the addition of nitrite [15]. Hence, antilisterial bacteriocin by PC *in situ* has been under focus as an alternative to control spoilage and stabilize the color of meat products. Moreover, the selection of bacteriocin-producing LAB as a potential protective culture within the autochthonous microbiota of fermented meat should be recommended, since autochthonous cultures may have better adaptability, faster growth and ensure prolonged shelf-life [15].

14.7.4 APPLICATIONS OF LABS IN FRUITS AND VEGETABLES

The bacteriocinogenic LAB are explored for biopreservation of minimally processed foods of plant origin (such as salads, sprouts, fruit juices, and sauerkraut). Especially salads offer the advantage of convenience and freshness, which is also favorable for survival and growth of *L. monocytogenes* [47]. The contamination occurs due to poor handling practices during the production of these foods, and microorganisms can grow under abnormal storage temperatures. In this context, bacteriocinogenic LAB as PC can serve as a hurdle to the growth of these pathogens. In fresh-cut ready-to-eat salads, nisin-Z producing *Lc. lactis* and bacteriocin producing *E. faecium* could inhibit *L. monocotogenes.* Similarly, *Lb. casei* was able to eliminate the post-processing contaminants coliforms and enterococci in ready-to-eat vegetables after three days of storage under refrigeration [49].

The use of bacteriocinogenic LAB in the fermentation of sauerkraut and other products is beneficial in enhancing the shelf-life of the product. Incorporation of bacteriocin-producing lactococcal strains with *Leu. mesenteroides* or natural microflora proved valuable for sauerkraut fermentation for controlling the spoilage caused by homo-fermentative *Lb. plantarum* [22, 47].

14.8 APPLICATIONS OF NON-BACTERIOCINOGENIC LAB AS PROTECTIVE CULTURE

During bacterial fermentation, organic acids are produced along with other metabolites that are also produced. The fermentation of organic acids viz., lactic, acetic, and propionic acids lower the pH of the food matrix, which plays a vital role in food preservation. Other antifungal metabolites are: H_2O_2, reuterin, phenyllactic, and hydroxyphenyllactic acids, diacetyl, acetoin, and fatty acids. On the other hand, LAB producing biocompounds other than bacteriocins, which exhibit synergistic effect, have been explored as non-bacteriocinogenic PC [33]. Fungal growth in foods is associated with the production of volatile compounds, bad odor, and release of mycotoxins, thus antifungal PC have gained importance. Sorbates, propionates, and benzoates have been used to overcome the problem of fungi, nevertheless, their usage is recently frowned upon by the consumers. Thus as an alternative antifungal protective agent, LAB has great potential to produce preservative-free foods without compromising shelf-life [5, 35].

Many researchers have isolated antifungal strains from different niches including fermented foods and have reported antifungal activity in yogurt, cheese, and bakery products. *Lactobacillus* species were evaluated with LAB in foods as antifungal PC and there are promising applications in dairy products [34]. For example, strains belonging to *Lb. alimentarius*, *Lb. sakei, Lb. fermentum, Lb. paracasei* or *Lb. casei* species demonstrated good inhibition of *Penicillium, Kluveromyces, Rhizopus, and Yersinia* spp. in yogurt [5]. Use of *Lb. reuteri, Lb. plantarum* and *Lb. rhamnosus* strains in cheddar and cottage cheese resulted in inhibition of *Penicillium* spp [21, 35]. The *Lb. sakei, Lb. plantarum, Lb. spicheri, Lb. brevis* and *Leu. Citrium* have shown *in situ* antifungal activity in bakery products, out of 270 LAB strains assessed for antifungal activity *in vitro* against *Penicillium, Cladosporium, and Wallemia* spp. [29].

Aflatoxins are used as a food safety hazard produced by *Aspergillus* spp. and the major ones are B1, B2, G1, G2; and M1 is considered as a human carcinogen. The use of protective culture is an attractive preventive approach to manage the risk of aflatoxins in foods [44]. Mainly *Lactobacillus, Bifidobacterium, and Propionibacterium* are reported to have the ability to bind Aflatoxins. The aflatoxin B1 in contaminated wheat flour during the bread-making process was found to be reduced by fermentation with yeast and *Lb. rhamnosus* [19].

The application of LAB caused 84–100% reduction of aflatoxin in bread with the extension of shelf-life up to 4 days [4]. While the addition of *Lb. plantarum* resulted in a decline of aflatoxin B1 concentration from 11 to 5.9 μg/kg during the storage of olives [26]. Similar observations have been noticed in milk *S. cerevisiae* and a pool of LAB strains 100% binding of aflatoxin M1 in milk for 60 minutes [13]. *Lb plantarum* was the highest aflatoxin M1 binding strain in yogurt during storage when used with *Str. thermophilus* and *Lb. bulgaricus* [20]. The aflatoxin binding ability of LAB is associated with factors, such as [2]: bacteria strain, food matrix, storage temperature and the storage period, etc. Furthermore, the binding is a surface phenomenon that is strongly correlated with the involvement of various metabolites released by LAB.

The mechanism of antifungal action of LAB is not well-known as compared to antibacterial peptides. Antifungal principle of LAB in foods may be attributed to acidification (lowering of pH), production of antifungal metabolites (antibiosis), and competition for specific nutrients [14]. However, the mechanism of antifungal action of a strain may not solely depend on the concentration of one type of antimicrobial. For instance, organic acids can diffuse into the cell through the cell membrane and can dissociate inside causing reduction of intracellular pH, disruption of membrane function leading to cell death. In addition, complex interactions between numerous metabolites released by LAB are believed to exert combined stress on fungal physiology with synergistic action causing greater inhibition. The interactive effect between the metabolites confirms the complexity of the antifungal mechanism [14].

Several methods have been proposed for optimization of antifungal activity in food products to enhance of production of antifungal metabolites and broaden antifungal spectrum [33]:

- Inducation of stress conditions to cultures;
- Use of co-cultures of antifungal strains;
- Supplementation with precursors, which trigger biosynthesis of metabolites (e.g., glycerol);
- Combination with other molecules, such as chitosan.

Even though so many antifungal strains have been identified, yet only a few cultures are made commercially available for utilization in food products.

14.9 PROTECTIVE CULTURES (PC) IN HURDLE TECHNOLOGY

Combinations of cultures for *in situ* production of bacteriocins and other hurdles have also been studied mainly to control *L. monocytogenes* or *S. aureus* in foods. Bacteriophage P100 and *Lb. sakei* together was successfully applied to prevent the progression of *L. monocytogenes* on cooked ham. This combined approach can also be a beneficial tactic to reduce the emergence of resistant pathogens [24].

The effectiveness of the combination is dependent on the product and target bacteria. Several authors have studied to develop PC to combine with other hurdles in food products. The successful combination should act synergistically, and the effect of the combination is more than an independent application [50]. The protective culture-based combination has shown to increase the effectiveness of HHP, active packaging, modified atmosphere packaging (MAP), and super chilling against pathogens and spoilage bacteria [46]. However, the selection of PC having compatibility with other hurdles need to be evaluated carefully.

14.10 FACTORS AFFECTING ACTIVITY OF PROTECTIVE CULTURES (PC)

Unlike conventional preservation, PC do not readily result in a successful outcome and the antagonistic activity is a dynamic process. The interactions between the production of bacteriocin and the microorganisms will depend on several intrinsic and extrinsic factors. The variation in any factor may ultimately influence the production of bacteriocin *in situ* and its activity. Table 14.3 indicates factors to regulate contamination and to ensure shelf-life and food safety in different food preparations. Some of them have contradictory effects or some have interrelated effects. The pH is the key factor, which influences the bacteriocin production; and food environment temperature influences the growth rate, biomass of bacteriocin producer and target microflora [25, 46].

TABLE 14.3 Factors Affecting the Protective Cultures in Foods

The Target Microbiota	The Bacteriocin Producing Strain	The Food Matrix
Bacteriocin sensitivity	Spectrum of activity	Processing factors
Growth rate	Rate of bacteriocin production	Storage temperature
Storage temperature	Stability of bacteriocin and producing trait	Initial microbial load
The protective effects of food constituents	Minimum growth temperature of producing strain	Composition of food and buffering capacity
Initial load of microflora	Interaction with food additives/ingredients	Interaction with food components
Microbial interactions	Adoption to the food environment	Synergistic effect between bacteriocin and metabolic end products
·	Survival during processing conditions	Solubility and distribution of bacteriocin
	Ability to cause spoilage of product	
	Any health benefits/hazards	

14.11 PROTECTIVE CULTURES (PC) AVAILABLE IN THE MARKET

The current regulations do not restrict the use of bacteriocin producing LAB in foods as these are considered food grade. The bacteriocins produced by LAB *in situ* need not be indicated on the product label [48]. Although numerous PC are identified, yet only few are commercialized for food applications. Selected examples of PC in the market are listed in Table 14.4.

TABLE 14.4 Protective Cultures Available in Market

Protective Culture	Company	Composition	Activity Spectrum	Recommended Application
Bactoferm™ F-LC	Chr. Hansen	*P. acidilactici, Lb. curvatus* and *Staph. xylosus.*	*L. monocytogenes*	Fermented sausages
Befesh™	Handary	*Lb. paracasei* *Prop. shermani*	Yeast and mold	Fermented milk products
Dairy Safe	CSK food enrichment	*Lc. lactis*	*Cl. tyrobutyricum*	Cheese
Delvo®Guard	DSM	*Lb. rhamnosus* and *Lb. sakei*	Yeast and mold	Dairy products
FreshQ®	Chr. Hansen	lactic acid bacteria	Yeast and mold	Yogurt, sour cream, cheese, kefir
Holdbac®	Dupont	*Prop. freudenreichii* subsp. *shermanii* JS and *Lb. rhamnosus* LC705 or *Lb. paracasei* SM20	Yeast and mold	Yogurt and cheese
Lyofast LPR A	SACCO	*Lb. rhamnosus* and *Lb. plantarum*	Yeast and mold	Cheese and fermented milk
SafePro®	Chr. Hansen	*Lb. curvatus*	Listeria	meat, salmon & salads

The process of production at the commercial level is like the starter cultures. The PC for commercial applications are produced as direct vat cultures. The strains are grown in batch fermenter followed by concentration and finally frozen for pellets or freeze-dried for powders, which is the most practical way of application.

14.12 SUMMARY

The bacteriocinogenic cultures are active only against Gram-positive bacteria; therefore inhibition of Gram-positive bacteria is doubtful. The control of Gram-negative bacteria can be enhanced by exploiting the synergies between the antimicrobials and other preservation techniques. The use of PC is very encouraging in fermented products with greater application in yogurt and cheese. Bacteriocins have a narrow spectrum of activity and inhibit only closely-related bacteria. An ideal protective culture should possess broad-spectrum activity. PC is effective to inhibit *Listeria* nevertheless the regrowth should be considered to design full-proof biopreservation. Although number of potential PC is being identified, yet there is a huge gap between *in vitro* activity and *in situ* success. Therefore, only a few PC are available in the market for food applications. Careful selection of LAB as protective culture and proper optimization is necessary for effectiveness in foods at low cost.

KEYWORDS

- **antimicrobials**
- **bacteriocin**
- **Embden-Meyerhof-Parnas**
- **food safety**
- **lactic acid bacteria**
- **protective culture**

REFERENCES

1. Adedokun, E. O., Rather, I. A., Bajpai, V. K., & Park, Y. H., (2016). Biocontrol efficacy of *Lactobacillus fermentum* YML014 against food spoilage molds using the tomato puree model. *Frontiers in Life Science*, *9*(1), 64–68.
2. Ahlberg, S. H., Joutsjoki, V., & Korhonen, H. J., (2015). Potential of lactic acid bacteria in aflatoxin risk mitigation. *International Journal of Food Microbiology*, *207*, 87–102.
3. Alvarez-Sieiro, P., Montalbán-López, M., Mu, D., & Kuipers, O. P., (2016). Bacteriocins of lactic acid bacteria: Extending the family. *Applied Microbiology and Biotechnology*, *100*(7), 2939–2951.
4. Asurmendi, P., Pascual, L., Dalcero, A., & Barberis, L., (2014). Incidence of lactic acid bacteria and *Aspergillus flavus* for potential antifungal activity of these bacteria. *Journal of Stored Products Research*, *56*, 33–37.

5. Aunsbjerg, S. D., Honoré, A. H., Marcussen, J., Ebrahimi, P., Vogensen, F. K., Benfeldt, C., Skov, T., & Knøchel, S., (2015). Contribution of volatiles to the antifungal effect of *Lactobacillus paracasei* in defined medium and yogurt. *International Journal of Food Microbiology, 194,* 46–53.
6. Ávila, M., Garde, S., Gaya, P., Medina, M., & Nuñez, M., (2006). Effect of high-pressure treatment and a bacteriocin-producing lactic culture on the proteolysis, texture, and taste of hispánicocheese. *Journal of Dairy Science, 89*(8), 2882–2893.
7. Aymerich, T., Rodríguez, M., Garriga, M., & Bover-Cid, S., (2019). Assessment of the bioprotective potential of lactic acid bacteria against *Listeria monocytogenes* on vacuum-packed cold-smoked salmon stored at 8°C. *Food Microbiology, 83,* 64–70.
8. Benech, R. O., Kheadr, E. E., Laridi, R., Lacroix, C., & Fliss, I., (2002). Inhibition of *Listeria innocua* in cheddar cheese by addition of nisin-Z in liposomes or by *in situ* production in mixed culture. *Applied and Environmental Microbiology, 68*(8), 3683–3690.
9. Bian, X., Muhammad, Z., Evivie, S. E., Luo, G. W., Xu, M., & Huo, G. C., (2016). Screening of antifungal potentials of *Lactobacillus helveticus* KLDS 1.8701 against spoilage microorganism and their effects on physicochemical properties and shelf-life of fermented soybean milk during preservation. *Food Control, 66,* 183–189.
10. Bogovič-Matijašić, B., Koman-Rajšp, M., Perko, B., & Rogelj, I., (2007). Inhibition of *Clostridium tyrobutyricum* in cheese by *Lactobacillus gasseri. International Dairy Journal, 17*(2), 157–166.
11. Carmen, M., Bengoechea, J., Bustos, I., Rodríguez, B., Requena, T., & Peláez, C., (2010). Control of late blowing in cheese by adding lacticin 3147-producing *Lactococcus lactis* IFPL 3593 to the starter. *International Dairy Journal, 20*(1), 18–24.
12. Comi, G., Andyanto, D., Manzano, M., & Iacumin, L., (2016). *Lactococcus lactis* and *Lactobacillus sakei* as bioprotective cultures to eliminate *Leuconostoc mesenteroides* spoilage and improve the shelf-life and sensorial characteristics of commercial cooked bacon. *Food Microbiology, 58,* 16–22.
13. Corassin, C. H., Bovo, F., Rosim, R. E., & Oliveira, C. A. F., (2013). Efficiency of *Saccharomyces cerevisiae* and lactic acid bacteria strains to bind aflatoxin M1 in UHT skim milk. *Food Control, 31*(1), 80–83.
14. Dagnas, S., Gauvry, E., Onno, B., & Membr, J. M., (2015). Quantifying effect of lactic, acetic, and propionic acids on growth of molds isolated from spoiled bakery products. *Journal of Food Protection, 78*(9), 1689–1698.
15. De Souza, B. M., Todorov, S. D., Ivanova, I., & Chobert, J. M., (2015). Improving safety of salami by application of bacteriocins produced by an autochthonous *Lactobacillus curvatus* isolate. *Food Microbiology, 46,* 254–262.
16. Delavenne, E., Ismail, R., Pawtowski, A., Mounier, J., Barbier, G., & Le-Blay, G., (2013). Assessment of lactobacilli strains as yogurt bioprotective cultures. *Food Control, 30*(1), 206–213.
17. Di Gioia, D., Mazzola, G., & Nikodinoska, I., (2016). Lactic acid bacteria as protective cultures in fermented pork meat to prevent *Clostridium* spp. growth. *International Journal of Food Microbiology, 235,* 53–59.
18. Elmalti, J., & Amarouch, H., (2008). Protective cultures used for the biopreservation of horse meat fermented sausage: Microbial and physicochemical characterization. *Journal of Food Safety, 28*(3), 324–345.

19. Elsanhoty, R. M., Ramadan, M. F., El-Gohery, S. S., Abol-Ela, M. F., & Azeke, M. A., (2013). Ability of selected microorganisms for removing aflatoxins *in vitro* and fate of aflatoxins in contaminated wheat during bread baking. *Food Control*, *33*(1), 287–292.
20. Elsanhoty, R. M., Salam, S. A., Ramadan, M. F., & Badr, F. H., (2014). Detoxification of aflatoxin M1 in yogurt using probiotics and lactic acid bacteria. *Food Control*, *43,* 129–134.
21. Fernandez, B., Vimont, A., Desfossés-Foucault, É., Daga, M., Arora, G., & Fliss, I., (2017). Antifungal activity of lactic and propionic acid bacteria and their potential as protective culture in cottage cheese. *Food Control*, *78,* 350–356.
22. Franz, C. M. A. P., Du Toit, M., Von, H. A., Schillinger, U., & Holzapfel, W. H., (1997). Production of nisin-like bacteriocins by *Lactococcus lactis* strains isolated from vegetables. *Journal of Basic Microbiology*, *37*(3), 187–196.
23. Gómez-Torres, N., Ávila, M., Gaya, P., & Garde, S., (2014). Prevention of late blowing defect by reuterin produced in cheese by a *Lactobacillus reuteri* adjunct. *Food Microbiology*, *42,* 82–88.
24. Holck, A., & Berg, J., (2009). Inhibition of *Listeria monocytogenes* in cooked ham by virulent bacteriophages and protective cultures. *Applied and Environmental Microbiology*, *75*(21), 6944.
25. Holzapfel, W. H., Geisen, R., & Schillinger, U., (1995). Biological preservation of foods with reference to protective cultures, bacteriocins and food-grade enzymes. *International Journal of Food Microbiology*, *24*(3), 343–362.
26. Kachouri, F., Ksontini, H., & Hamdi, M., (2014). Removal of aflatoxin B1 and inhibition of *Aspergillus flavus* growth by the use of *Lactobacillus plantarum* on olives. *Journal of Food Protection*, *77*(10), 1760–1767.
27. Katla, T., Møretrø, T., Aasen, I. M., Holck, A., Axelsson, L., & Naterstad, K., (2001). Inhibition of *Listeria monocytogenes* in cold smoked salmon by addition of sakacin P and/or live *Lactobacillus sakei* cultures. *Food Microbiology*, *18*(4), 431–439.
28. Kondrotiene, K., Kasnauskyte, N., Serniene, L., & Gölz, G., (2018). Characterization and application of newly isolated nisin producing *Lactococcus lactis* strains for control of *Listeria monocytogenes* growth in fresh cheese. *LWT-Food Science and Technology*, *87* 507–514.
29. Le-Lay, C., Mounier, J., Vasseur, V., Weill, A., Le Blay, G., Barbier, G., & Cotton, E., (2016). *In vitro* and *in situ* screening of lactic acid bacteria and propionibacteria antifungal activities against bakery product spoilage molds. *Food Control*, *60,* 247–255.
30. Lee, E. H., Khan, I., & Oh, D. H., (2018). Evaluation of the efficacy of nisin-loaded chitosan nanoparticles against foodborne pathogens in orange juice. *Journal of Food Science and Technology*, *55*(3), 1127–1133.
31. Leroi, F., Arbey, N., Joffraud, J. J., & Chevalier, F., (1996). Effect of inoculation with lactic acid bacteria on extending the shelf-life of vacuum-packed cold smoked salmon. *International Journal of Food Science and Technology*, *31*(6), 497–504.
32. Leroi, F., Cornet, J., Chevalier, F., Cardinal, M., & Coeuret, G., (2015). Selection of bioprotective cultures for preventing cold-smoked salmon spoilage. *International Journal of Food Microbiology*, *213* 79–87.
33. Leyva-Salas, M., Thierry, A., Lemaître, M., Garric, G., & Harel-Oger, M., (2018). Antifungal activity of lactic acid bacteria combinations in dairy mimicking models

and their potential as bioprotective cultures in pilot scale applications. *Frontiers in Microbiology*, *9* 1787.
34. Li, H., Liu, L., Zhang, S., Uluko, H., Cui, W., & Lv, J., (2013). Potential use of *Lactobacillus casei*AST18 as a bioprotective culture in yogurt. *Food Control*, *34*(2), 675–680.
35. Lynch, K. M., Pawlowska, A. M., Brosnan, B., & Coffey, A., (2014). Application of *Lactobacillus amylovorus* as an antifungal adjunct to extend the shelf-life of cheddar cheese. *International Dairy Journal*, *34*(1), 167–173.
36. Maisnier-Patin, S., Deschamps, N., Tatini, S. R., & Richard, J., (1992). Inhibition of *Listeria monocytogenes* in camembert cheese made with a nisin-producing starter. *Lait*, *72*(3), 249–263.
37. Martínez-Cuesta, M. C., Requena, T., & Peláez, C., (2006). Cell membrane damage induced by lacticin 3147 enhances aldehyde formation in *Lactococcus lactis* IFPL730. *International Journal of Food Microbiology*, *109*(3), 198–204.
38. Nikodinoska, I., Baffoni, L., Di Gioia, D., & Manso, B., (2019). Protective cultures against foodborne pathogens in a nitrite reduced fermented meat product. *LWT-Food Science* and *Technology*, *101* 293–299.
39. Pei, J., Yue, T., & Jin, W., (2016). Application of bacteriocin RC20975 in apple juice. *Food Science and Technology International*, *23*(2), 166–173.
40. Pei, J., Yue, T., & Yuan, Y., (2014). Control of *Alicyclobacillus acidoterrestris* in fruit juices by a newly discovered bacteriocin. *World Journal of Microbiology and Biotechnology*, *30*(3), 855–863.
41. Quattrini, M., Bernardi, C., Stuknytė, M., & Masotti, F., (2018). Functional characterization of *Lactobacillus plantarum* ITEM 17215: A potential biocontrol agent of fungi with plant growth promoting traits, able to enhance the nutritional value of cereal products. *Food Research International*, *106,* 936–944.
42. Reis, J. A., Paula, A. T., Casarotti, S. N., & Penna, A. L. B., (2012). Lactic acid bacteria antimicrobial compounds: Characteristics and applications. *Food Engineering Reviews*, *4*(2), 124–140.
43. Russo, P., Fares, C., Longo, A., Spano, G., & Capozzi, V., (2017). *Lactobacillus plantarum* with broad antifungal activity as a protective starter culture for bread production. *Foods*, *6*(12), 110.
44. Saladino, F., Luz, C., Manyes, L., Fernández-Franzón, M., & Meca, G., (2016). *In vitro* antifungal activity of lactic acid bacteria against mycotoxigenic fungi and their application in loaf bread shelf-life improvement. *Food Control*, *67,* 273–277.
45. Skinner, G. E., Solomon, H. M., & Fingerhut, G. A., (1999). Prevention of *Clostridium botulinum* type-A, proteolytic *b* and *e* toxin formation in refrigerated pea soup by *Lactobacillus plantarum* ATCC 8014. *Journal of Food Science*, *64*(4), 724–727.
46. Stratakos, A. C., Linton, M., Tessema, G. T., Skjerdal, T., Patterson, M. F., & Koidis, A., (2016). Effect of high pressure processing in combination with *Weissella viridescens* as a protective culture against *Listeria monocytogenes* in ready-to-eat salads of different pH. *Food Control*, *61* 6–12.
47. Trias, R., Badosa, E., Montesinos, E., & Bañeras, L., (2008). Bioprotective *Leuconostoc* strains against *Listeria monocytogenes* in fresh fruits and vegetables. *International Journal of Food Microbiology*, *127*(1), 91–98.
48. Varsha, K. K., & Nampoothiri, K. M., (2016). Appraisal of lactic acid bacteria as protective cultures. *Food Control*, *69* 61–64.

49. Vescovo, M., Orsi, C., Scolari, G., & Torriani, S., (1995). Inhibitory effect of selected lactic acid bacteria on microflora associated with ready-to-use vegetables. *Letters in Applied Microbiology*, *21*(2), 121–125.
50. Wiernasz, N., Cornet, J., Cardinal, M., Pilet, M. F., Passerini, D., & Leroi, F., (2017). Lactic acid bacteria selection for biopreservation as a part of hurdle technology approach applied on seafood. *Frontiers in Marine Science*, *4*, 119.
51. Zaman, M. Z., Abu-Bakar, F., & Jinap, S., (2011). Novel starter cultures to inhibit biogenic amines accumulation during fish sauce fermentation. *International Journal of Food Microbiology*, *145*(1), 84–91.

INDEX

β

β-glucan, 98, 99, 103, 128
β-glucanase, 98, 99, 103, 105
β-sheet, 111, 119, 128

A

Absorptive phase, 6
Acetic acid, 39, 76–78, 291
Acetoglycerides, 266
Acidification, 74, 96, 253, 295, 296, 299
Acquired
  immune deficiency syndrome, 9
  resistance, 170
    class II bacteriocins, 171
    class III bacteriocins, 172
    cyclic bacteriocins, 172
    lantibiotics, 170
Acrolein, 83
Active compound, 56
Additive effect, 25, 36, 37, 80, 241
Aerobic
  bacteria, 39
  fungi, 87
Aeromonas strains, 147
Agaricus bisporus, 150
Alcoholic extracts, 58
Aldehyde, 24
Aloe
  barbadensis, 60
  vera, 60
Amalgamation, 12
Amino
  acids, 56, 79, 97, 111, 119, 126, 128, 164, 216, 289, 296
  group, 168
Aminotransferase enzyme, 81
Amorphous polymeric material, 264
Amperometric biosensor, 281
Amphipathic molecules, 128
Amphipathical α-helix structure, 119
Amphipathicity, 119, 122
Amylopectin, 198
Anaerobic
  approach, 13
  conditions, 83, 87, 149
  degradation, 87
  fermentation, 13
Anionic surface, 119
Antagonistic activity, 74, 81, 300
Antibacterial
  action, 37, 290
  activity, 24, 25, 75, 98, 103, 129
  agents, 95, 99
  factor, 4
  spectrum, 95, 96
  system, 96, 99
Antibiotic
  resistant
    bacteria (ARB), 24, 96, 144–146
    strains, 143
  treatments, 147, 156
Anti-cryptosporidium, 8
Antifungal
  activity, 78, 80, 82, 84, 89, 128, 298, 299
  agents, 75, 87, 89
  carboxylic acids, 79
  compounds, 75, 76, 84, 85, 87, 90, 290
  peptides (AFPs), 77, 79, 80, 128
  properties, 75, 80
  protective agent, 298
Anti-infective agents, 100
Anti-listerial effects, 36, 39
Antimicrobial, 22, 25, 30, 34, 36, 37, 39, 40, 90, 119, 120, 253, 256–258, 287, 288, 290, 295, 302
  activities, 60, 62, 63, 75, 82, 95, 98, 118, 119, 122, 125–128, 163, 258, 288
  additives, 25
  agents, 80, 85, 96, 100, 125, 152, 249, 254, 257, 258
  combinations, 38

compounds, 75, 102, 204, 288
effects, 7, 83, 96, 230, 231, 256, 257
enzyme, 95, 96, 98, 100, 104, 105
bacteriophage lysins, 99
chitinase, 99
glucose oxidase (GOX), 97
lactoperoxidase (LP), 4, 96, 97, 102
lysozyme (LYZ), 98
fractions, 3, 14
metabolites, 74, 87, 89, 161, 163, 288
packaging, 250, 252, 256, 258, 259, 269
peptides (AMPs), 79, 111–113, 117–129, 168, 176
antimicrobial attributes, 127
antioxidant activities, 126
antitumor activity, 127
beneficial symbiotic interaction, 125
bonding based classification, 121
database (APD), 111, 112, 117, 118
innate immunity activation, 125
intracellular active AMPs, 124
membrane active AMPs, 122
role in probiosis association, 126
sources, 113
structural classification, 118
polymers, 257
potential, 24, 80
properties, 250, 253
resistance, 156
sequence database (AMSdb), 117
spectrum, 37, 38
substances, 74, 96, 98, 100, 241
Antimycotic properties, 82
Antioxidant, 3, 13, 53, 56, 58, 126, 228, 253, 257
activity, 58, 59, 61, 63
Anti-oxidative
activity, 53
status, 61
Anti-protozoans activities, 127
Anti-tumor activity, 5
Aptamers, 280, 281
Arbitrary units (AU), 166
Aromatherapy, 22
Aromatic elements, 188
Ascorbic acid (AC), 57, 239, 240, 242
Aspergillus
flavus, 80, 81, 236
niger, 33, 87, 98, 231, 237
Atherosclerosis, 126
Atmospheric
plasma, 195, 199
pressure plasma (APP), 187, 190, 205
jet (APPJ), 200
Atomic fragmentation, 194
Autoimmune diseases, 5
Autolysin, 170
Autooxidation, 58

B

Bacillus
cereus, 36, 37, 113, 168, 172, 236, 258
clausii, 99
subtilis, 36, 99, 113, 149, 170
Bacteria cell membrane, 35
Bacterial
blight, 150, 151
cells, 23, 34, 39, 83, 97, 98, 147, 169, 239
endospore, 33
inactivation, 242
loads, 13
translocation, 4
Bactericidal effect, 163, 230
Bactericinogenic culture, 166
Bacteriocin, 25, 33, 36, 37, 40, 79, 89, 118, 127, 161, 163–176, 204, 257, 288–293, 295–302
activity mode, 164
immunity, 165
potential use, 293
resistance development, 161, 163, 165–167, 170, 173, 176
acquired resistance, 170
innate resistance, 167
strategies, 172
sensitivity, 166
Bacteriocinogenic
cultures, 161, 163, 173, 302
strains, 161, 163, 165, 167
Bacteriophage, 36, 39, 95, 99, 105, 146, 147, 161, 163
lysin, 95, 99, 104, 105
Bacteriostatic effect, 163
Bacteriovorax, 148, 153
Barrel-Stave model, 122, 123, 129
B-cell, 5
Bdelloplasts, 143

Bdellovibrio, 143, 145–154, 156
and like organisms (BALO), 148, 149, 151, 153, 154
application, 149–151
aquaculture potential, 147
genome, 145
prey encounter mechanism, 146
potential role
disease management, 152
Benzoic acid, 75, 76, 78
Bifido-bacteria, 60
Bioactive
components, 3, 4, 11, 12, 57, 89, 265
compounds, 53, 54, 57, 228, 289
constituents, 8
immune constituents, 3, 13
ingredients, 54, 56, 265
molecules, 22, 64
packaging, 249, 265
peptide, 80, 126
sequence, 118
Bioactivity, 12
Bioavailable flavonoids, 265
Biochemical pathways, 37, 75, 289
Biocidal activities, 89
Biocomponents, 24
Biocompounds, 75, 200, 298
Biocontrol agents, 156
Bioconversion, 75, 78, 79, 81, 83
Biodegradable
films, 267
matrix, 85
packaging, 265
plastics, 252
polymeric materials, 264
Biofilm, 34, 99, 118, 143, 144, 147, 151–154
Bio-functional
components, 54
parameters, 198
Biological sensor, 268
Bioluminescence, 277
Biomolecular interaction, 278, 282
Biomolecules, 3, 6, 75, 278
Biopesticides, 104
Bioplastics, 264
classification, 264
Biopolymer, 257, 266, 267
Biopreservation, 22, 74, 78, 89, 90, 95, 165, 188, 287, 288, 290, 297, 302
Bioprotectant, 87–89
Bioprotection, 290
Bioprotective
compounds, 79
cultures, 75, 85, 89, 161, 163
properties, 86
Bioreceptor, 268
Biosensor, 251, 253, 268, 269, 275, 277, 278, 280–283
classification, 279
biorecognition molecule, 279
devices, 277
operation principle, 278
requirements, 277
technology history, 278
types, 277
Biotechnology, 89, 100
Biphasic model, 234
Bovine colostrum (BC), 3–14, 171
concentrates (BCCs), 8
preservation and storage, 9
chemical preservation, 11
colostrum silage, 13
freezing and freeze drying/lyophilization, 11
heat treatment, 10
high pressure processing (HPP), 12
microfiltration (MF) treatment, 12
pasteurization, 10
spray drying, 11
protective role, 7
therapeutic role, 8
Broad
lytic spectrum, 100
spectrum activity, 163
Butyl hydroquinone (BHQ), 33

C

Cancer chemotherapy, 127
Candida albicans, 80, 84
Carbohydrate, 87, 128, 233
chains, 264
Carbon
dioxide (CO2), 23, 82, 100, 254, 259, 262, 263
moiety, 82
Carboxylic acid, 78

Cardiovascular
diseases (CVD), 54, 58, 59, 64
system, 9
Carotene, 4, 5, 10, 57
Carotenoids, 53, 54
Carpet model, 123, 124, 129
Carvacrol addition, 33
Catalase (CAT), 83, 84, 102, 103, 200, 289
Catalyzation, 198
Cationic antimicrobial peptides (CAMPs), 168, 169, 176
Cell
based biosensors, 269
death, 23, 78, 111, 112, 121, 123, 164, 168, 299
envelope, 79, 164, 165, 167–169, 171, 172
free supernatants, 86
lines, 154, 155
membrane, 23, 31, 32, 38, 77, 84, 112, 119, 121–124, 128, 164, 165, 168, 231, 299
nucleus, 197
nutrients, 150
proteins, 97, 104
structure, 232
surface protein receptors, 147
wall, 32, 33, 37, 75, 79, 99, 103, 128, 168, 170, 172
composition, 33
Cellular
contents, 23, 154
interactions, 196
membranes, 197
metabolism, 89
molecules, 23
oxidative stresses, 126
proteins, 83
wall permeation, 5
Center for Food Safety and Applied Nutrition (CFSAN), 174
Chelating agents, 98
Chemical
pesticides, 104
reactors, 191
Chemiluminescence, 277
Chemotactic receptor proteins, 146
Chemotaxis, 5
Chiral molecules, 80
Chitinase, 99, 103, 105
Chitooligomers, 103
Chitosan, 34, 89, 257, 258, 266, 299
Chologenic death, 230
Chromosomes, 165, 197
Chrysophrys major, 114, 115
Cinnamaldehyde, 23, 24, 30, 31, 39, 40
Cinnamic acid, 36, 78, 79
Citric acids, 39
Citrus extracts, 38
Clostridium
perfringens, 30, 104
tyrobutyricum, 103
Cold plasma, 144, 187, 191, 196, 197, 199–202, 211, 227
Colletotrichum gloeosporiodes, 32
Colony-forming unit (CFU), 144, 148, 152, 153, 156, 282
Colostral treatment, 12
Colostrum, 3–14
post-thermal treatment, 10
silage, 13, 14
Commercial scale, 23, 37, 202, 288
Conductometric biosensor, 269
Controlled atmospheric storage (CAS), 262, 263
Corona discharges (CDs), 191, 205
Critical temperature
based indicators (CTI), 260, 269
indicators, 260
time-based indicators (CTTI), 260
Cryptosporidiosis, 9
Cryptosporidium parvum, 8
Cuminum cyminum, 28, 36, 62
Cyclic
dipeptides (CDPs), 80
lipoproteins, 128
peptides, 77, 90
Cyclobutane thiamine dimers, 230
Cyclotides database, 118
Cyprinus carpio, 114
Cystic fibrosis, 154
Cytochrome oxidase, 289
Cytokines, 3, 8, 14, 155
Cytoplasm, 38, 78, 79, 123
Cytoplasmic membrane, 97, 170, 171

Cytotoxic
actions, 5
products, 95, 96
Cytotoxicity, 81, 127

D

Dairy products
herbs application, 58
cheese, 61
fat-rich dairy products, 58
fermented milk, 60
sandesh, 61
Death kinetics, 25
Debaryomyces hansenii, 85, 86
Decimal reduction
dose, 25
time, 25
Decoy cells, 154
Degree of unsaturation, 82
Dehydration, 9, 83, 253
Density waves, 282
Deoxyribonucleic acid (DNA), 33, 83, 112, 119, 124, 127, 175, 197, 230, 232, 269, 280, 289
Depolarization, 123, 232
Dermaseptin-S4 (DRS-S4), 128
Desaturation, 8
Destructurized starch (DS), 264
Dextrose equivalent (DE), 57
Diagnostic enzyme, 98
Diarrhea, 6, 8, 10, 14, 162
Dielectric
barrier discharge, 187, 191
plasma needle barrier discharge (DBDs), 187, 190, 191, 200, 202
Dietary
fibers, 54
nucleotides, 8
Diffraction, 277
Direct current (DC), 190, 192, 230
D-limonene, 30, 37
D-value, 25, 26

E

Edible packaging, 249, 251, 265, 266, 269
Electric pulses, 31
Electrochemical
biosensor, 269
reactions, 191
Electromagnetic
field, 191
waves, 188
Electrons, 97, 190, 191, 193, 195, 282
Electroporation, 31, 192
Electrostatic interaction, 122
Embden-Meyerhof-Parnas (EMP), 290, 302
Emulsification activity, 56
Endolysins, 99, 100, 104, 105
Endospores, 198, 277
Endo-thermic reaction, 277, 278
Endotoxemia, 8
Endotoxins, 8
Enterococcus faecalis, 113, 171, 293
Enterotoxigenic *Escherichia coli* (ETEC), 7, 8
Environment Protection Agency (EPA), 173, 174
Enzyme-linked immune sorbent assay (ELISA), 280, 283
Epithelial
cells, 7, 119, 125, 143, 155
lining, 6
Escherichia coli (E. coli), 6, 7, 12, 25–40, 81, 83, 98, 113, 125, 143, 149, 151, 152, 155, 156, 215, 230, 234, 236–238, 241, 255, 258, 259, 276, 281, 282, 293
Essential oil (EO), 21–26, 30–40, 59, 204, 257
antibacterial activity, 23
combined approach, 25
composition, 23
Ethylene, 201, 220, 253, 257, 259, 263
vinyl alcohol (EVA), 257
copolymer (EVOH), 220
Ethylenediamine tetraacetic acid (EDTA), 98, 100, 101, 259
Eugenol, 23, 24, 36, 39, 40
Eukaryotes, 117
Eukaryotic cells, 163
Eurotium rubrum, 80
Extended
AMPs, 120, 128
spectrum ß-lactamase (ESBL), 155, 156
Extracellular compounds, 78
Extra-chromosomal material, 165

F

Fatty acid, 8, 40, 53, 73, 78, 82, 89, 170–172, 290, 298
peroxidation, 126
Federal
Food, Drug, and Cosmetic Act (FFDCA), 173, 174
Insecticide, Fungicide, and Rodenticide Act (FIFRA), 174
Meat Inspection Act (FMIA), 174
Fermentation, 13, 74, 75, 78, 81, 86–88, 171, 188, 257, 289, 291, 295, 297, 298
Ferric myoglobin, 218
Ferulic acid, 63
First
in first-out (FIFO), 261
order microbial kinetics, 35
Flagella, 146
Fluorescence, 277
Food
Food and Agriculture Organization (FAO), 84, 144, 162
Food and Drug Administration (FDA), 57, 74, 103, 173, 174, 227, 229
grade microorganisms, 38
matrix, 37, 96, 176, 213, 232, 254, 257, 280, 293, 297, 298
packaging optimization, 267
pathogens, 283
preservation, 21, 25, 26, 36, 38, 61, 96, 161, 162, 187–189, 192–195, 199, 204, 205, 229, 232, 258, 259, 297
Safety, 21, 22, 30, 40, 143, 144, 146, 161–163, 173, 174, 176, 196, 203, 218, 241, 251–253, 258, 260, 262, 269, 275, 276, 283, 287, 288, 290, 298, 300, 302
Safety and Inspection Services (FSIS), 173, 218
Safety and Standards Authority of India, 276
systems
antifungal LAB, 84, 86–88
antimicrobial enzymes, 100, 102–104
Foodborne
bacteria, 21
diseases (FBD), 162, 176, 268, 276
illnesses, 21, 162
pathogens, 30, 187, 205, 287
Functional
foods, 9, 53, 54, 58, 64
groups, 24, 173
Fungal
cell, 82
walls, 95, 96
protoplast, 103
Fungicides, 257
Fungistatic agent, 78
Fusarium
culmorum, 86, 88
strains, 87

G

Gamma-proteobacteria, 145
Gastroenteritis, 14
Gastrointestinal (GI), 4, 6–9, 126, 175, 288
tract (GIT), 4, 126, 175, 288
Gelatin, 57, 126
Generally regarded as safe (GRAS), 22, 74, 98, 103, 165, 173, 174, 290
Genetic engineering, 104
Germination, 30, 33, 189, 277, 296
Global food security, 144
Globalization, 162
Glucose oxidase (GOX), 97, 98, 101–103, 105, 259, 279
glucose system, 98
Glutamate decarboxylase (GAD), 168, 176
Glutathione S-transferase (GST), 200
Gram
negative, 8, 83, 96, 98, 99, 124, 143–148, 150, 152–156, 164, 215, 235, 302
bacteria, 83, 124, 143, 147, 150, 152–154, 156, 164, 302
genera, 143
microbes, 8
microbial pathogens, 8
organisms, 145
pathogens, 156
positive, 96, 149, 154, 164, 215
bacteria, 83, 98, 164, 171, 235, 302
pathogens, 38
strains, 99
Guava extracts, 39
Gum Arabic, 57
Gut
associated lymphoid tissues (GALT), 7
microbiota, 126, 127, 129, 149

H

Heat
denaturation, 98, 219
sensitive
enzyme, 279
nutrients, 228
treatment, 10, 12, 14, 21, 30, 194, 215, 253, 295
Helicobacter pylori, 6, 7
Hemolytic
activity, 119, 128
uremic syndrome, 7
Herbal
bioactive, 53, 60, 64
compounds, 56
crop production, 55
extracts, 55, 56, 63
medicine, 53, 55, 64
importance and scope, 55
supplementation, 55
Heterofermentation, 75, 171
Heterolactic fermentative, 290
High
hydrostatic pressure (HHP), 12, 13, 25, 27, 28, 31–33, 187, 194, 211–221, 227, 228, 300
effects, 215
products processing, 213
pressure processing (HPP), 12, 14
Homeostasis, 78, 126
Homo-fermentation, 75
Homogenization, 9
Homolactic fermentation, 290
Hospitalizations, 21
Host cell plasma membranes, 150
Hostindependent (HI), 88, 147
Human
cell lines, 81, 143, 155
cytomegalovirus (HCMV), 128
epithelial colorectal adenocarcinoma cells, 127
lymphocyte proliferation, 5
Humoral immune response, 5
Hurdle technology, 40, 161, 163, 173, 241, 242, 299
Hydrocarbon, 23, 24
Hydrocinnamic acid, 36
Hydrocolloids, 266
Hydrodistillation, 23
Hydrogen
bonding, 280
peroxide, 77, 83, 96, 97, 200, 279
Hydrogenation, 199
Hydrolysates, 79, 126
Hydrolysis, 62, 99, 127, 217, 261, 290
Hydrolytic
enzyme, 98
genes, 145
Hydrophobic
amino acids, 119
compounds, 23
interactions, 119
nature, 199
Hydrophobicity, 37, 79, 98, 119, 128, 201
Hydrostatic pressure, 127
Hydroxy propionaldehyde (HPA), 83
Hydroxyl group, 24, 82
Hyperimmune bovine colostrum (HBC), 7

I

Immune
cells, 277, 280
factors, 4
milk, 3, 4, 14
modulators, 5, 111
Immunesensors, 281
Immunodeficiency virus, 6
Immunoglobulin, 3, 4, 6, 7, 9–13
IgA, 4, 7, 10
IgG, 4, 8, 10–13
IgM, 10, 11
Immunomodulators, 5, 112
Immunosensors, 280
Immunotherapeutic
agent, 4, 14
roles, 3
In vitro, 5, 8, 148, 149, 172, 175, 235, 236, 241, 281, 293, 295, 298, 302
cell proliferation, 8
Inflammation, 4, 149
Infrared (IR), 211, 229, 231
Innate resistance, 167
bacterial cell envelope variations, 168
bacteriocin degradation, 167
growth conditions resistance, 167
immunity mimicry, 167

Inoculants, 87
Insulin-like growth factor (IGF), 5, 6, 10, 11
Intelligent packaging, 249, 251, 252, 255, 256, 259, 261, 266, 269
Intense
  light pulses (ILP), 228–230, 232, 233, 239, 240
  pulsed light (IPL), 228, 235
Interferon (IFN), 5, 6
Interleukins (IL), 6, 155
Intermolecular binding, 257
Intracellular
  components, 31, 35, 82, 123
  metabolites, 112
  molecules, 111, 128
  peptidases, 79
  redox state, 83
  signal molecules, 80
Isostatic
  principle, 212
  theory, 195

L

Lachrymal glands, 96
Lactalbumins, 5
Lactic acid bacteria (LAB), 13, 29, 36, 73–90, 101, 103, 164, 217, 257, 287–299, 301, 302
  antifungal compounds, 75
    cyclic dipetides, 80
    fatty acids, 82
    organic acids, 75, 77, 78
    other antifungal compounds, 83
    phenyllactic acid (PLA), 80, 81, 89, 264, 266, 290
    proteinaceous antifungal compounds, 79
    reuterin, 83
    volatile compounds, 81
  applications
    dairy products, 295
    fruits and vegetables, 297
    marine foods, 296
    meat products, 296
  protective cultures (PC), 287–290, 292, 295–302
    antimicrobial substances production, 289
    competitive exclusion, 289
    principles combination, 289
Lactobacillus, 60, 62, 77, 79, 87, 113, 119, 127, 292, 296, 298
  acidophilus, 60, 62
  bulgaricus, 60
  paracasei, 60, 292
Lactoferricin (Lf), 3, 4, 8, 11, 14, 80, 119, 120, 127
Lactoperoxidase (LP), 4, 96, 105
  system (LPS), 8, 96, 97, 100, 102
Lactose metabolism, 289
Lantibiotics, 113, 119, 121, 164, 167, 170
Large scale dairy production units, 3
Le-Chatelier principle, 212, 221
Leuconostoc mesenteroides, 80
Leukocytes, 5, 120
Limit of detection (LOD), 277, 281, 282
Linear
  correlation, 234
  fashion, 59
  peptides, 121
Lipid, 11, 39, 82, 97, 122, 123, 261, 264, 266
  bilayer, 82, 123, 171
  membrane, 112, 127
  monolayers, 170
  oxidation, 217, 253
  peroxidation, 126
  peroxide, 200
Lipopolysaccharide (LPS), 8, 98
Lipoteichoic acids (LTAs), 168, 170
Listeria
  innocua, 36, 237, 293, 296
  monocytogenes, 36, 62, 81, 113, 166, 167, 218, 232, 234, 236–238, 293
Log-linear model, 234
Lymphokines, 3, 5, 14
Lyophilization, 3, 10, 14
Lysins, 99, 100, 118
Lysostaphin, 113, 172
Lysozyme (LYZ), 4, 98, 100–103, 105, 115, 117, 257, 259
Lytic enzyme, 103

M

Macrophages, 5, 119
Malondialdehyde (MDA), 197
Maltodextrin, 56, 57
Mannose phosphotransferase system (Man-PTS), 164, 171

Manothermosonication, 241
Matrix metalloproteinase, 125
Membrane
destruction, 154
disruption, 78, 124, 129
fluidity, 84, 122, 171, 172
Mentha longifolia, 36
Messenger ribonucleic acid, 79
Metabolic compounds, 78
Metabolites leakage, 121
Metallo-ß-lactamase enzymes, 155
Metasequoia glyptostroboides, 28, 36
Microbes, 7, 74, 96, 111, 119, 125, 150, 189, 192, 193, 195–198, 200–202, 204, 215, 227, 230–232, 235, 237, 239, 241, 250, 258, 275
Microbial
cells, 189, 196, 231, 235
characteristics, 232
decontamination, 205, 236
destruction, 35, 229, 232–234, 239, 241
elimination, 236
growth, 124, 163, 240, 249, 255, 256, 296
inactivation, 27, 230, 232–235, 239, 241
photochemical mechanism, 230
photophysical mechanism, 232
photothermal mechanism, 231
infection, 125, 128
invasion, 125
membrane, 119
quality, 221, 255, 259, 261
spoilage, 144, 162, 250, 251, 253, 260, 275, 288
static effects, 257
Microbicidal protein, 124
Microbiological safety, 33
Microcapsules, 56
Microdevices, 269
Micro-encapsulation immobilization, 265
Microfiltration (MF), 3, 12, 13, 228
Microflora, 6, 7, 79, 81, 89, 99, 215, 219, 237, 263, 289, 297, 300
Microplasma arrays, 199
Microscopic ordering, 195
Microsensors, 268
Microwave-assisted extraction, 23
Mimic physiological shearstress, 154
Minimal inhibitory concentration (MIC), 37, 38, 78, 80, 84, 166, 254
Minimum bactericidal concentration (MBC), 39
Modified atmosphere packaging (MAP), 34, 40, 63, 100, 259, 262, 263, 269, 300
Molecular
components, 25
structure, 97
weight (MW), 23, 80, 164, 216, 289, 290
Monoterpene, 23
Morbidity, 10
Mucosal
lining, 6
membranes, 7
surfaces, 99
Multiple
drug resistant (MDR), 80, 155, 156
immune-enhancing factors, 5
Mycotoxins, 74, 298

N

N-acetyl-glycosamine, 98
N-acetylmuramic acid, 98
Nano packaging system, 264
Nanobioactive packaging, 265
Nano-composites, 264
Nano-crystalline, 264
Nano-emulsions, 30, 35, 37, 266
Nano-encapsulations, 265
Nano-modulated lattice films, 264
Nano-packaging material, 265
Nanostructured multilayer films, 264
Narrow spectrum activity, 163
Natural killer (NK), 5, 8, 114
Necrotic lesions, 151
Neonatal animals, 5
Neuro toxicities, 156
Neutrophils, 119
N-glycosylation sites, 96
Nisin, 34, 36, 37, 100–102, 165–171, 240, 241, 257, 259, 290, 296, 297
resistant strains, 167
Nitrogen-fixation endosymbionts, 125
Nodule specific cysteine rich (NCR), 125
Non-antimicrobial amalgamations, 6
Non-bacteriocinogenic LAB applications, 297
Non-covalent interactions, 78, 121
Non-equilibrium, 191

Non-food applications, 264, 265
Non-protein nitrogen (NPN), 8, 13
Non-proteinnitrogen (NPN), 8
Non-thermal
components, 195
methodologies, 188
plasma (NTP), 187, 189–192, 195, 197–202, 204, 205
applications, 195, 196
atmospheric pressure plasma (APP), 190
corona discharges (CDS), 190
dielectric barrier discharges (DBDS), 191
effect, 201
impact, 200
processing methods, 192
radio frequencies (RFS), 191
treatment, 33
Novel packaging systems
types, 252
active packaging, 253
antimicrobial packaging, 256
bioplastic (biodegradable) packaging, 263
fresh-check indicators, 261
intelligent packaging, 255
modified/controlled atmospheric packaging, 262
time temperature indicators (TTIS), 258
Nucleases, 145
Nucleic acid (NAs), 78, 97, 119, 146, 230, 279, 281, 283, 289
Nucleobases, 8
Nucleosides, 8, 84, 279
Nutraceuticals, 3, 7, 9, 55, 64

O

Ocimum basilicum, 33
Ohmic heating (OH), 25, 35, 36, 40, 77, 80, 117, 197
Omega-3 fatty acids, 54
Oncorhynchus mykiss, 114, 217
Optical
biosensor, 280, 282
density, 277
Oral
cavity, 153
provision, 8
Organic
acids, 25, 31, 38, 73, 75, 77–79, 87, 89, 254, 257, 289, 297, 299
molecules, 197
Organoleptic
properties, 74, 188
qualities, 187
Osmolarity, 168
Oxidation-reduction reaction, 278
Oxidative stress, 32, 75, 77, 83, 89, 126
Oxidoreductase enzymes, 96
Ozone liberation, 191

P

Paenibacillus polymyxa, 113, 167
Passive immunity, 3, 4, 13
Pasteurization, 3, 10, 11, 32, 190, 193, 202, 228, 275, 276
Pathogenic
bacteria, 6, 36, 37, 39, 63, 102, 105, 147, 153, 254, 255, 257, 259, 275, 296
bacterial strains, 152
fungi, 99, 103
microbes, 10, 191, 275
microbiota, 75
microorganisms, 11, 22, 37, 99, 163, 214, 218, 228, 257
Pathogens, 4, 6, 8, 21, 30, 33, 34, 36, 38, 40, 62, 77, 79, 81, 95, 98–100, 119, 124–128, 143, 144, 147, 149, 151–156, 161–163, 165, 193, 203, 256, 257, 276, 283, 287, 288, 290, 296, 297, 299, 300
Penicillium
chrysogenum, 33
roqueforti, 80
verrucosum, 81
Peptides, 5, 36, 53, 62, 79, 80, 111, 112, 114, 117–126, 128, 163, 164, 290, 299
Peptidoglycan, 98, 99, 146, 164, 168, 172
Peracetic acid (PAA), 144, 151
Permeabilization, 124, 198
Peroxidase, 84, 200
Peroxyl radicals, 202
Pesticides, 204, 276, 283
PH, 13, 23, 31, 32, 35, 38, 39, 63, 77, 78, 82, 97, 173, 261, 262, 281, 288, 297, 299, 300
Phage lysin enzymes, 100
Phagocytic activity, 5

Pharmaceuticals, 129
Phenolic, 202
  components, 200
  compounds, 24, 37, 58
Phenols, 23, 24
Phenotype, 170
Phenyllactic acid, 81
Phenylpropanoids, 23
Phenylpyruvic acid (PPA), 81
Phosphate buffer saline (PBS), 282
Phospholipids, 111, 122, 168
Phosphorescence, 277
Photochemical reactions, 33
Photothermal
  effect, 231
  mechanism, 230, 242
Physicochemical
  attributes, 56, 61, 118
  characteristics, 102, 296
Physico-optical characteristics, 278
Phytochemicals, 55, 265
Phytopathogens, 144, 146, 151
Phytosterols, 53, 64
Piezoelectric biosensors, 282
Plasma
  discharge technology, 201
  ionization, 190
  membrane, 78, 112, 128
  oscillations, 188
  sterilization, 195, 201
  technology impact, 202
Plasmids, 165
Plastic resistance, 147
Plethora, 3, 4, 6, 7
Pneumocandins, 128
Pneumonia, 10, 14, 152, 155
Polar head group, 123
Polyamide (PA), 36, 113, 164, 167, 220
Polyanhydrides, 257
Polyethylene (PE), 220, 257
  terephthalate (PET), 220
Polymer, 99, 252, 257, 261
  materials, 252
Polymeric
  material matrix, 257
  matrices, 253
  plastic materials, 263
Polymerization, 56, 261
Polypeptide chains, 121
Polyphenoloxidase, 200, 240
Polyphenols, 53, 54, 56, 57, 64
Polysaccharides, 56, 125, 265, 266
Polyunsaturated fatty acids (PUFA), 8, 217
Polyvinyl alcohol (PVOH), 257
Poration, 112, 192
Potent
  antioxidative peptides, 126
  enzymes, 155
Potential
  agents, 127
  bactericidal activity, 102
Potentiometric biosensors, 269
Poultry products inspection act (PPIA), 174
Prebiotics, 54, 265
Predation, 146, 148, 150–156
Predator flagellums, 146
Predatory bacteria
  application, 155
  biocontrol potential, 146
Pressurization, 32
Primary carbon metabolism, 75
Proactive elements, 204
Probiotic
  agent, 147
  bacteria, 60
Prokaryotes, 117
Proline-rich polypeptides (PRP), 3, 5, 14
Prophylaxis, 3, 6, 8, 14
Propionic acid, 76–78, 259, 297
Protective
  culture (PC), 88, 89, 287, 288, 291, 292, 294, 297–302
    activity affecting factors, 300
    isolation, 292
    LAB availability, 291
  enzymes, 37
Protein
  data bank (PDB), 120
  denaturation, 11, 12
  films, 257, 266
Proteinaceous compounds, 77
Proteolysis, 62, 79, 296
Proteolytic
  action, 13
  activity, 62
  agent, 124

degradation, 119
enzymes, 79
Proton gradient interference, 75–77, 89
Protozoa, 83, 125, 127, 128
Pseudomonas
aeruginosa, 113, 127, 128, 152
tolaasii, 150, 151
Pulsed
electric field (PEF), 25, 27, 28, 30, 31, 40, 187, 192, 211, 227, 228, 240, 241
light (PL), 34, 227–230, 232–237, 239–242
light technology (PLT), 227–237, 239–242
advantages, 241
disadvantages, 241
effect, 239
efficacy, 235
factors affecting microbial inactivation, 232
microbial inactivation mechanism, 230
principle, 229

Q

Quartz crystal microbalance, 282
Quasi-neutral electrified gas, 188
Quaternary ammonium compounds, 151
Quick response time, 278

R

Radiant energy, 233
Radio
frequencies (RFs), 191
frequency identification (RFID), 255, 256
sensitivity, 33, 34
Raw colostrum, 10
Reaction kinetics, 234, 241, 242, 261
Reactive oxygen species (ROS), 75, 125, 126, 197, 202
Real-time
detailing, 251
information, 249, 255, 256
monitoring, 278
Receptor molecules, 165
Refractive index, 22, 277, 282
Regulatory agencies, 176
Relative
density, 22
humidity, 10
Rennet coagulation time (RCT), 53
Reuterin, 73, 77, 83, 89, 90, 100, 290, 296, 298
Rhizomes, 22
Rhizopus mucilaginosa, 86
Ribonuclease activity, 83
Ribonucleic acid (RNA), 79, 124, 145, 230, 280, 293
Ribonucleotides, 8

S

Salicylic acid, 79
Salmonella
enteric, 39
enterica, 35, 149, 151
enteritidis, 38, 62
typhimurium, 27, 101, 125, 236, 293
Scanning electron
micrographs, 152
microscope imaging (SEMI), 150, 156
microscopy, 154
Seafood, 84, 103, 148, 217, 221, 296
Secondary carbon metabolism, 75, 79
Self-immunity (SI), 165
Seminal plasmin, 124
Septicemia, 14
Serotypes, 153
Serum proteins, 10
Sigmoidal model, 235
Single-cell planktonic bacteria, 152
Solvent-extraction, 23
Sonolysis, 35
Spoilage bacteria, 38, 39, 77, 300
Sporozoite-host cell adhesion, 9
Spray
dried bovine colostrums (SDBC), 11
drying, 3, 11, 12, 14, 57
Staphylococcus aureus, 6, 62, 81, 113, 128, 149, 236, 257, 258, 293
Stationary-phase cells, 167
Steam-distillation, 23
Steric hindrance, 154
Streptococcus
pneumoniae, 127
thermophilus, 60, 62, 292
Sturgeon farming, 147
Succinic acid, 75
Sulfur dioxide, 163, 254

Supercritical carbon dioxide extraction, 23
Suppress T-cells, 5
Surface
  active agent, 57
  molecular size, 257
  plasma resonance (SPR), 269, 279, 282
  plasmon (SPs), 277, 282
    resonance, 282, 283
Synergistic effect, 25, 31–34, 36–38, 298

T

Tachyplesin, 119
Taxonomic identification, 174
T-cells, 5, 8
Teichoic acids (TAs), 168, 172
Tensile strength, 252, 264, 267
Terminalia arjuna, 59
Terpene compounds, 23
Terpenoids, 23, 54
Tetramic acid, 83
Therapeutic
  agents, 8, 156
  effects, 129
  potential, 112
  use of AMPs, 129
Thermal
  intensity, 26
  plasma, 191, 195, 199, 202
  stability, 53
Thermochromic inks, 249, 256
Thermoplastic, 264
  starch (TPS), 264, 267
Thermosonication, 35, 227, 240, 241
Thiocyanate, 84, 96, 97, 102
Thiol-reactive compounds, 31
Thioredoxin reductase, 32
Time
  based temperature indicators (TTIs), 249, 251, 253, 255, 256, 259–261
  temperature indicators (TTIs), 251, 269
Tissue growth factors (TGF), 3, 5, 13
Toroidal
  model, 123, 129
  pore model, 123
Toxicity, 84, 95, 96, 127, 281
Toxin identification indicators (TII), 255
Transcriptional activator, 171
Transducers
  types, 281
    amperometric transducer, 281
    conductometric transducer, 282
    electrochemical-based transducers, 281
    potentiometric transducer, 281
Transforming growth factor ß (TGF ß), 5
Transposons, 165
Tryptic soy broth (TSB), 39, 154, 292
Tryptophan, 120
Tumorigenes, 127
Tyrosine, 81

U

Ultra high temperature (UHT), 63, 228
Ultrasonic
  application, 9
  extraction, 23
Ultra-sonication, 193
Ultrasound treatment, 35
Ultraviolet (UV), 28, 29, 33, 34, 144, 191, 198, 202, 228–231, 240, 241
United States
  Department of Agriculture (USDA), 162, 173, 174, 218
  Food and Drug Administration (US-FDA), 57, 74, 103
Upper respiratory tract infection (URTI), 9

V

Vacuoles, 119, 232
Value addition, 54, 64
Vanillic acid, 78
Vegetative cells, 192, 235
Viscosity, 11, 12, 57, 216, 233, 237, 282
Volatile compounds, 56, 57, 78, 82, 298

W

Water vapor permeation, 252
Weibull model, 235
World Health Organization (WHO), 262

Z

Zataria multiflora, 34, 36